Cambridge IGCSE® & O Level
Complete
Physics
Fourth Edition

Stephen Pople
Anna Harris
Naseemunissa Azam
Elliot Sarkodie-Addo
Helen Roff

OXFORD
UNIVERSITY PRESS

Great Clarendon Street, Oxford, OX2 6DP, United Kingdom

Oxford University Press is a department of the University of Oxford. It furthers the University's objective of excellence in research, scholarship, and education by publishing worldwide. Oxford is a registered trade mark of Oxford University Press in the UK and in certain other countries

© Oxford University Press 2021

The moral rights of the authors have been asserted

First published in 2021

All rights reserved. No part of this publication may be reproduced, stored in a retrieval system, or transmitted, in any form or by any means, without the prior permission in writing of Oxford University Press, or as expressly permitted by law, by licence or under terms agreed with the appropriate reprographics rights organization. Enquiries concerning reproduction outside the scope of the above should be sent to the Rights Department, Oxford University Press, at the address above.

You must not circulate this work in any other form and you must impose this same condition on any acquirer

British Library Cataloguing in Publication Data
Data available

978-1-38-200594-4 (standard)
10 9 8 7 6 5

978-1-38-200593-7 (enhanced)
10 9 8 7 6 5

Paper used in the production of this book is a natural, recyclable product made from wood grown in sustainable forests. The manufacturing process conforms to the environmental regulations of the country of origin.

Printed in China by Golden Cup

Acknowledgements

IGCSE® is the registered trademark of Cambridge International Examinations. All answers have been written by the authors. In examination, the way marks are awarded may be different.

Stephen Pople would like to thank Susan Pople and Dr Darren Lewis for their help, with special thanks also to Zachary and Thea Hopley for keeping his spirits up.

The publisher and authors would like to thank the following for permission to use photographs and other copyright material:

Cover: Willyam Bradberry/Shutterstock.

Photos: p11: VaclavHajduch/Shutterstock; **p14 & 22:** sciencephotos/Alamy Stock Photo; **p16:** Sergey Lapin/Shutterstock; **p18:** NASA; p23l: janrysavy/iStockphoto; p23r: JPL/University of Arizona/NASA; p23bl: Trekholidays/iStockphoto; p23br: ElementalImaging/iStockphoto; **p27:** David Hallett/Stringer/AFP/Getty Image; **p28:** Keith Kent/ Science Photo Library; **p32:** Design Pics Inc/Alamy Stock Photo; **p34:** Charles M. Duke Jr./NASA; **p39:** Fly_Fast/iStockphoto; **p40:** EHStock/iStockphoto; **p42:** Takeshi Takahara/Science Photo Library; p43t: Pixbox77/Shutterstock; p43l: PCN Photography/Alamy Stock Photo; p43r: Trubavin/Shutterstock; **p48:** imageBROKER/Alamy Stock Photo; **p54:** Kqlsm/Shutterstock; **p59:** zebra0209/Shutterstock; **p63:** agefotostock/Alamy Stock Photo; **p67:** Gang Liu/Shutterstock; p69l: Andrew Buckin/Shutterstock; p69m: Norlito Gumapac/Fotolia; p69r: Peter Gould/OUP; **p70:** OAR/National Undersea Research Program (NURP); Woods Hole Oceanographic Inst./NOAA; **p72:** NASA/JPL/UCSD/JSC; **p74:** Philippe Plailly/Science Photo Library; **p80:** Action Plus Sports Images/Alamy Stock Photo; p88t: Juri Bizgajmer/Dreamstime; p88b: Michael1959/iStockphoto; **p90:** Gary Parker/Science Photo Library; **p93:** David R. Frazier Photolibrary, Inc./Alamy Stock Photo; **p99:** Paul Drabot/Shutterstock; **p100:** Trekholidays/iStockphoto; **p107:** Jean-Francois Monier/Stringer/AFP/Getty Images; **p109:** Mike Birkhead Associates/Photodisc/Getty Images; p113t: World History Archive/Alamy Stock Photo; p113m: V travels/Shutterstock; p113b: TigerStocks/Shutterstock; **p114:** Peter Gould/OUP; p115l: Justin Pumfrey/The Image Bank/Getty Images; p115m: Kertlis/iStockphoto; p115r: Lisegagne/iStockphoto; **p119:** Andersen-Ross/Getty Images; **p124:** Mario7/Shutterstock; **p128:** Janine Wiedel Photolibrary/Alamy Stock Photo; p129l: Antony Jones/Contributor/Getty Images; p129r: Avpics/Alamy Stock Photo; **p130:** NOAA Central Library; OAR/ERL/National Severe Storms Laboratory (NSSL)/NOAA Photo Library; **p134:** Merlin Tuttle/Science Photo Library; **p135:** Monkey Business/Fotolia; **p139:** tropicalpixsingapore/iStockphoto; p140t: Busypix/iStockphoto; p140b: Peter Menzel/Science Photo Library; **p141:** NASA/Science Photo Library; **p142:** Brasiliao/Shutterstock; **p146:** Peter Gould/OUP; p149r: David M. Martin, MD/Science Photo Library; p149l &163: David Parker/Science Photo Library; **p151:** Roydee/iStockphoto; **p160:** Jorisvo/Shutterstock; p161t: Krzyssagit/Dreamstime; p161b: Tomazl/iStockphoto; **p163:** ADragan/Shutterstock; **p167:** Craigie Aitchison/Fotolia; **p168:** Peter Menzel/Science Photo Library; p170&172: Peter Menzel/Science Photo Library; p174 - **p179:** Peter Gould/OUP; **p184:** Akbar Baloch/Reuters; **p192:** helloijan/Shutterstock; **p193:** Peter Gould/OUP; **p197:** Los Alamos National Laboratory/Science Photo Library; **p199:** Sciencephotos/Alamy Stock Photo; **p200:** Peter Gould/OUP; **p201:** Bryan & Cherry Alexander Photography/Arcticphoto; **p203:** Sciencephotos /Alamy Stock Photo; **p204:** Pollyana Ventura/E+/Getty Images; **p205:** Peter Gould/OUP; **p207:** Sciencephotos/Alamy Stock Photo; **p209:** Peter Gould/OUP; **p211:** Ajay Bhaskar/Shutterstock; **p213:** dcdebs/iStockphoto; p215r: Natali Goryachaya/Dreamstime; p215l & p216l: Peter Gould/OUP; p216r: Tyler Olson/Shutterstock; **p218:** JMDZ/Fotolia; **p221:** Stephen Minkler/Dreamstime; **p230:** David Parker/Science Photo Library; **p231:** Science Photo Library; **p237:** Martin Bond/Science Photo Library; **p238:** US Department Of Energy/Science Photo Library; **p239:** Gelpi/Shutterstock; **p240:** BSIP/Universal Images Group/Getty Images; **p241:** siam.pukkato/Shutterstock; **p243:** Ted Kinsman /Science Source /Science Photo Library; **p244:** © 2007 CERN; **p249:** Event Horizon Telescope collaboration; **p251:** julienTello/iStockphoto; p254t: NASA/Science Photo Library; p254l: Science Photo Library - Victor Habbick Visions/Brand X Pictures/Getty Images; p255l: Jerry Lodriguss/Science Photo Library; p255r: David Parker/Science Photo Library; **p257:** Science Photo Library; p258r: Audrius Birbilas/Shutterstock; p257b: David Parker/Science Photo Library; **p259:** Celestial Image Co/Science Photo Library; **p262:** David Nunuk/Science Photo Library; **p267:** tankbmb/iStockphoto; p268t: LawrenceSawyer/iStockphoto; p268b: Transport of Delight/Alamy Stock Photo; p269t: Alorusalorus/Dreamstime; p269b: TOSP/Shutterstock; **p270:** Sputnik/Science Photo Library; p271t: Prof. Peter Fowler/Science Photo Library; p271b: CERN/Science Photo Library; p272t: Bryan & Cherry Alexander Photography/Arcticphoto; p272b: Switas/iStockphoto; p273t: Dorling Kindersley/UIG/Science Photo Library; p273b: Roman Krochuk/Shutterstock; **p274:** Science Photo Library; p275t: Royal Astronomical Society/Science Photo Library; p275b: NASA; **p277:** Peter Menzel/Science Photo Library; p278 & **p279:** Peter Gould/OUP.

Artwork by: GreenGate Publishing Services, QBS Learning & OUP

Every effort has been made to contact copyright holders of material reproduced in this book. Any omissions will be rectified in subsequent printings if notice is given to the publisher.

This Student Book refers to the Cambridge IGCSE Physics (0625) and Cambridge O Level Physics (5054) Syllabuses published by Cambridge Assessment International Education.

This work has been developed independently from and is not endorsed by or otherwise connected with Cambridge Assessment International Education.

Introduction

If you are studying physics for Cambridge IGCSE®, then this book is designed for you. It explains the concepts that you will meet, and should help you with your practical work. It is mostly written in double-page units which we have called **spreads**. These are grouped into sections.

Sections 1 to 11 The main areas of physics are covered here. At the end of each of these sections there is a revision summary giving the main topics covered in each spread.

History of key ideas Section 12 describes how scientists have developed their understanding of physics over the years.

Practical physics Section 13 tells you how to plan and carry out experiments and interpret the results. It includes suggestions for investigations, and guidance on taking practical tests.

Mathematics for physics Section 14 summarizes the mathematical skills you will need when studying physics for Cambridge IGCSE.

Examination questions There are practice examination questions at the end of each section (1 to 11). In addition, Section 15 contains a collection of some alternative-to-practical questions.

Reference section Section 16 includes essential equations, units of measurement, circuit symbols, answers to questions, and an index.

Core syllabus content	Supplement syllabus content
If you are following the Core syllabus content, you can ignore any material with a red line beside it.	For this, you need all the material on the white pages, including the supplement material marked with a red line.

The **Enhanced Online Book** supports this student book by offering high-quality digital resources that help to build scientific and examination skills in preparation for the high-stakes IGCSE assessment. If you purchase access to the digital course, you will find a wealth of additional resources to help you with your studies and revision:

- A worksheet and interactive quiz for every unit
- On Your Marks activities to help you achieve your best
- Glossary quizzes to consolidate your understanding of scientific terminology
- Full practice papers with mark schemes

Each person has their own way of working, but the following tips might help you to get the most from this book:

- Use the contents page — this will provide information on large topics
- Use the index — this will allow you to use a single word to direct you to pages where you can find out more.
- Use the questions — this is the best way of checking whether you have learned and understood the material on each spread.

Questions are to be found on most units and within or at the end of each section. Harder questions are identified by the blue circle.

Stephen Pople

Contents

*Watch for this symbol, below and throughout the book. It indicates spreads or parts of spreads that have been included to provide extension material to set physics in a broader context.

For information about the link between spreads and the syllabus, see pages vii–x.

Syllabus and spreads vii

1 Measurements and units

1.1 Numbers and units 12
1.2 A system of units 14
1.3 Measuring length and time 16
1.4 Volume and density 18
1.5 Measuring volume and density 20
1.6 More about mass and density 22
Check-up 24

2 Forces and motion

2.1 Speed, velocity, and acceleration 28
2.2 Motion graphs 30
2.3 Recording motion 32
2.4 Free fall 34
2.5 More motion graphs 36
2.6 Forces in balance 38
2.7 Force, mass, and acceleration 40
2.8 Friction 42
2.9 Force, weight, and gravity 44
2.10 Action and reaction* 46
2.11 Momentum (1) 48
2.12 Momentum (2) 50
2.13 More about vectors 52
2.14 Moving in circles 54
Check-up 56

3 Forces and pressure

3.1 Forces and turning effects 60
3.2 Centre of gravity 62
3.3 More about moments 64
3.4 Stretching and compressing 66
3.5 Pressure 68
3.6 Pressure in liquids 70
3.7 Pressure from the air* 72
3.8 Gas pressure and volume 74
Check-up 76

4 Forces and energy

4.1 Work and energy 80
4.2 Energy transfers 82
4.3 Calculating PE and KE 84
4.4 Efficiency and power 86
4.5 Energy for electricity (1) 88
4.6 Energy for electricity (2) 90
4.7 Energy resources 92
4.8 How the world gets its energy 94
Check-up 96

5 Thermal effects

5.1 Moving particles 100
5.2 Temperature 102
5.3 Expanding solids and liquids 104
5.4 Heating gases 106
5.5 Thermal conduction 108
5.6 Convection 110
5.7 Thermal radiation 112
5.8 Liquids and vapours 114
5.9 Specific heat capacity 116
5.10 Latent heat 118
Check-up 120

6 Waves and sounds

6.1	Transverse and longitudinal waves	124
6.2	Wave effects	126
6.3	Sound waves	128
6.4	Speed of sound and echoes	130
6.5	Characteristics of sound waves	132
6.6	Ultrasound	134
	Check-up	136

7 Rays and waves

7.1	Light rays and waves	140
7.2	Reflection in plane mirrors (1)	142
7.3	Reflection in plane mirrors (2)	144
7.4	Refraction of light	146
7.5	Total internal reflection	148
7.6	Refraction calculations	150
7.7	Lenses (1)	152
7.8	Lenses (2)	154
7.9	More lenses in action	156
7.10	Electromagnetic waves (1)	158
7.11	Electromagnetic waves (2)	160
7.12	Sending signals	162
	Check-up	164

8 Electricity

8.1	Electric charge (1)	168
8.2	Electric charge (2)	170
8.3	Electric fields	172
8.4	Current in a simple circuit	174
8.5	Potential difference	176
8.6	Resistance (1)	178
8.7	Resistance (2)	180
8.8	More about resistance factors	182
8.9	Series and parallel circuits (1)	184
8.10	Series and parallel circuits (2)	186
8.11	More on components	188
8.12	Electrical energy and power	190
8.13	Living with electricity	192
	Check-up	194

9 Magnets and currents

9.1	Magnets	198
9.2	Magnetic fields	200
9.3	Magnetic effect of a current	202
9.4	Electromagnets	204
9.5	Magnetic force on a current	206
9.6	Electric motors	208
9.7	Electromagnetic induction	210
9.8	More about induced currents	212
9.9	Generators	214
9.10	Coils and transformers (1)	216
9.11	Coils and transformers (2)	218
9.12	Power across the country	220
	Check-up	222

10 Atoms and radioactivity

10.1	Inside atoms	226
10.2	Nuclear radiation (1)	228
10.3	Nuclear radiation (2)	230
10.4	Radioactive decay (1)	232
10.5	Radioactive decay (2)	234
10.6	Nuclear energy	236
10.7	Fusion future	238
10.8	Using radioactivity	240
10.9	Atoms and particles (1)	242
10.10	Atoms and particles (2)*	244
	Check-up	246

11 The Earth in space

11.1	Sun, Earth, and Moon	250
11.2	The Solar System (1)	252
11.3	The Solar System (2)	254
11.4	Objects in orbit	256
11.5	Sun, stars, and galaxies (1)	258
11.6	Sun, stars, and galaxies (2)	260
11.7	The expanding Universe	262
	Check-up	264

12 History of key ideas

12.1	Force, motion, and energy*	268
12.2	Rays, waves, and particles*	270
12.3	Magnetism and electricity*	272
12.4	The Earth and beyond*	274
	Key developments in physics	276

13 Practical physics

13.1	Working safely	278
13.2	Planning and preparing	280
13.3	Measuring and recording	282
13.4	Dealing with data	284
13.5	Evaluating and improving	285
13.6	Some experimental investigations	286
13.7	Taking a practical test	290
	Check-up	291

14 Mathematics for physics

The essential mathematics	294

15 IGCSE practice questions

Multichoice questions (Core)	298
Multichoice questions (Extended)	300
IGCSE theory questions	302
IGCSE alternative-to-practical questions	312

16 Reference

Useful equations	316
Units and elements	318
Electrical symbols and codes	319
Answers	320
Index	333

www.oxfordsecondary.com/complete-igcse-science

Syllabus and spreads

Below, is an outline of the Cambridge IGCSE syllabus as it stood at the time of publication, along with details of where each topic is covered in the book. Before constructing a teaching or revision programme, please check with the latest version of the syllabus/specification for any changes.

IGCSE syllabus section			Spread
1		**Motion, Forces and Energy**	
	1.1	Physical quantities and measurement techniques	1.3 2.1 2.13
	1.2	Motion	2.1 2.2 2.3 2.4 2.5 2.6
	1.3	Mass and Weight	1.6 2.9
	1.4	Density	1.4 1.5 1.6
	1.5	Forces	2.6 2.7 2.8 2.14 3.1 3.2 3.4
	1.6	Momentum	2.11 2.12
	1.7	Energy, Work and Power	4.1 4.2 4.3 4.4 4.5 4.6 4.7 4.8
	1.8	Pressure	3.5 3.6
2		**Thermal Physics**	
	2.1	Kinetic particle model of matter	3.8 5.1 5.2 5.4
	2.2	Thermal properties and temperature	5.2 5.3 5.4 5.8 5.9 5.10
	2.3	Transfer of thermal energy	5.5 5.6 5.7

IGCSE syllabus section			Spread
3		**Waves**	
	3.1	General properties of waves	6.1 6.2
	3.2	Light	7.1 7.2 7.3 7.4 7.5 7.7 7.8 7.9 7.12
	3.3	Electromagnetic Spectrum	7.10 7.11 7.12
	3.4	Sound	6.1 6.3 6.4 6.5 6.6
4		**Electricity and Magnetism**	
	4.1	Simple phenomena of magnetism	9.1 9.2
	4.2	Electrical quantities	8.1 8.2 8.3 8.4 8.5 8.6 8.7 8.8 8.11 8.12 8.13 9.9
	4.3	Electric circuits	8.4 8.5 8.6 8.7 8.9 8.10 8.11
	4.4	Electrical safety	8.13
	4.5	Electromagnetic effects	9.3 9.4 9.5 9.6 9.7 9.8 9.9 9.10 9.11 9.12 10.2

IGCSE syllabus section		Spread
5	**Nuclear physics**	
5.1	The nuclear model of the atom	10.1
		10.2
		10.4
		10.6
		10.7
		10.9
5.2	Radioactivity	10.2
		10.3
		10.4
		10.5
		10.8
6	**Space physics**	
6.1	Earth and the Solar System	11.1
		11.2
		11.3
		11.4
		11.5
6.2	Stars and the universe	11.1
		11.5
		11.6
		11.7

Assessment for IGCSE

The IGCSE examination will include questions that test you in three different ways. These are called Assessment Objectives (AO for short). How these different AOs are tested in the examination is explained in the table below:

Assessment Objective	What the syllabus calls these objectives	What this means in the examination
AO1	Knowledge with understanding	Questions which mainly test your recall (and understanding) of what you have learned. About 50% of the marks in the examination are for AO1.
AO2	Handling information and problem solving	Using what you have learned in unfamiliar situations. These questions often ask you to examine data in graphs or tables, or to carry out calculations. About 30% of the marks are for AO2.
AO3	Experimental skills and investigations	These are tested on the Practical Paper or the Alternative to Practical (20% of the total marks). However, the skills you develop in practising for these papers may be valuable in handling questions on the theory papers.

The end-of-section questions in this book include examples of those testing AO1, AO2 and AO3. Your teacher will help you to attempt questions of all types. You can see from the above table that it will not be enough to try only 'recall' questions.

All candidates take three papers.

The make-up of each assessment programme is shown below:

Core assessment

Questions are based on Core content.

Paper 1: Multiple Choice (Core), 45 mins	Paper 3: Theory (Core), 1 hour 15 mins
There are a total of 40 marks available, worth 30% of your iGCSE. The paper consists of multiple-choice questions.	There are a total of 80 marks available, worth 50% of your IGCSE. The paper consists of compulsory short-answer and structured questions.

Extended assessment

Questions are based on the Core and Supplement subject content.

Paper 2: Multiple Choice (Extended), 45 mins	Paper 4: Theory (Extended), 1 hour 15 mins
There are a total of 40 marks available, worth 30% of your IGCSE. The paper consists of multiple-choice questions.	There are a total of 80 marks available, worth 50% of your IGCSE. The paper consists of compulsory short-answer and structured questions.

Practical assessment

Students take either Paper 5 or Paper 6.

Paper 5: Practical Tests, 1 hour 15 mins	Paper 6: Alternative to Practical, 1 hour 15 mins
There are a total of 40 marks available, worth 20% of your IGCSE. You will be required to do experiments in a lab as part of the assessment.	There are a total of 40 marks available, worth 20% of your IGCSE. You will NOT be required to do experiments in a lab as part of the assessment.

1 Measurements and units

- PHYSICAL QUANTITIES
- UNITS AND PREFIXES
- SCIENTIFIC NOTATION
- SI UNITS
- MASS
- TIME
- LENGTH
- VOLUME
- DENSITY

An astronomical clock in Prague, in the Czech Republic. As well as giving the time, the clock also shows the positions of the Sun and Moon relative to the constellations of the zodiac. Until about fifty years ago, scientists had to rely on mechanical clocks, such as the one above, to measure time. Today, they have access to atomic clocks whose timekeeping varies by less than a second in a million years.

1.1 Numbers and units

Objectives: to know that a physical quantity is a number and unit - to use prefixes to make units bigger or smaller - to understand scientific notation.

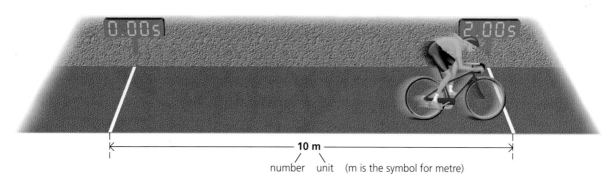

10 m
number unit (m is the symbol for metre)

When you make a measurement, you might get a result like the one above: a distance of 10 m. The complete measurement is called a **physical quantity**. It is made up of two parts: a number and a unit.

10 m really means 10 × m (ten times metre), just as in algebra, $10x$ means $10 \times x$ (ten times x). You can treat the m just like a symbol in an algebraic equation. This is important when combining units.

Combining units

In the diagram above, the girl cycles 10 metres in 2 s. So she travels 5 metres every second. Her *speed* is 5 metres per second. To work out the speed, you divide the distance travelled by the time taken, like this:

$$\text{speed} = \frac{10 \text{ m}}{2 \text{ s}} \quad \text{(s is the symbol for second)}$$

As m and s can be treated as algebraic symbols:

$$\text{speed} = \frac{10}{2} \cdot \frac{\text{m}}{\text{s}} = 5 \frac{\text{m}}{\text{s}}$$

To save space, $5 \frac{\text{m}}{\text{s}}$ is usually written as 5m/s.

So m/s is the unit of speed.

Rights and wrongs

This equation is correct: $\quad \text{speed} = \frac{10 \text{ m}}{2 \text{ s}} = 5 \text{ m/s}$

This equation is incorrect: $\quad \text{speed} = \frac{10}{2} = 5 \text{ m/s}$

It is incorrect because the m and s have been left out. 10 divided by 2 equals 5, and not 5 m/s.

Strictly speaking, units should be included at *all* stages of a calculation, not just at the end. However, in this book, the 'incorrect' type of equation will sometimes be used so that you can follow the arithmetic without units which make the calculation look more complicated.

Advanced units

5 m/s is a space-saving way of writing $5 \frac{\text{m}}{\text{s}}$.

But $5 \frac{\text{m}}{\text{s}}$ equals $5 \text{ m} \frac{1}{\text{s}}$.

Also, $\frac{1}{\text{s}}$ can be written as s^{-1}.

So the speed can be written as 5 m s^{-1}.

This method of showing units is more common in advanced work.

Tables and graphs

You may see table headings or graph axes labelled like this:

$\frac{\text{distance}}{\text{m}}$ or distance/m

That is because the values shown are just numbers, without units. So:

If distance = 10 m

Then $\frac{\text{distance}}{\text{m}} = 10$

MEASUREMENTS AND UNITS

Bigger and smaller

You can make a unit bigger or smaller by putting an extra symbol, called a prefix, in front. (Below, W stands for watt, a unit of power.)

prefix	meaning		example
G (giga)	1 000 000 000	(10^9)	GW (gigawatt)
M (mega)	1 000 000	(10^6)	MW (megawatt)
k (kilo)	1000	(10^3)	km (kilometre)
d (deci)	$\frac{1}{10}$	(10^{-1})	dm (decimetre)
c (centi)	$\frac{1}{100}$	(10^{-2})	cm (centimetre)
m (milli)	$\frac{1}{1000}$	(10^{-3})	mm (millimetre)
μ (micro)	$\frac{1}{1 000 000}$	(10^{-6})	μW (microwatt)
n (nano)	$\frac{1}{1 000 000 000}$	(10^{-9})	nm (nanometre)

Powers of 10

$1000 = 10 \times 10 \times 10 = 10^3$
$100 = 10 \times 10 = 10^2$
$0.1 = \frac{1}{10} = 10^{-1}$
$0.01 = \frac{1}{100} = \frac{1}{10^2} = 10^{-2}$
$0.001 = \frac{1}{1000} = \frac{1}{10^3} = 10^{-3}$

'milli' means 'thousandth', *not* 'millionth'

* You would not normally be tested on micro, nano or giga in a Cambridge IGCSE examination (see also yellow panel at the start of the next spread, 1.2).

Scientific notation

An atlas says that the population of Iceland is this:

320 000

There are two problems with giving the number in this form. Writing lots of zeros isn't very convenient. Also, you don't know which zeros are accurate. Most are only there to show you that it is a six-figure number. These problems are avoided if the number is written using powers of ten:

3.2×10^5 ($10^5 = 10 \times 10 \times 10 \times 10 \times 10 = 100 000$)

'3.2×10^5' tells you that the figures 3 and 2 are important. The number is being given to *two significant figures*. If the population were known more accurately, to three significant figures, it might be written like this:

3.20×10^5

Numbers written using powers of ten are in **scientific notation** or **standard form**. The examples on the right are to one significant figure.

decimal	fraction	scientific notation
500		5×10^2
0.5	$\frac{5}{10}$	5×10^{-1}
0.05	$\frac{5}{100}$	5×10^{-2}
0.005	$\frac{5}{1000}$	5×10^{-3}

1. How many grams are there in 1 kilogram?
2. How many millimetres are there in 1 metre?
3. How many microseconds are there in 1 second?
4. This equation is used to work out the area of a rectangle: area = length × width.
   If a rectangle measures X m by 2 m, calculate its area, and include the units in your calculation.
5. Write down the following in km:
   2000 m   200 m 2×10^4 m
6. Write down the following in s:
   5000 ms 5×10^7 μs
7. Using scientific notation, write down the following to two significant figures:
   1500 m   1 500 000 m   0.15 m   0.015 m

Related topics: SI units **1.2**; speed **2.1**; significant figures **13.3**

13

1.2 A system of units

Objectives: - to know what SI units are – to know the units used for mass, length, and time.

Mass: lb, oz, g, kg, cwt, ton

Length: cm, yd, m, mile, ft, mm, km

Time: s, hour, day, month, year, ms

E A line down the side of the text means that the material is only required for Extended Level.

* An asterisk indicates extension material, provided to set physics in a broader context. You would not normally be tested on this in a CAIE IGCSE examination.

There are many different units – including those above. But in scientific work, life is much easier if everyone uses a common system of units.

SI units

Most scientists use **SI units** (full name: Le Système International d'Unités). The basic SI units for measuring mass, time, and length are the kilogram, the second, and the metre. From these **base units** come a whole range of units for measuring volume, speed, force, energy, and other quantities.

Other SI base units include the ampere (for measuring electric current) and the kelvin (for measuring temperature).

Mass

Mass is a measure of the quantity of matter in an object. It has two effects:
- All objects are attracted to the Earth. The greater the mass of an object, the stronger is the Earth's gravitational pull on it.
- All objects resist being made to go faster, slower, or in a different direction. The greater the mass, the greater the resistance to change in motion.

▲ The mass of an object can be found using a **balance** like this. The balance really detects the gravitational pull on the object on the pan, but the scale is marked to show the mass.

The SI base unit of mass is the **kilogram** (symbol **kg**). At one time, the standard kilogram was a block of platinum alloy stored in Paris. However, there is now a more accurate but more complicated definition involving an electromagnetic balance. Other units based on the kilogram are shown below.

mass	comparison with base unit	scientific notation	approximate size
1 tonne (t)	1000 kg	10^3 kg	medium-sized car
1 kilogram (kg)	1 kg		bag of sugar
1 gram (g)	1 g	$\frac{1}{1\,000}$ kg	10^{-3} kg	banknote
1 milligram (mg)	$\frac{1}{1\,000}$ g	$\frac{1}{1\,000\,000}$ kg	10^{-6} kg	human hair

Note: the SI base unit of mass is the **kilogram**, not the gram

MEASUREMENTS AND UNITS

Time

The SI base unit of time is the **second** (symbol **s**). Here are some shorter units based on the second:

1 millisecond (ms) = $\frac{1}{1000}$ s = 10^{-3} s

1 microsecond (μs) = $\frac{1}{1\,000\,000}$ s = 10^{-6} s

1 nanosecond (ns) = $\frac{1}{1\,000\,000\,000}$ s = 10^{-9} s

To keep time, clocks and watches need something that beats at a steady rate. Some old clocks used the swings of a pendulum. Modern digital watches count the vibrations made by a tiny quartz crystal.

> The second was originally defined as $\frac{1}{60 \times 60 \times 24}$ of a day, one day being the time it takes the Earth to rotate once. But the Earth's rotation is not quite constant. So, for accuracy, the second is now defined in terms of something that never changes: the frequency of an oscillation which can occur in the nucleus of a caesium atom.

Length

The SI base unit of length is the **metre** (symbol **m**). At one time, the standard metre was the distance between two marks on a metal bar kept at the Office of Weights and Measures in Paris. A more accurate standard is now used, based on the speed of light, as explained on the right.

> By definition, one metre is the distance travelled by light in a vacuum in $\frac{1}{299\,792\,458}$ of a second.

There are larger and smaller units of length based on the metre:

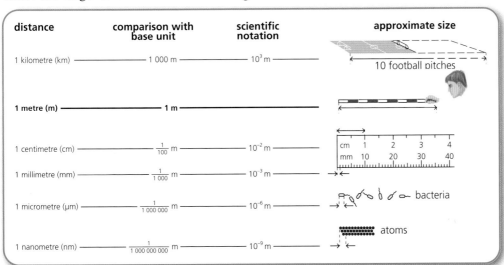

1. What is the SI unit of length?
2. What is the SI unit of mass?
3. What is the SI unit of time?
4. What do the following symbols stand for?
   g   mg   t   μm   ms
5. Write down the value of
   a 1564 mm in m   b 1750 g in kg
   c 26 t in kg   d 62 μs in s
   e 3.65×10^4 g in kg   f 6.16×10^{-7} mm in m
6. The 500 pages of a book have a mass of 2.50 kg. What is the mass of each page **a** in kg **b** in mg?
7. km   μg   μm   t   nm   kg   m
   ms   s   mg   ns   μs   g   mm

   Arrange the above units in three columns as below. The units in each column should be in order, with the largest at the top.

	mass	length	time
largest unit →			

Related topics: numbers and units **1.1**; mass **2.7**

15

1.3 Measuring length and time

Objectives: to know how to measure length with a rule and time with a stopwatch – to describe some methods of improving accuracy.

Measuring length

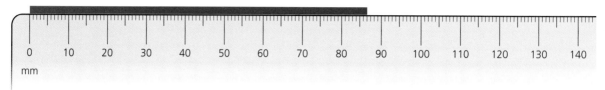

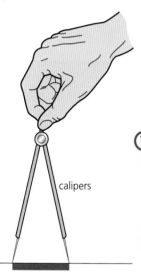

▲ If the rule cannot be placed next to the object being measured, calipers can be used.

Lengths from a few millimetres up to a metre can be measured using a **rule**, as shown above. When using the rule, the scale should be placed right next to the object being measured. If this is not possible, **calipers** can be used, as shown on the left. The calipers are set so that their points exactly match the ends of the object. Then they are moved across to a rule to make the measurement.

(E) Lengths of several metres can be measured using a **tape** with a scale on it.

Accurately measuring small objects is more difficult, but there are ways around the problem. Say, for example, you wanted to find the thickness of a sheet of A4 paper.

Use a ruler to measure the thickness of a 500 sheet pack: 49 mm

Dividing 49 mm by 500 gives the thickness of one sheet: 0.098 mm

Measuring length with light

Surveyors don't need a tape to measure the dimensions of a room. They can use a laser tape measure instead. Despite its name, no tape is involved. The surveyor places the instrument against one wall, points it at the opposite wall, presses a button, and reads the distance on the display.

There are various systems, but in one type, the instrument fires a pulse of laser light at the opposite wall, picks up the reflection, measures the time delay between the outgoing and returning pulses and uses this to calculate the distance.

Light travels at a speed close to 300 000 000 metres per second. So, for example, if the pulse had to travel 30 metres out and back, it would take 100 nanoseconds. If this were the time measured, the display would show a distance of 15 metres. (In this example, the numbers have been simplified. Typically, the instrument is accurate to within 3 mm.)

MEASUREMENTS AND UNITS

Measuring time

Time intervals of many seconds or minutes can be measured using a **stopclock** or a **stopwatch**. Some instruments have an **analogue** display, with a needle ('hand') moving round a circular scale. Others have a **digital** display, which shows a number. There are buttons for starting the timing, stopping it, and resetting the instrument to zero.

With a hand-operated stopclock or stopwatch, making accurate measurements of short time intervals (a few seconds or less) can be difficult. This is because of the time it takes you to react when you have to press the button. Fortunately, in some experiments, there is a simple way of overcoming the problem. Here is an example:

Zero error
You have to allow for this on many measuring instruments. For example, bathroom scales might give a reading of 46.2 kg when someone stands on them, but 0.1 kg when they step off and the expected reading is zero. In this case, the zero error is 0.1 kg and the corrected measurement is 46.1 kg.

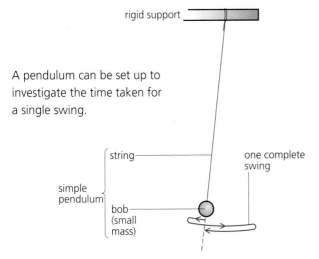

A pendulum can be set up to investigate the time taken for a single swing.

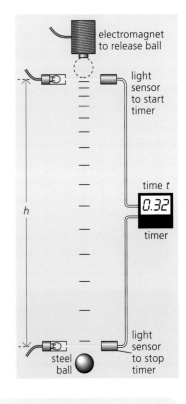

▶ Measuring the time t it takes for a steel ball to fall a distance h.

The pendulum above takes about two seconds to make one complete swing. Provided the swings are small, every swing takes the same time. This time is called its **period**. You can find it accurately by measuring the time for 25 swings, and then dividing the result by 25. For example:

Time for 25 swings = 55 seconds

So: time for 1 swing = 55/25 seconds = 2.2 seconds

Another method of improving accuracy is to use automatic timing, as shown in the example on the right. Here, the time taken for a small object to fall a short distance is being measured. The timer is started automatically when the ball cuts one light beam and stopped when it cuts another.

1. On the opposite page, there is a diagram of a rule.
   a. What is the reading on its scale?
   b. The rule has not be drawn to its true size. What is the length of the red line as printed?
2. A student measures the time taken for 20 swings of a pendulum. He finds that the time taken in 46 seconds.
   a. What time does the pendulum take for one swing?
   b. How could the student have found the time for one swing more accurately?
3. A student wants to find the thickness of one page of this book.
   a. Explain how she might do this accurately.
   b. Measure this book and then find your own value for the thickness of one page.
4. a. What is meant by *zero error*?
   b. Give an example of when you would have to allow for it.

Related topics: units of length and time 1.2; timing a falling object 2.4

1.4 Volume and density

Objectives: to know the equation linking mass, volume, and density, and how to use the equation to solve problems.

Volume

The quantity of space an object takes up is called its volume.

The SI unit of volume is the **cubic metre** (m^3). However, this is rather large for everyday work, so other units are often used for convenience, as shown in the diagrams below:

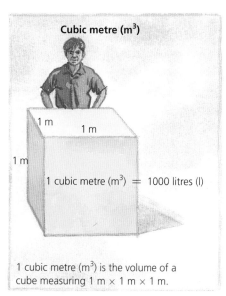

Cubic metre (m^3)

1 cubic metre (m^3) = 1000 litres (l)

1 cubic metre (m^3) is the volume of a cube measuring 1 m × 1 m × 1 m.

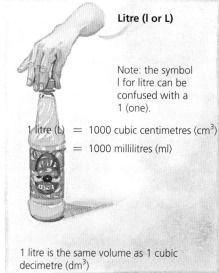

Litre (l or L)

Note: the symbol l for litre can be confused with a 1 (one).

1 litre (l) = 1000 cubic centimetres (cm^3)
= 1000 millilitres (ml)

1 litre is the same volume as 1 cubic decimetre (dm^3).

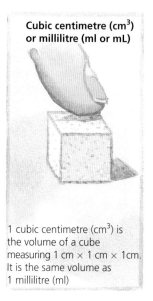

Cubic centimetre (cm^3) or millilitre (ml or mL)

1 cubic centimetre (cm^3) is the volume of a cube measuring 1 cm × 1 cm × 1cm. It is the same volume as 1 millilitre (ml).

Density

Is lead heavier than water? Not necessarily. It depends on the volumes of lead and water being compared. However, lead is more dense than water: it has more kilograms packed into every cubic metre.

The **density** of a material is calculated like this:

$$\text{density} = \frac{\text{mass}}{\text{volume}}$$

In the case of water:

a mass of 1000 kg of water has a volume of 1 m^3

a mass of 2000 kg of water has a volume of 2 m^3

a mass of 3000 kg of water has a volume of 3 m^3, and so on.

Using any of these sets of figures in the above equation, the density of water works out to be 1000 kg/m^3.

If masses are measured in grams (g) and volumes in cubic centimetres (cm^3), it is simpler to calculate densities in g/cm^3. Converting to kg/m^3 is easy:

$1 \text{ g/cm}^3 = 1000 \text{ kg/m}^3$

The density of water is 1 g/cm^3. This simple value is no accident. The kilogram (1000 g) was originally supposed to be the mass of 1000 cm^3 of

▲ The glowing gas in the tail of a comet stretches for millions of kilometres behind the comet's core. The density of the gas is less than a kilogram per cubic *kilometre*.

water (pure, and at 4 °C). However, a very slight error was made in the early measurement, so this is no longer used as a definition of the kilogram.

substance	density kg/m³	density g/cm³	substance	density kg/m³	density g/cm³
air	1.3	0.0013	granite	2700	2.7
expanded polystyrene	14	0.014	aluminium	2700	2.7
wood (beech)	750	0.75	steel (stainless)	7800	7.8
petrol	800	0.80	copper	8900	8.9
ice (0 °C)	920	0.92	lead	11 400	11.4
polythene	950	0.95	mercury	13 600	13.6
water (4 °C)	1000	1.0	gold	19 300	19.3
concrete	2400	2.4	platinum	21 500	21.5
glass (varies)	2500	2.5	osmium	22 600	22.6

The densities of solids and liquids vary slightly with temperature. Most substances get a little bigger when heated. The increase in volume reduces the density. The densities of gases can vary enormously depending on how compressed they are.

The rare metal osmium is the densest substance found on Earth. If this book were made of osmium, it would weigh as much as a heavy suitcase.

Density calculations

The equation linking density, mass, and volume can be written in symbols:

$$\rho = \frac{m}{V}$$ where ρ = density, m = mass, and V = volume

This equation can be rearranged to give: $V = \frac{m}{\rho}$ and $m = V\rho$

These are useful if the density is known, but the volume or mass is to be calculated. On the right is a method of finding all three equations.

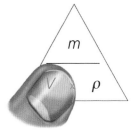
▲ Cover V in the triangle and you can see what V is equal to. It works for m and ρ as well.

Example Using density data from the table above, calculate the mass of steel having the same volume as 5400 kg of aluminium.

First, calculate the volume of 5400 kg of aluminium. In this case, ρ is 2700 kg/m³, m is 5400 kg, and V is to be found. So:

$$V = \frac{m}{\rho} = \frac{5400 \text{ kg}}{2700 \text{ kg/m}^3} = 2 \text{ m}^3$$

This is also the volume of the steel. Therefore, for the steel, ρ is 7800 kg/m³, V is 2 m³, and m is to be found. So:

$$m = V\rho = 7800 \text{ kg/m}^3 \times 2 \text{ m}^3 = 15\,600 \text{ kg}$$

So the mass of steel is 15 600 kg.

In the density equation, the symbol ρ is the Greek letter 'rho'.

1. How many cm³ are there in 1 m³?
2. How many cm³ are there in 1 litre?
3. How many ml are there in 1 m³?
4. A tankful of liquid has a volume of 0.2 m³. What is the volume in **a** litres **b** cm³ **c** ml?
5. Aluminium has a density of 2700 kg/m³.
   a What is the density in g/cm³?
   b What is the mass of 20 cm³ of aluminium?
   c What is the volume of 27 g of aluminium?

Use the information in the table of densities at the top of the page to answer the following:

6. What material, of mass 39 g, has a volume of 5 cm³?
7. What is the mass of air in a room measuring 5 m × 2 m × 3 m?
8. What is the volume of a storage tank which will hold 3200 kg of petrol?
9. What mass of lead has the same volume as 1600 kg of petrol?

Related topics: pressure in liquids 3.6

1.5 Measuring volume and density

Objectives: to know how to measure the volume of a solid or liquid – to know how to measure the density of a solid or liquid.

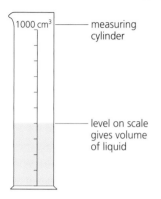

▲ Measuring the volume of a liquid

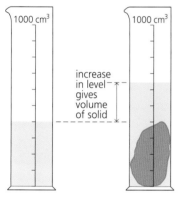

▲ Measuring the volume of a small solid

Measuring volume

Liquid A volume of about a litre or so can be measured using a measuring cylinder. When the liquid is poured into the cylinder, the level on the scale gives the volume.

Most measuring cylinders have scales marked in millilitres (ml), or cubic centimetres (cm^3).

Regular solid If an object has a simple shape, its volume can be calculated. For example:

  volume of a rectangular block = length × width × height
  volume of a cylinder = π × $radius^2$ × height

Irregular solid If the shape is too awkward for the volume to be calculated, the solid can be lowered into a partly filled measuring cylinder as shown on the left. The *rise* in level on the volume scale gives the volume of the solid.

If the solid floats, it can be weighed down with a lump of metal. The total volume is found. The volume of the metal is measured in a separate experiment and then subtracted from this total.

Using a displacement can If the solid is too big for a measuring cylinder, its volume can be found using a displacement can, shown below left. First, the can is filled up to the level of the spout (this is done by overfilling it, and then waiting for the surplus water to run out). Then the solid is slowly lowered into the water. The solid is now taking up space once occupied by the water – in other words, it has displaced its own volume of water. The displaced water is collected in a beaker and emptied into a measuring cylinder.

The displacement method, so the story goes, was discovered by accident, by Archimedes. You can find out how on the opposite page.

Measuring density

The density of a material can be found by calculation, once the volume and mass have been measured. The mass of a small solid or of a liquid can be measured using a balance. However, in the case of a liquid, you must remember to allow for the mass of its container.

Here are some readings from an experiment to find the density of a liquid:

volume of liquid in measuring cylinder	= 400 cm^3	(A)
mass of measuring cylinder	= 240 g	(B)
mass of measuring cylinder with liquid in	= 560 g	(C)

Therefore: mass of liquid = 560 g − 240 g = 320 g   (C − B)

Therefore density of liquid = $\dfrac{\text{mass}}{\text{volume}} = \dfrac{320 \text{ g}}{400 \text{ cm}^3} = 0.8 \text{ g/cm}^3$

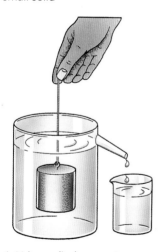

▲ Using a displacement can. Provided the can is filled to the spout at the start, the volume of water collected in the beaker is equal to the volume of the object lowered into the can.

MEASUREMENTS AND UNITS

Checking the density of a liquid*

A quick method of finding the density of a liquid it to use a small float called a **hydrometer**. There is an example on the right. It is based on the idea that a floating object floats higher up in a denser liquid. You can read more about floating, sinking, and the link with density in the next spread, 1.6.

The scale on a hydrometer normally indicates the *relative* density (or 'specific gravity') of the liquid: that is the density compared with water (1000 kg/m^3). A reading of 1.05 means that the density of the liquid is 1050 kg/m^3.

Density checks like this are important in some production processes. For example, creamy milk is slightly less dense than skimmed milk, and strong beer is slightly less dense than weak beer.

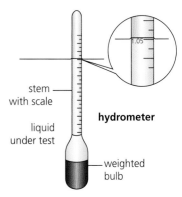

Archimedes and the crown

Archimedes, a Greek mathematician, lived in Syracuse (now in Sicily) around 250 BCE. He made important discoveries about levers and liquids, but is probably best remembered for his clever solution to a problem set him by the King of Syracuse.

The King had given his goldsmith some gold to make a crown. But when the crown was delivered, the King was suspicious. Perhaps the goldsmith had stolen some of the gold and mixed in cheaper silver instead. The King asked Archimedes to test the crown.

Archimedes knew that the crown was the correct mass. He also knew that silver was less dense than gold. So a crown with silver in it would have a greater volume than it should have. But how could he measure the volume? Stepping into his bath one day, so the story goes, Archimedes noticed the rise in water level. Here was the answer! He was so excited that he lept from his bath and ran naked through the streets, shouting "Eureka!", which means "I have found it!".

Later, Archimedes put the crown in a container of water and measured the rise in level. Then he did the same with an equal mass of pure gold. The rise in level was different. So the crown could not have been pure gold.

	crown A	crown B	crown C
mass/g	3750	3750	3750
volume/cm^3	357	194	315

density: gold 19.3 g/cm^3; silver 10.5 g/cm^3

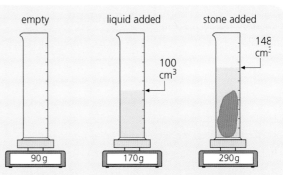

1 Use the information above to decide which crown is gold, which is silver, and which is a mixture.

2 Use the information above to calculate:
   a the mass, volume, and density of the liquid
   b the mass, volume, and density of the stone.

Related topics: volume and density **1.4**

1.6 More about mass and density

Objectives: to describe what a beam balance is used for – to predict, by comparing densities, whether something will float or sink in a liquid (or gas).

Density essentials

$$\text{density} = \frac{\text{mass}}{\text{volume}}$$

Comparing masses

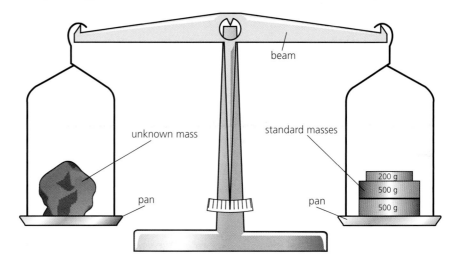

▶ A simple beam balance

The device above is called a **beam balance**. It is the simplest, and probably the oldest, way of finding the mass of something. You put the object in one pan, then add standard masses to the other pan until the beam balances in a level position. If you have to add 1.2 kg of standard masses, as in the diagram, then you know that the object also has a mass of 1.2 kg.

The balance is really comparing weights rather than masses. Weight is the downward pull of gravity. The beam balances when the downward pull on one pan is equal to the downward pull on the other. However, masses can be compared because of the way gravity acts on them. If the objects in the two pans have the same weight, they must also have the same mass.

When using a balance like the one above, you might say that you were 'weighing' something. However, 1.2 kg is the mass of the object, not its weight. Weight is a force, measured in force units called **newtons**. For more on this, and the difference between mass and weight, see spreads 2.7 and 2.9.

A more modern type of balance is shown on the left.

▲ A more modern type of balance. It detects the gravitational pull on the object on the pan, but gives its reading in units of mass.

1. On the Moon, the force of gravity on an object is only about one sixth of its value on Earth. Decide whether each of the following would give an accurate measurement of mass if used on the Moon.
   a A beam balance like the one in the diagram at the top of the page.
   b A balance like the one in the photograph above.

2. A balloon like the one on the opposite page contains 2000 m³ of air. When the air is cold, its density is 1.3 kg/m³. When heated, the air expands so that some is pushed out of the hole at the bottom, and the density falls to 1.1 kg/m³. Calculate the following.
   a The mass of air in the balloon when cold.
   b The mass of air in the balloon when hot.
   c The mass of air lost from the balloon during heating.

MEASUREMENTS AND UNITS

Planet density

The density of a planet increases towards the centre. However, the average density can be found by dividing the total mass by the total volume. The mass of a planet affects its gravitational pull and, therefore, the orbit of any moon circling it. The mass can be calculated from this. The volume can be calculated once the diameter is known.

The average density gives clues about a planet's structure:

Earth
Average density 5520 kg/m³

This is about double the density of the rocks near the surface, so the Earth must have a high density core – probably mainly iron.

Jupiter
Average density 1330 kg/m³

The low average density is one reason why scientists think that Jupiter is a sphere mostly of hydrogen and helium gas, with a small, rocky core.

not to scale

Float or sink?

You can tell whether a material will float or sink by comparing its density with that of the surrounding liquid (or gas). If it is less dense, it will float; if it is more dense, it will sink. For example, wood is less dense than water, so it floats; steel is more dense, so it sinks.

Oil is less dense than water, so it floats on water.

Density differences aren't the cause of floating or sinking, just a way of predicting which will occur. Floating is made possible by an upward force produced whenever an object is immersed in a liquid (or gas). To feel this force, try pushing an empty bottle down into water.

▲ Hot air is less dense than cold air, so a hot-air balloon will rise upwards – provided the fabric, gas cylinders, basket, and passengers do not increase the average density by too much.

▲ Ice is less dense than water in its liquid form, so icebergs float.

Related topics: mass **1.2**; volume and density **1.4–1.5**; force **2.6**; mass and weight **2.9**; convection **5.6**; densities of planets **11.2**

23

Check-up on measurements and units

Further questions

1 Copy and complete the table shown below:

measurement	unit	symbol
length	?	?
?	kilogram	?
?	?	s

[6]

2 Write down the number of
  A mg in 1 g
  B g in 1 kg
  C mg in 1 kg
  D mm in 4 km
  E cm in 5 km [5]

3 Write down the values of
  a 300 cm, in m
  b 500 g, in kg
  c 1500 m, in km
  d 250 ms, in s
  e 0.5 s, in ms
  f 0.75 km, in m
  g 2.5 kg, in g
  h 0.8 m, in mm [8]

4 The volume of a rectangular block can be calculated using this equation:
  volume = length × width × height
Using this information, copy and complete the table below. [4]

length/cm	width/cm	height/cm	volume of rectangular block/cm³
2	3	4	?
5	5	?	100
6	?	5	300
?	10	10	50

5 In each of the following pairs, which quantity is the larger?
  a 2 km or 2500 m?
  b 2 m or 1500 mm?
  c 2 tonnes or 3000 kg?
  d 2 litres or 300 cm³? [4]

6 Which of the following statements is/are correct?
  A One milligram equals one million grams.
  B One thousand milligrams equals one gram.
  C One million milligrams equals one gram.
  D One million milligrams equals one kilogram. [2]

7

m	g/cm³	m³	km	cm³
kg	ms	ml	kg/m³	s

Which of the above are
  a units of mass?
  b units of length
  c units of volume?
  d units of time?
  e units of density? [10]

8 Which block is made of the densest material?

block	mass/g	length/cm	breadth/cm	height/cm
A	480	5	4	4
B	360	10	4	3
C	800	10	5	2
D	600	5	4	3

[1]

9 The mass of a measuring cylinder and its contents are measured before and after putting a stone in it.

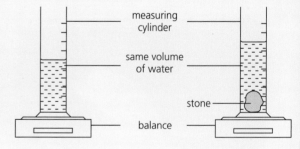

Which of the following could you calculate using measurements taken from the apparatus above?
  A the density of the liquid only
  B the density of the stone only
  C the densities of the liquid and the stone [2]

10 A plastic bag filled with air has a volume of 0.008 m³. When air in the bag is squeezed into a rigid container, the mass of the container (with air) increases from 0.02 kg to 0.03 kg. Use the formula

$$\text{density} = \frac{\text{mass}}{\text{volume}}$$

to calculate the density of the air in the bag. [2]

11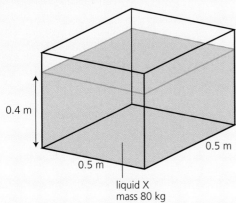
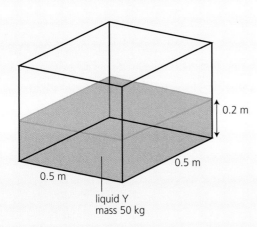

In the diagram above, the tanks contain two different liquids, X and Y.
a What is the volume of each liquid in m³? [2]
b If you had 1 m³ of the liquid X, what would its mass be? [2]
c What is the density of liquid X? [2]
d What is the density of liquid Y? [2]

12 Use the table of data on p19 (Spread 1.4) to answer the following:
a Which of the solids in the table will float in water? Explain how you made your decision. [5]
b Which solids in the table will float in petrol? [2]
c Petrol and water don't mix. If some water is tipped into petrol, what would you expect to happen? Explain your answer. [2]

13 The table shows the density of various substances.

substance	density/ g/cm³
copper	8.9
iron	7.9
kerosene	0.87
mercury	13.6
water	1.0

Consider the following statements:
A 1 cm³ of mercury has a greater mass than 1 cm³ of any other substance in this table – true or false?
B 1 cm³ of water has a smaller mass than 1 cm³ of any other substance in this table – true or false?
C 1 g of iron has a smaller volume than 1 g of copper – true or false?
D 1 g of mercury has a greater mass than 1 g of copper – true or false? [2]

14 A student decides to measure the period of a pendulum (the period is the time taken for one complete swing). Using a stopwatch, he finds that eight complete swings take 7.4 seconds. With his calculator, he then uses this data to work out the time for one swing. The number shown on his calculator is 0.925.
a Is it acceptable for the student to claim that the period of the pendulum is 0.925 seconds? Explain your answer. [2]
b How could the student measure the period more accurately? [2]
c Later, another student finds that 100 complete swings take 92.8 seconds. From these measurements, what is the period of the pendulum? [2]

MEASUREMENTS AND UNITS

Use the list below when you revise for your IGCSE examination. The spread number, in brackets, tells you where to find more information.

Revision checklist

Core Level
- [ ] How to use units. (1.1)
- [ ] Making bigger or smaller units using prefixes. (1.1)
- [ ] Writing numbers in scientific (standard) notation. (1.1)
- [ ] Significant figures. (1.1)
- [ ] SI units, including the metre, kilogram, and second. (1.2)
- [ ] How to measure lengths. (1.3)
- [ ] How to measure short intervals of time. (1.3)
- [ ] How to find the period of a simple pendulum. (1.3)
- [ ] Units for measuring volume. (1.4)
- [ ] How density is defined. (1.4)
- [ ] Using the equation linking density, mass, and volume. (1.4)
- [ ] Finding the volume of a regular solid. (1.5)
- [ ] Using a measuring cylinder to find the volume of a liquid. (1.5)
- [ ] Measuring the density of liquid. (1.5)
- [ ] Measuring the density of a regular solid. (1.5)
- [ ] How to use a displacement can. (1.5)
- [ ] Measuring the density of an irregular solid. (1.5)
- [ ] How to compare masses with a beam balance. (1.6)

Extended Level
As for Core Level, plus the following:
- [ ] Use density data to predict whether a material will sink or float. (1.6)

2
Forces and motion

- SPEED AND VELOCITY
- ACCELERATION
- FREE FALL
- FORCE AND MASS
- FRICTION
- GRAVITY
- MOMENTUM
- VECTORS AND SCALARS
- CIRCULAR MOTION

A bungee jumper leaps more than 180 metres from the top of the Sky Tower in Auckland, New Zealand. With nothing to oppose his fall, he would hit the ground at a speed of 60 metres per second. However, his fall is slowed by the resistance of the air rushing past him, and eventually stopped by the pull of the bungee rope. Side ropes are also being used to stop him crashing into the tower.

2.1 Speed, velocity, and acceleration

Objectives: to know how speed, velocity, acceleration and deceleration are defined – to explain what vectors are.

▲ *Thrust* supersonic car travelling faster than sound. For speed records, cars are timed over a measured distance (either one kilometre or one mile). The speed is worked out from the average of two runs – down the course and then back again – so that the effects of wind are cancelled out.

Travel times
time taken to travel 1 kilometre (1000 m)

Runner — 150 s
Grand Prix car — 10 s
Passenger jet — 4 s
Sound — 3 s
International Space Station — 0.13 s

Speed

If a car travels between two points on a road, its average speed can be calculated like this:

$$\text{average speed} = \frac{\text{distance moved}}{\text{time taken}} \qquad \text{In symbols: } v = \frac{s}{t}$$

If distance is measured in metres (m) and time in seconds (s), speed is measured in metres per second (m/s). For example: if a car moves 90 m in 3 s, its average speed is 30 m/s.

On most journeys, the speed of a car varies, so the actual speed at any moment is usually different from the average speed. To find an actual speed, you need to discover how far the car moves in the shortest time you can measure. For example, if a car moves 0.20 metres in 0.01 s:

$$\text{speed} = \frac{0.20 \text{ m}}{0.01 \text{ s}} = 20 \text{ m/s}$$

Velocity

Velocity means the speed of something *and* its direction of travel. For example, a cyclist might have a velocity of 10 m/s due east. On paper, this velocity can be shown using an arrow:

10 m/s →

For motion in a straight line you can use a + or − to indicate direction. For example:

+10 m/s (velocity of 10 m/s *to the right*)

−10 m/s (velocity of 10 m/s *to the left*)

Note: +10 m/s may be written without the +, just as 10 m/s.

Ⓔ Quantities, such as velocity, which have a direction as well as a magnitude (size) are called **vectors**.

FORCES AND MOTION

Acceleration

Something is accelerating if its velocity is *changing*. Acceleration is calculated like this:

$$\text{average acceleration} = \frac{\text{change in velocity}}{\text{time taken}}$$

In symbols: $$a = \frac{\Delta v}{\Delta t}$$

The symbol Δ stands for 'change in'.

For example, if a car increases its velocity from zero to 12 m/s in 4 s:

average acceleration = 12/4 = 3 m/s² (omitting some units for simplicity)

Note that acceleration is measured in metres per second² (m/s²).

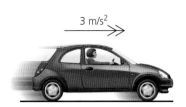

time	velocity
0 s	0 m/s
1 s	3 m/s
2 s	6 m/s
3 s	9 m/s
4 s	12 m/s

The velocity of this car is increasing by 3 m/s every second. The car has a steady acceleration of 3 m/s².

Acceleration is a vector. It can be shown using an arrow (usually double-headed). Alternatively, a + or − sign can be used to indicate whether the velocity is increasing or decreasing. For example:

+3 m/s² (velocity *increasing* by 3 m/s every second)

−3 m/s² (velocity *decreasing* by 3 m/s every second)

A *negative* acceleration is called a **deceleration** or a **retardation**.

A *uniform* acceleration means a constant (steady) acceleration.

Solving a problem

Example The car on the right passes post A with a velocity of 12 m/s. If it has a steady acceleration of 3 m/s², what is its velocity 5 s later, at B?

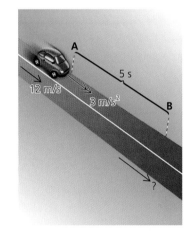

The car is gaining 3 m/s of velocity every second. So in 5 s, it gains an extra 15 m/s on top of its original 12 m/s. Therefore its final velocity is 27 m/s. Note that the result is worked out like this:

final velocity = original velocity + extra velocity

So: final velocity = original velocity + (acceleration × time)

The above equation also works for retardation. If a car has a retardation of 3 m/s², you treat this as an acceleration of −3 m/s².

1. A car travels 600 m in 30 s. What is its average speed? Why is its actual speed usually different from its average speed?
2. How is velocity different from speed?
3. A car has a steady speed of 8 m/s.
   a. How far does the car travel in 8 s?
   b. How long does the car take to travel 160 m?
4. Calculate the average speed of each thing in the chart of travel times on the opposite page.
5. A car has an acceleration of +2 m/s². What does this tell you about the velocity of the car? What is meant by an acceleration of −2 m/s²?
6. A car takes 8 s to increase its velocity from 10 m/s to 30 m/s. What is its average acceration?
7. A motor cycle, travelling at 20 m/s, takes 5 s to stop. What is its average retardation?
8. An aircraft on its take-off run has a steady acceleration of 3 m/s².
   a. What velocity does the aircraft gain in 4 s?
   b. If the aircraft passes one post on the runway at a velocity of 20 m/s, what is its velocity 8 s later?
9. A truck travelling at 25 m/s puts its brakes on for 4 s. This produces a retardation of 2 m/s². What does the truck's velocity drop to?

2.2 Motion graphs

Objectives: to interpret distance-time and speed-time graphs – to know how to calculate speed, acceleration, and distance travelled from such graphs.

Distance–time graphs

Graphs can be useful when studying motion. Below, a car is travelling along a straight road, away from a marker post. The car's distance from the post is measured every second. The charts and graphs show four different examples of what the car's motion might be.

On a graph, the line's rise on the vertical scale divided by its rise on the horizontal scale is called the **gradient**, as shown on the left. With a distance–time graph, the gradient tells you how much extra distance is travelled every second. So:

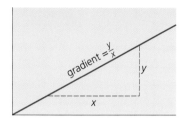

▲ On a straight line graph like this, the gradient has the same value wherever you measure y and x.

> On a distance–time graph, the gradient of the line is numerically equal to the speed.

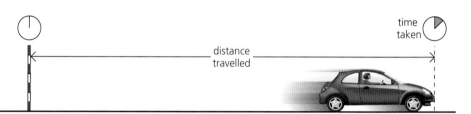

A Car travelling at **steady speed**

time/ s	0	1	2	3	4	5
distance/ m	0	10	20	30	40	50

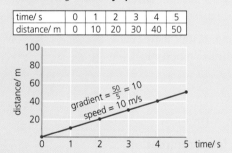

The line rises 10 m on the distance scale for every 1 s on the time scale.

B Car travelling at **higher steady speed**

time/ s	0	1	2	3	4	5
distance/ m	0	20	40	60	80	100

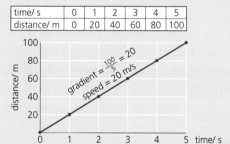

The line is steeper than before. It rises 20 m on the distance scale for every 1 s on the time scale.

C Car **accelerating**

time/ s	0	1	2	3	4	5
distance/ m	0	10	25	45	70	100

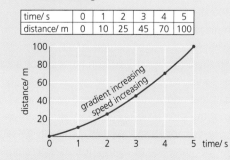

The speed rises. So the car travels further each second than the one before, and the line curves upwards.

D Car **stopped**

time/ s	0	1	2	3	4	5
distance/ m	50	50	50	50	50	50

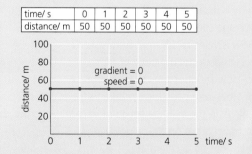

The car is parked 50 m from the post, so this distance stays the same.

FORCES AND MOTION

Speed–time graphs

Each speed–time graph below is for a car travelling along a straight road. The gradient tells you how much extra speed is gained every second. So:

> On a speed–time graph, the gradient of the line is numerically equal to the acceleration.

In graph E, the car travels at a steady 15 m/s for 5 s, so the distance travelled is 75 m. The area of the shaded rectangle, calculated using the scale numbers, is also 75. This principle works for more complicated graph lines as well. In graph F, the area of the shaded triangle, $\frac{1}{2} \times$ base $\times$ height, equals 50. So the distance travelled is 50 metres.

> On a speed–time graph, the area under the line is numerically equal to the distance travelled.

!> **Velocity–time graphs**
Velocity is speed in a particular direction.
Where there is no change in the direction of motion, a velocity–time graph looks the same as a speed–time graph.

E Car travelling at **steady speed**

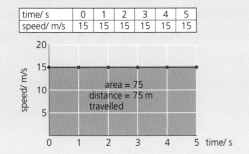

The speed stays the same, so the line stays at the same level.

F Car with **steady acceleration**

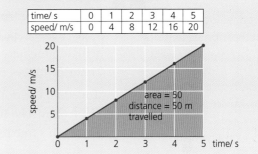

As the car gains speed, the line rises 4 m/s on the speed scale for every 1 s on the time scale.

Q

1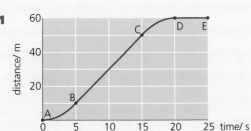

The distance–time graph above is for a motor cycle travelling along a straight road.
  a What is the motor cycle doing between points D and E on the graph?
  b Between which points is it accelerating?
  c Between which points is its speed steady?
  d What is this steady speed?
  e What is the distance travelled between A and D?
  f What is the average speed between A and D?

2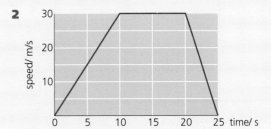

The speed–time graph above is for another motor cycle travelling along the same road.
  a What is the motor cycle's maximum speed?
  b What is the acceleration during the first 10 s?
  c What is its deceleration during the last 5 s?
  d What distance is travelled during the first 10 s?
  e What is the total distance travelled?
  f What is the time taken for the whole journey?
  g What is the average speed for the whole journey?

Related topics: speed, velocity, and acceleration 2.1

2.3 Recording motion

Objective: to explain how, in the lab, the acceleration of a trolley can be found by measuring the distances between dots on ticker tape.

Using ticker-tape

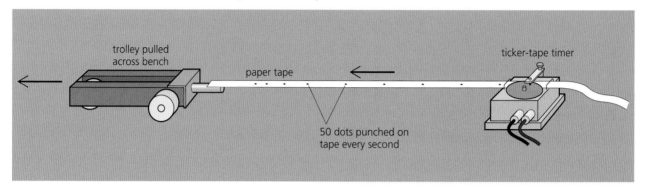

Speed, velocity, and acceleration essentials

$$\text{speed}^* = \frac{\text{distance moved}}{\text{time taken}}$$

velocity is speed in a particular direction

$$\text{acceleration}^* = \frac{\text{change in velocity}}{\text{time taken}}$$

*average

In the laboratory, motion can be investigated using a trolley like the one above. As the trolley travels along the bench, it pulls a length of paper tape (ticker-tape) behind it. The tape passes through a ticker-tape timer which punches carbon dots on the tape at regular intervals. A typical timer produces 50 dots every second.

Together, the dots on the tape form a complete record of the motion of the trolley. The further apart the dots, the faster the trolley is moving. Here are some examples:

steady speed — distance between dots stays the same

higher steady speed — distance between dots greater than before

acceleration — distance between dots increases

acceleration ———— then ———— retardation

▼ Motion can also be recorded photographically. These images of the Sun were taken at regular intervals, at midsummer, in Alaska. Even at midnight, the Sun is still above the horizon.

Calculations from tape

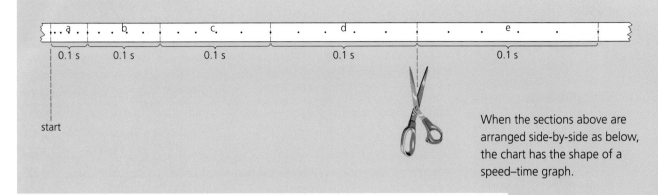

When the sections above are arranged side-by-side as below, the chart has the shape of a speed–time graph.

The ticker-tape record above is for a trolley with steady acceleration.

The tape has been marked off in sections 5 dot-spaces long. One dot-space is the distance travelled by the trolley in 1/50 second (0.02 s). So 5 dot-spaces is the distance travelled in 1/10 second (0.1 s).

If the tape is chopped up into its 5 dot-space sections, and the sections put side-by-side in order, the result is a chart like the one on the right. The chart is the shape of a speed–time graph. The lengths of the sections represent speeds because the trolley travels further in each 0.1 s as its speed increases. Side-by-side, the sections provide a time scale because each section starts 0.1 s after the one before.

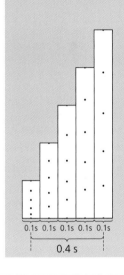

The acceleration of the trolley can be found from measurements on the tape. Do questions 2 and 3 below to discover how.

1

Describe the motion of the trolley that produced the ticker-tape record above.

2 The dots on the tape below were made by a ticker-tape timer producing 50 dots per second.
  a Count the number of dot-spaces between A and B. Then calculate the time it took the tape to move from A to B.
  b Using a ruler, measure the distance from A to B in mm. Then calculate the average speed of the trolley between A and B, in mm/s.
  c Measure the distance from C to D, then calculate the average speed of the trolley between C and D.
  d Section CD was completed exactly one second after section AB. Calculate the acceleration of the trolley in mm/s^2.

3 Look at the chart above.
  a Using a ruler, measure the distance travelled by the trolley in the first 0.1 s recorded on the tape.
  b Calculate the trolley's average speed during this first 0.1 s.
  c Measure the distance travelled by the trolley in the last 0.1 s recorded on the tape.
  d Calculate the average speed during this last 0.1 s.
  e Calculate the gain in speed during the 0.4 s.
  f Calculate the acceleration of the trolley in mm/s^2.

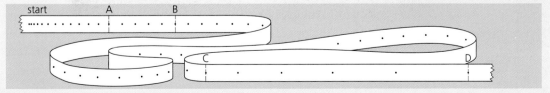

Related topics: speed, velocity, and acceleration 2.1; motion graphs 2.2 and 2.5

2.4 Free fall

Objectives: to describe how objects accelerate when dropped (with no air resistance) - to explain what is meant by the acceleration of free fall.

The acceleration of free fall, *g*

If you drop a lead weight and a feather, both fall downwards because of gravity. However, the feather is slowed much more by the air.

The diagram on the left shows what would happen if there were no air resistance. Both objects would fall with the same downward acceleration: 9.8 m/s². This is called the **acceleration of free fall**. It is the same for *all* objects falling near the Earth's surface, light and heavy alike.

The acceleration of free fall is represented by the symbol **g**. Its value varies slightly from one place on the Earth's surface to another, because the Earth's gravitational pull varies. However, the variation is less than 1%. Moving away from the Earth and out into space, *g* decreases.

Note that the value of *g* near the Earth's surface is close to 10 m/s². This simple figure is accurate enough for many calculations, and will be the one used in this book.

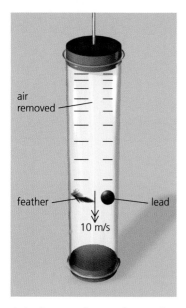

▲ In the experiment above, all the air has been removed from the tube. Without air resistance, a light object falls with the same acceleration as a heavy one.

▲ On the Moon, the acceleration of free fall is only 1.6 m/s². And as there is no atmosphere, a feather would fall with the same acceleration as a lead weight.

Measuring *g**

An experiment to find *g* is shown on the left. The principle is to measure the time taken for a steel ball to drop through a known height, and to calculate the acceleration from this. Air resistance has little effect on a small, heavy ball falling only a short distance, so the ball's acceleration is effectively *g*.

The ball is dropped by cutting the power to the electromagnet. The electronic timer is automatically switched on when the ball passes through the upper light beam, and switched off when it passes through the lower beam. If the height of the fall is *h* and the time taken is *t*, then *g* can be calculated using this equation (derived from other equations):

$$g = \frac{2h}{t^2}$$

▲ Experiment to measure *g*

FORCES AND MOTION

E Up and down

In the following example, assume that g is 10 m/s², and that there is no air resistance.

The ball on the right is thrown upwards with a velocity of 30 m/s. The diagram shows the velocity of the ball every second as it rises to its highest point and then falls back to where it started.

As an *upward* velocity of 30 m/s is the same as a *downward* velocity of −30 m/s, the motion of the ball can be described like this:

At 0 s....   the downward velocity is − m/s
After 1 s.... the downward velocity is −20 m/s
After 2 s.... the downward velocity is −10 m/s   } 10 m/s is being added to the downward velocity every second
After 3 s.... the downward velocity is   0 m/s
After 4 s.... the downward velocity is +10 m/s
After 5 s.... the downward velocity is +20 m/s
After 6 s.... the downward velocity is +30 m/s

Whether the ball is travelling up or down, it is gaining downward velocity at the rate of 10 m/s per second. So it always has a downward acceleration of 10 m/s², which is g. Even when the ball is moving upwards, or is stationary at its highest point, it still has downward acceleration.

Below, you can see a velocity–time graph for the motion.

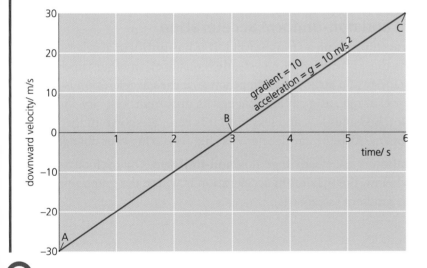

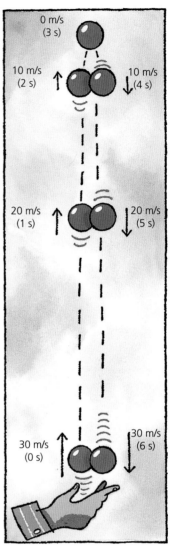

▲ A ball in flight. As g is 10 m/s², the ball's velocity changes by 10 m/s every second.

◀ The velocity–time graph for the ball's motion is shown on the left.

Q

Assume that g = 10 m/s² and that there is no air resistance.

1 A stone is dropped from rest. What is its speed
   a after 1 s   b after 2 s   c after 5 s?

2 A stone is thrown downwards at 20 m/s. What is its speed
   a after 1 s   b after 2 s   c after 5 s?

3 A stone is thrown upwards at 20 m/s. What is its speed
   a after 1 s   b after 2 s   c after 5 s?

4 This question is about the three points, A, B, and C, on the graph above left.
   a In which direction is the ball moving at point C?
   b At which point is the ball stationary?
   c At which point is the ball at its maximum height?
   d What is the ball's acceleration at point C?
   e What is the ball's acceleration at point A?
   f What is the ball's acceleration at point B?
   g At which point does the ball have the same speed as when it was thrown?

Related topics: acceleration 2.1; motion graphs 2.2 and 2.5; gravitational force 2.9

2.5 More motion graphs

Objective: to know how to tell from the shape of a speed-time graph whether the acceleration is uniform (steady) or non-uniform (varying).

Motion graph essentials

Here are four examples of velocity–time graphs for a car travelling along a straight line:

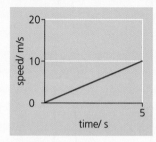 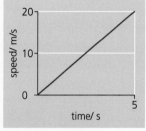 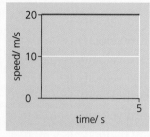 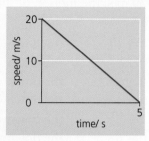

Steady acceleration of 2 m/s²
The speed of the car increases by 2 m/s every second.
The initial speed is zero, so the car is starting from rest.

Steady acceleration of 4 m/s²
The speed of the car increases by 4 m/s every second.
The initial speed is zero, so the car is starting from rest.

Zero acceleration
The car has a steady speed of 20 m/s.

Steady retardation (deceleration) of 4 m/s²
The speed of the car decreases by 4 m/s². In other words: the acceleration is –4 m/s².
The final speed is zero, so the car comes to rest.

Uniform and non-uniform acceleration

A car is travelling along a straight road. If it has **uniform** acceleration, this means that its acceleration is steady (constant). In other words, it is gaining velocity at a steady rate. In practice, a car's acceleration is rarely steady. For example, as a car approaches its maximum velocity, the acceleration becomes less and less until it is zero, as shown in the example below. Also the car decelerates slightly during gear changes.

If acceleration is not steady then it is **non-uniform**. On a velocity–time graph, as below, the maximum acceleration is where the graph line has its highest gradient (steepness).

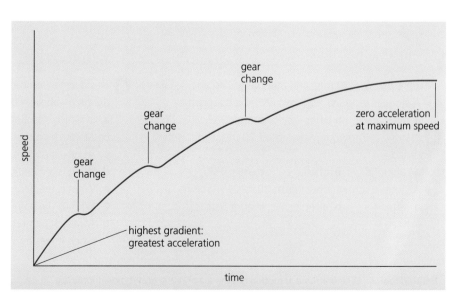

FORCES AND MOTION

Here are more examples of uniform and non-uniform acceleration:

A stone is dropped from a great height. With no air resistance, the velocity–time graph for the stone would be as shown below left. The acceleration would be uniform. It would be 10 m/s², the acceleration of free fall, g.

E In practice, there is air resistance on the stone. This affects its motion, producing non-uniform acceleration, as shown below right. At the instant the stone is dropped, it has no velocity. This means that its initial acceleration is g because there is not yet any air resistance on it. However, as the velocity increases, air resistance also increases. Eventually, the air resistance is so great that the velocity reaches a maximum and the acceleration falls to zero.

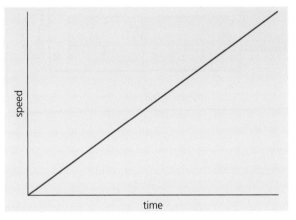

▲ Uniform acceleration of a falling stone with no air resistance acting.

▲ Non-uniform acceleration of a falling stone with air resistance acting.

On a speed–time graph, the area under the line is numerically equal to the distance travelled. This applies whether the motion is uniform or non-uniform – in other words, whether the graph line is straight or curved. With a straight-line graph, the area can be calculated. With a curved-line graph, this may not be possible, although an estimate can be made by counting squares. When doing this, remember that the area must be worked out using the scale numbers on the axis. It isn't the 'real' area on the paper.

1 A boat moves off from its mooring in a straight line. A speed–time graph for its motion is shown on the right. The graph has been divided into sections, AB, BC, CD, and DE. Over which section (or sections) of the graph does the boat
   a have its greatest speed?
   b have its greatest acceleration?
   c have retardation?
   d have uniform acceleration or retardation?
   e have non-uniform acceleration or retardation?
   f travel the greatest distance?
2 Sketch a speed–time graph for a beach-ball falling from a great height. How will this graph differ from that for a falling stone, shown above right?

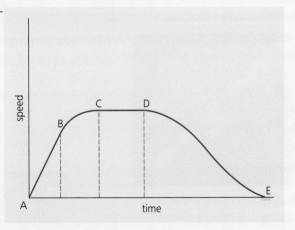

Related topics: speed, velocity, and acceleration 2.1; motion graphs 2.2; g and free fall 2.4

2.6 Forces in balance

Objectives: to know how force is measured – to describe the motion an object if there is no force, or balanced forces, on it – to explain what terminal velocity is.

Typical forces in newtons

force to switch on a
  bathroom light....... 10 N

force to pull
  open a drinks can...... 20 N

force to lift
  a heavy suitcase..... 200 N

force from a large
  jet engine.... 250 000 N

A **force** is a push or a pull, exerted by one object on another. It has direction as well as magnitude (size), so it is a vector.

The SI unit of force is the **newton (N)**. Small forces can be measured using a spring balance like the one below. The greater the force, the more the spring is stretched and the higher the reading on the scale:

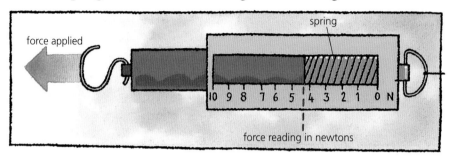

Common forces

Here are some examples of forces:

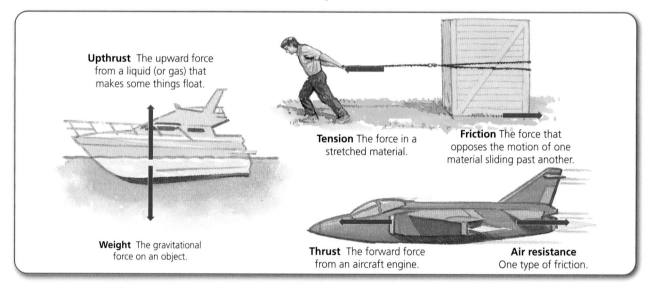

Upthrust The upward force from a liquid (or gas) that makes some things float.

Weight The gravitational force on an object.

Tension The force in a stretched material.

Friction The force that opposes the motion of one material sliding past another.

Thrust The forward force from an aircraft engine.

Air resistance One type of friction.

Motion without force

On Earth, unpowered vehicles soon come to rest because of friction. But with no friction, gravity, or other external force on it, a moving object will keep moving for ever – at a steady speed in a straight line. It doesn't need a force to keep it moving.

This idea is summed up in a law first put forward by Sir Isaac Newton in 1687:

▲ Deep in space with no forces to slow it, a moving object will keep moving for ever.

> If no external force is acting on it, an object will
> – if stationary, remain stationary
> – if moving, keep moving at a steady speed in a straight line.

This is known as **Newton's first law of motion**.

FORCES AND MOTION

Balanced forces

An object may have several forces on it. But if the forces are in balance, they cancel each other out. Then, the object behaves as if there is no force on it at all. Here are some examples:

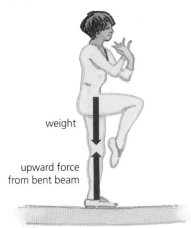
Stationary gymnast

Skater with steady velocity

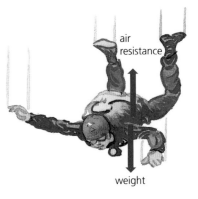
Skydiver with steady velocity

With balanced forces on it, an object is *either* at rest, *or* moving at a steady velocity (steady speed in a straight line). That follows from Newton's first law.

E Terminal velocity

When a skydiver falls from a hovering helicopter, as her speed increases, the air resistance on her also increases. Eventually, it is enough to balance her weight, and she gains no more speed. She is at her **terminal velocity**. Typically, this is about 60 m/s, though the actual value depends on air conditions, as well as the size, shape, and weight of the skydiver.

When the skydiver opens her parachute, the extra area of material increases the air resistance. She loses speed rapidly until the forces are again in balance, at a greatly reduced terminal velocity.

If air resistance balances her weight, why doesn't a skydiver stay still? If she wasn't moving, there wouldn't be any air resistance. And with only her weight acting, she would gain velocity.

Surely, if she is travelling downwards, her weight must be greater than the air resistance? Only if is she is gaining velocity. At a steady velocity, the forces must be in balance. That follows from Newton's first law.

▲ If a skydiver is falling at a steady velocity, the forces on her are balanced: her weight downwards is exactly matched by the air resistance upwards.

1. What is the SI unit of force?
2. What does Newton's first law of motion tell you about the forces on an object that is **a** stationary **b** moving at a steady velocity?
3. The parachutist on the right is descending at a steady velocity.
   a What name is given to this velocity?
   b Copy the diagram. Mark in and label another force acting.
   c How does this force compare with the weight?
   d If the parachutist used a larger parachute, how would this affect the steady velocity reached? Explain why.

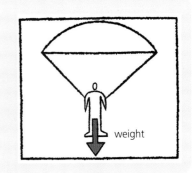

Related topics: friction and moving vehicles 2.8; weight and mass 2.9

2.7 Force, mass, and acceleration

Objectives: to know and use the equation linking force, mass, and acceleration – to find the resultant of two forces acting in a straight line.

Inertia and mass

▶ Once a massive ship like this is moving, it is extremely difficult to stop.

If an object is at rest, it takes a force to make it move. If it is moving, it takes a force to make it go faster, slower, or in a different direction. So all objects resist a change in velocity – even if the velocity is zero. This resistance to change in velocity is called **inertia**. The more mass something has, the more inertia it has.

Velocity is speed in a particular direction.

Any change in velocity is an acceleration. So the more mass something has, the more difficult it is to make it accelerate.

Resultant force

In the diagram on the left, the two forces are unbalanced. Together, they are equivalent to a single force. This is called the **resultant force**.

These two forces...

3 N ← → 5 N

are equivalent to a single force of (5−3) N....

→ 2 N

This is the **resultant force**.

If forces are balanced, the resultant force is zero and there is no acceleration. Any other resultant force causes an acceleration – in the same direction as the resultant force.

ⒺLinking force, mass, and acceleration

There is a link between the resultant force acting, the mass, and the acceleration produced. For example:

If this resultant force...	acts on this mass...	then this is the acceleration...
1 N	1 kg	1 m/s^2
2 N	2 kg	1 m/s^2
4 N	2 kg	2 m/s^2
6 N	2 kg	3 m/s^2

In all cases, the following equation applies:

$$\text{resultant force} = \text{mass} \times \text{acceleration}$$

Symbols and units
F = force, in newtons (N)
m = mass, in kilograms (kg)
a = acceleration, in metres/second^2 (m/s^2)

In symbols: $$F = ma$$

This relationship between force, mass, and acceleration is sometimes called **Newton's second law of motion**.

FORCES AND MOTION

Example What is the acceleration of the model car on the right?

First, work out the resultant force on the car. A force of 18 N to the *right* combined with a force of 10 N to the *left* is equivalent to a force of (18 − 10) N to the *right*. So the resultant force is 8 N.

Next, work out the acceleration when $F = 8$ N and $m = 2$ kg:

$F = ma$

So: $8 = 2a$ (omitting units for simplicity)

Rearranged, this gives $a = 4$. So the car's acceleration is 4 m/s².

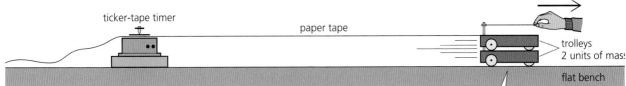

Finding the link

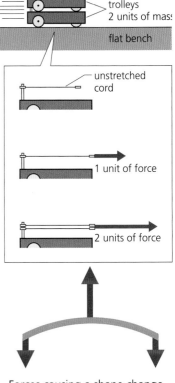

The link between force, mass, and acceleration can be found experimentally using the equipment above. Different forces are applied to the trolley by pulling it along with one, two, or three elastic cords, stretched to the same length each time. During each run, the ticker-tape timer marks a series of dots on the paper tape. The acceleration can be calculated from the spacing of the dots. To vary the mass, one, two, or three trolleys are used in a stack.

Defining the newton

A 1 N resultant force acting on 1 kg produces an acceleration of 1 m/s². This simple result is no accident. It arises from the way the newton is defined:

> 1 newton is the force required to give a mass of 1 kilogram an acceleration of 1 m/s².

Further effects of forces

Forces do not only affect motion. If two or more forces act on something, they change its shape or volume (or both). The effect is slight with hard objects, but can be very noticeable with flexible ones, as shown on the right.

Forces causing a shape change

1 a What equation links resultant force, mass, and acceleration?
  b Use this equation to calculate the resultant force on each of the stones shown below.

2 a What is the resultant force on the car below?
  b What is the car's acceleration?
  c If the total frictional force rises to 1500 N, what happens to the car?

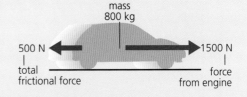

Related topics: mass 1.2; acceleration 2.1; using ticker-tape 2.3; balanced forces 2.6; stretching and compressing 3.4

41

2.8 Friction

Objectives: to describe what friction is and when it causes heating – to give different examples of friction and describe when it is and isn't useful.

reducing friction
roller bearing + grease

brake pad   tyre gripping road
using friction

▲ This wheel is mounted on roller bearings to reduce friction.

▼ Air resistance is a form of dynamic friction. When a car is travelling fast, it is the largest of all the frictional forces opposing motion.
Air resistance wastes energy, so less air resistance means better fuel consumption. Car bodies are specially shaped to smooth the air flow past them and reduce air resistance. A low frontal area also helps.

Friction is the force that tries to stop materials sliding across each other. There is friction between your hands when you rub them together, and friction between your shoes and the ground when you walk along. Friction prevents machinery from moving freely and heats up its moving parts. To reduce friction, wheels are mounted on ball or roller bearings, with oil or grease to make the moving surfaces slippery.

Friction is not always a nuisance. It gives shoes and tyres grip on the ground, and it is used in most braking systems. On a bicycle, for example, rubber blocks are pressed against the wheels to slow them down.

Two kinds of friction

When the block below is pulled gently, friction stops it moving. As the force is increased, the friction rises until the block is about to slip. This is the starting or **static** friction. With a greater downward force on the block, the static friction is higher. Once the block starts to slide, the friction drops: moving or **dynamic** friction is less than static friction.

 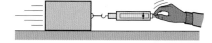

static friction is greater than. ..   .. dynamic friction

Dynamic friction heats materials up. When something is moved against the force of friction, its energy of motion (called kinetic energy) is converted into thermal energy (heat). Brakes and other machinery must be designed so that they get rid of this thermal energy. Otherwise their moving parts may become so hot that they seize up.

FORCES AND MOTION

Drag

Objects experience friction when they move through a liquid or a gas. This force is called drag. **Drag** (also know as air resistance) acts on an aircraft as it moves through the air. Drag acts on a boat as it moves across water. And if you drop a pebble into deep water, drag slows its descent.

Friction highs and lows

As the Earth moves through space, it runs into small bits of material, also orbiting the Sun. These mostly range in size from grains of sand to small pebbles, and they can hit the atmosphere at speeds of up to 70 km/s (150 000 mph). Frictional heating makes them burn up, causing a streak of light called a meteor (or 'shooting star'), as on the right. Sometimes, the burning produces a fireball.

Below are more examples of friction in action.

▲ A curling stone slides across the ice towards a target. To make the stone travel further, the sweepers brush vigorously in front of it with brooms. Friction from the brooms has a heating effect which melts some of the ice. The melting layer reduces friction under the stone.

▲ The top of a surfboard is often given a wax coating. Tiny bumps of wax increase friction by sticking to the surfer's feet. However, the underside of a surfboard has a smooth, glassy surface so that it can slide across the water with as little friction as possible.

1. In a car, friction is essential in some parts, but needs to be reduced in others. Give *two* examples of where friction is
   a essential   b needs to be reduced.
2. Why are car bodies designed so that air resistance is reduced as much as possible?
3. Comparing the *top* and *bottom* of a surfboard:
   a On which surface does the friction need to be high? Explain why.
   b On which surface does the friction need to be low? Explain why.
4. Write down whether, in each of the following examples, the friction has a heating effect:
   a The soles of your shoes gripping the ground when you are standing on a slope.
   b A crate being dragged across the ground.

Related topics: speed **2.01**; thermal energy **4.1**; energy transfers **4.2**

2.9 Force, weight, and gravity

Objectives: – to understand the connection between weight and mass – to explain what gravitational field strength is and how it can vary from one place to another.

▲ Near the Earth's surface, a 1 kg mass has a gravitational force on it of about 10 newtons. This is its weight.

Gravitational force

If you hang an object from a spring balance, you measure a downward pull from the Earth. This pull is called a **gravitational force**.

No one is sure what causes gravitational force, but here are some of its main features:

- *All* masses attract each other.
- The greater the masses, the stronger the force.
- The closer the masses, the stronger the force.

The pull between small masses is extremely weak. It is less than 10^{-7} N between you and this book! But the Earth is so massive that its gravitational pull is strong enough to hold most things firmly on the ground.

Weight

Weight is another name for the Earth's gravitational force on an object. Like other forces, it is measured in newtons (N).

Near the Earth's surface, an object of mass 1 kg has a weight of 9.8 N, though 10 N is accurate enough for many calculations and will be used in this book. Greater masses have greater weights. Here are some examples:

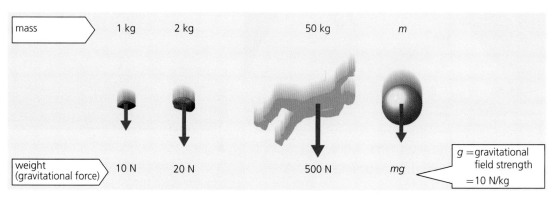

mass	1 kg	2 kg	50 kg	m
weight (gravitational force)	10 N	20 N	500 N	mg

g = gravitational field strength = 10 N/kg

Symbols and units

W = weight, in newtons (N)
m = mass, in kilograms (kg)
g = gravitational field strength, 10 N/kg near the Earth's surface

Gravitational field strength, g

A **gravitational field** is a region in which a mass experiences a force due to gravitational attraction. The Earth has a gravitational field around it. Near the surface, this exerts a force of 10 newtons on each kilogram of mass: the Earth's **gravitational field strength** is 10 newtons per kilogram (N/kg).

Gravitational field strength is represented by the symbol **g**. So:

$$\text{weight} = \text{mass} \times g \qquad (g = 10 \text{ N/kg})$$

In symbols: $W = mg$

In everyday language, we often use the word 'weight' when it should be 'mass'. Even balances, which detect weight, are normally marked in mass units. But the person in the diagram above doesn't 'weigh' 50 kilograms. He has a *mass* of 50 kilograms and a *weight* of 500 newtons.

FORCES AND MOTION

Example What is the acceleration of the rocket on the right?

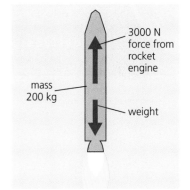

To find the acceleration, you need to know the resultant force on the rocket. And to find that, you need to know the rocket's weight:

$$\text{weight} = mg = 200 \text{ kg} \times 10 \text{ N/kg} = 2000 \text{ N}$$

So: resultant force (upwards) = 3000 N - 2000 N = 1000 N

But: resultant force = mass × acceleration
So: 1000 N = 200 kg × acceleration

Rearranged, this gives: acceleration = 5 m/s²

Changing weight, fixed mass

On the Moon, your weight (in newtons) would be less than on Earth, because the Moon's gravitational field is weaker.

Even on Earth, your weight can vary slightly from place to place, because the Earth's gravitational field strength varies. Moving away from the Earth, your weight decreases. If you could go deep into space, and be free of any gravitational pull, your weight would be zero.

Whether on the Earth, on the Moon, or deep in space, your body always has the same resistance to a change in motion. So your mass (in kg) doesn't change – at least, not under normal circumstances. But...

According to Einstein's theory of relativity, mass *can* change. For example, it increases when an object gains speed. However, the change is far too small to detect at speeds much below the speed of light. For all practical purposes, you can assume that mass is constant.

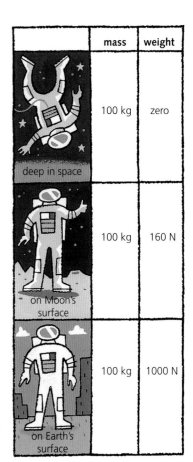

Two meanings for *g*

In the diagram opposite, the acceleration of each object can be worked out using the equation force = mass × acceleration. For example, the 2 kg mass has a 20 N force on it, so its acceleration is 10 m/s².

You get the same result for all the other objects. In each case, the acceleration works out at 10 m/s², or *g* (where *g* is the Earth's gravitational field strength, 10 N/kg).

So *g* has *two* meanings. In both cases, **g** is a vector:
- *g* is the gravitational field strength (10 newtons per kilogram).
- *g* is the acceleration of free fall (10 metres per second²).

Assume that $g = 10$ N/kg and there is no air resistance.

1 The rocks above are falling near the Earth's surface.
  a What is the weight of each rock?
  b What is the gravitational field strength?
  c What is the acceleration of each rock?

2 A spacecraft travels from Earth to Mars, where the gravitational field strength near the surface is 3.7 N/kg. The spacecraft is carrying a probe which has a mass of 100 kg when measured on Earth.
  a What is the probe's weight on Earth?
  b What is the probe's mass in space?
  c What is the probe's mass on Mars?
  d What is the probe's weight on Mars?
  e When the probe is falling, near the surface of Mars, what is its acceleration?

Related topics: kg 1.1; vectors 2.1 and 2.13; resultant force and acceleration 2.7; energy and mass 10.6

2.10 Action and reaction*

Objectives: – to understand that forces always occur in pairs, and the effects that this can have.

Action–reaction pairs

A single force cannot exist by itself. Forces are always pushes or pulls between *two* objects. So they always occur in pairs.

The experiment below shows the effect of a pair of forces. To begin with, the two trolleys are stationary. One of them contains a spring-loaded piston which shoots out when a release pin is hit.

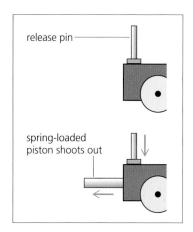

When the piston is released, the trolleys shoot off in opposite directions. Although the piston comes from one trolley only, two equal but opposite forces are produced, one acting on each trolley. The paired forces are known as the **action** and the **reaction**, but it doesn't matter which you call which. One cannot exist without the other.

Here are some more examples of action–reaction pairs:

If forces always occur in pairs, why don't they cancel each other out?
The forces in each pair act on *different* objects, not the same object.

If a skydiver is pulled downwards, why isn't the Earth pulled upwards?
It is! But the Earth is so massive that the upward force on it has far too small an effect for any movement to be detected.

FORCES AND MOTION

Newton's third law of motion

Isaac Newton was the first person to point out that every force has an equal but opposite partner acting on a different object. This idea is summed up by **Newton's third law of motion**:

> If object A exerts a force on object B, then object B will exert an equal but opposite force on object A.

Here is another way of stating the same law:

> To every action there is an equal but opposite reaction.

Rockets and jets

Rockets use the action–reaction principle. A rocket engine gets thrust in one direction by pushing out a huge mass of gas very quickly in the opposite direction. The gas is produced by burning fuel and oxygen. These are either stored as cold liquids, or the fuel may be stored in chemical compounds which have been compressed into solid pellets.

How can a rocket accelerate through space if there is nothing for it to push against? It *does* have something to push against – the huge mass of gas from its burning fuel and oxygen. Fuel and oxygen make up over 90% of the mass of a fully loaded rocket.

Jet engines also get thrust by pushing out a huge mass of gas. But the gas is mostly air that has been drawn in at the front:

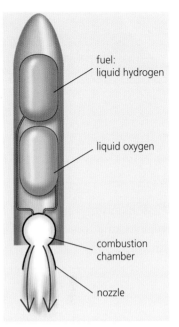

▲ A rocket engine. In the combustion chamber, a huge mass of hot gas expands and rushes out of the nozzle. The gas is produced by burning fuel and oxygen.

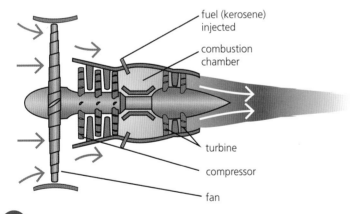

◀ A jet engine. The big fan at the front pushes out a huge mass of air. However, some of the air doesn't come straight out. It is compressed and used to burn fuel in a combustion chamber. As the hot exhaust gas expands, it rushes out of the engine, pushing round a turbine as it goes. The spinning turbine drives the fan and the compressor.

Q

1. The person on the right weighs 500 N. The diagram shows the force of his feet pressing on the ground.
   a. Copy the diagram. Label the size of the force (in newtons).
   b. Draw in the force that the ground exerts on the person's feet. Label the size of this force.
2. When a gun is fired, it exerts a forward force on the bullet. Why does the gun recoil backwards?
3. In the diagram on the opposite page, the forces on the runner and on the ground are equal. Why does the runner move forwards, yet the ground apparently does not move backwards?

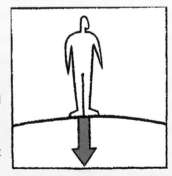

Related topics: force 2.6; gravitational force 2.9

2.11 Momentum (1)

Objectives: to know how to calculate momentum, and impulse – to know the link between force and a change in momentum

▶ Momentum = mass × velocity, and this truck has lots of it.

People say that a heavy vehicle travelling fast has lots of **momentum**. However, momentum has an exact scientific definition:

> momentum = mass × velocity   in symbols: $p=mv$

For example, if a model car has a mass of 2 kg and a velocity of 3 m/s, its momentum = mass × velocity = 2 kg × 3 m/s = 6 kg m/s

Like velocity, momentum is a vector, so a + or a − is often used to indicate its direction. For example:

> momentum of car moving to the *right* = +6 kg m/s
> momentum of car moving to the *left* = −6 kg m/s

Linking force and momentum: Newton's second law of motion

With a resultant force on it, an object will accelerate. Therefore, its velocity will change, and so will its momentum. The force and the momentum change are linked by this equation:

$$\text{resultant force} = \frac{\text{change in momentum}}{\text{time}} = \frac{\Delta p}{\Delta t}$$

or: resultant force = rate of change of momentum

The link between a resultant force and the rate of change of momentum it produces is known as **Newton's second law of motion**.

The above equation is really another way of saying that 'force = mass × acceleration'. The panel on the left explains why.

Impulse

From the previous equation, it follows that:

> resultant force × time = change in momentum

The quantity 'force × time' is called an **impulse**.

Newton noted that, when the same force acted for the same time on different masses, a large mass would gain less velocity than a smaller one, but the change in 'mass × velocity' was the same in every case. It was this observation that led to the concept of momentum and the second law.

Two versions of the same law

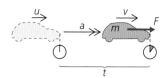

A resultant force F acts on an object of mass m for a time t. As a result, its velocity increases from u to v, its acceleration over this time being a. From Newton's second law of motion:

resultant force
$= \dfrac{\text{change in momentum}}{\text{time}}$

So: $F = \dfrac{mv - mu}{t}$

$= m\left(\dfrac{v - u}{t}\right)$

But: $a = \left(\dfrac{v - u}{t}\right)$

So: $F = ma$

In words:
resultant force = mass × acceleration

FORCES AND MOTION

E Solving problems

Example 1 A model car of mass 2 kg is travelling in a straight line. If its velocity increases from 3 m/s to 9 m/s in 4 s, what is the resultant force on it?

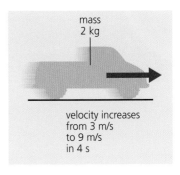

velocity increases from 3 m/s to 9 m/s in 4 s

To begin with:
momentum = mass × velocity = 2 kg × 3 m/s = 6 kg m/s

4 seconds later:
momentum = mass × velocity = 2 kg × 9 m/s = 18 kg m/s

So: change in momentum = 12 kg m/s

But: resultant force = $\dfrac{\text{change in momentum}}{\text{time}} = \dfrac{12 \text{ kg m/s}}{4 \text{ s}}$

So: resultant force = 3 N

The problem can also be solved by working out the car's acceleration and then using the equation: resultant force = mass × acceleration.

Example 2* A small rocket pushes out 2 kg of exhaust gas every second at a velocity of 100 m/s. What thrust (force) is produced by the engine?

By Newton's third law of motion, the forward force on the engine is equal to the backward force pushing out the exhaust gas. That force can be calculated by finding the rate of change of momentum of the gas:

In 1 second, 2 kg of gas increases its velocity from 0 to 100 m/s.

So: change in momentum = mass × velocity change
= 2 kg × 100 m/s = 200 kg m/s

force on gas = $\dfrac{\text{change in momentum}}{\text{time}} = \dfrac{200 \text{ kg m/s}}{1 \text{ s}}$

So: thrust = 200 N

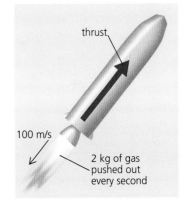

Q

1. What equation is used to calculate momentum?
2. What equation links the resultant force with the change in momentum it produces?
3. When a resultant force acts for 3 seconds on the trolley below, its velocity increases to 6 m/s.
   a. What is the momentum of the trolley before the force acts?
   b. What is the momentum after the force has acted?
   c. What is the change in momentum?
   d. What is the change in momentum every second?
   e. What is the resultant force on the trolley?
   Now you will calculate the resultant force on the trolley using different steps:
   f. What is the trolley's change in velocity?
   g. What is the trolley's acceleration?
   h. What equation links force, mass, and acceleration?
   i. What is the resultant force on the trolley?
4.* A jet engine pushes out 50 kg of gas (mainly air) every second, at a velocity of 150 m/s.
   a. What thrust (force) does the engine produce?
   b. If the engine pushed out twice the mass of gas at half the velocity, what would the thrust be?

Related topics: velocity, acceleration as vectors **2.1**; force, mass, acceleration, Newton's 2nd law **2.8**; Newton's 3rd law **2.10**; momentum and molecules **5.4**

2.12 Momentum (2)

Objectives: – to know that momentum is conserved – to do calculations on the motion of objects that are either colliding or being pushed apart.

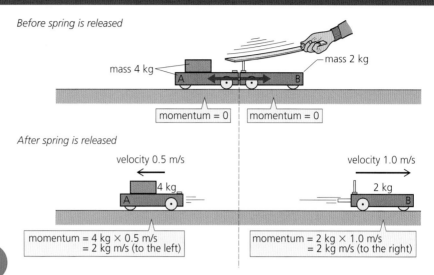

Before spring is released

After spring is released

Velocity and momentum essentials

Velocity is speed in a particular direction.

momentum = mass × velocity
(kg m/s)   (kg)   (m/s)

Velocity and momentum are vectors. They have direction as well as magnitude (size). Their direction can be shown using an arrow, or a + or −.

To begin with, the trolleys above are stationary. But when a spring-loaded piston is released between them, they shoot off in opposite directions. Their velocities can be measured using ticker-tape timers.

When the trolleys shoot apart, the trolley with least mass has most velocity. The diagram shows typical mass and velocity values. These illustrate a rule which applies in all such experiments:

mass × velocity to the left = mass × velocity to the right
(trolley A)   (trolley B)

This result is to be expected. From Newton's third law of motion, the forces on the two trolleys are equal but opposite. Also, the forces act for the same time. So they should cause equal but opposite changes in momentum (as force × time = change in momentum).

Conservation of momentum

With the mass and velocity values above, the total momentum of the trolleys before and after separation can be found. As momentum is a vector, its direction must be allowed for. In the following calculations, a momentum gain to the *right* is counted as *positive* (+):

Before the spring is released: total momentum of trolleys = 0

After the spring is released:
momentum of trolley A = mass × velocity = 4 kg × −0.5 m/s = −2 kg m/s
momentum of trolley B = mass × velocity = 2 kg × 1.0 m/s = +2 kg m/s
So: total momentum of trolleys = 0

So the total momentum (zero) is unchanged by the release of the spring. This is an example of the **law of conservation of momentum**:

> When two or more objects act on each other, their total momentum remains constant, provided no external forces are acting.

FORCES AND MOTION

Collision problem

Before the collision

After the collision

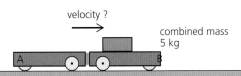

Example When the two trolleys above collide, they stick together. What is their velocity after the collision?

According to the law of conservation of momentum, the total momentum of the trolleys is the same after the collision as before:

Before the collision:
momentum of trolley A = mass × velocity = 1 kg × 2 m/s = +2 kg m/s
momentum of trolley B = mass × velocity = 4 kg × −3 m/s = −12 kg m/s
So: total momentum of trolleys A and B = −10 kg m/s

After the collision:
total momentum of trolleys A and B = −10 kg m/s   (as above)
So: combined mass × velocity = −10 kg m/s
So: 5 kg × velocity = −10 kg m/s
So: velocity of trolleys = −2 m/s

Therefore the trolleys have a velocity of 2 m/s to the left.

Momentum and energy

Moving objects have kinetic energy (see spread 4.1). In a collision, some of that energy may be changed into other forms.

If a collision is elastic, the total kinetic energy of the moving objects is the same after the collision as before. In other words, there is 'perfect bounce'. However, most collisions are not like this. The total kinetic energy is less after the collision than before. In such cases, the 'missing' energy is converted into thermal energy (heat).

1. A trolley of mass 2 kg rests next to a trolley of mass 3 kg on a flat bench. When a spring is released between the trolleys, and they are pushed apart, the 2 kg trolley travels to the left at 6 m/s.
Before separation:
a What is the total momentum of the trolleys?
After separation:
b What is the total momentum of the trolleys?
c What is the momentum of the 2 kg trolley?
d What is the momentum of the 3 kg trolley?
e What is the velocity of the 3 kg trolley?

2. A 16 kg mass travelling to the right at 5 m/s collides with a 4 kg mass travelling to the left, also at 5 m/s. When the masses collide, they stick together and move along the same line as before.
Before the collision:
a What is the momentum of the 16 kg mass?
b What is the momentum of the 4 kg mass?
c What is the total momentum of the masses?
After the collision:
d What is the total momentum of the masses?
e What is the velocity of the masses?

Related topics: velocity and vectors **2.1**; using ticker-tape **2.3**; Newton's 3rd law **2.10**; kinetic energy **4.1-4.3**

2.13 More about vectors

Objectives: – to know how vectors are different from scalars – to know how to find the resultant of two vectors at right-angles.

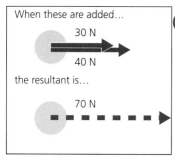

When these are added...
30 N
40 N
the resultant is...
70 N

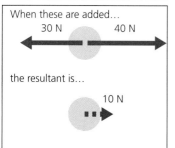

When these are added...
30 N   40 N
the resultant is...
10 N

(E) Vectors and scalars

Quantities such as force, which have a direction as well as a magnitude (size), are called **vectors**.

Two vectors acting at a point can be replaced by a single vector with the same effect. This is their **resultant**. On the left, you can see how to find it in two simple cases. Finding the resultant of two or more vectors is called *adding* the vectors.

Quantities such as mass and volume, which have magnitude but no direction, are called **scalars**. Adding scalars is easy. A mass of 30 kg added to a mass of 40 kg always gives a mass of 70 kg.

Adding vectors: the parallelogram rule

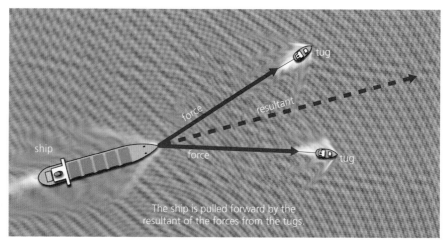

The ship is pulled forward by the resultant of the forces from the tugs.

Why the rule works

To see why the parallelogram rule works, consider this simple example using displacement vectors:

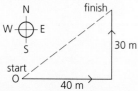

Above, someone starts at O, walks 40 m east, then 30 m north. From Pythagoras' theorem, the person must end up 50 m from O.

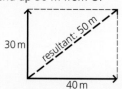

Above, the journey has been shown as the sum of two displacement vectors. When the parallelogram is drawn, its diagonal gives the correct displacement.

(E) The **parallelogram rule** is a method of finding the resultant in situations like the one above, where the vectors are not in line. It works like this:

To find the resultant of two vectors (for example, forces of 30 N and 40 N acting at a point O, as in the diagram below):

1. On paper, draw two lines from O to represent the vectors. The directions must be accurate, and the length of each line must be in proportion to the magnitude of each vector.
2. Draw in two more lines to complete a parallelogram.
3. Draw in the diagonal from O and measure its length. The diagonal represents the resultant in both magnitude and direction. (Below, for example, the resultant is a force of 60 N at 26° to the horizontal.)

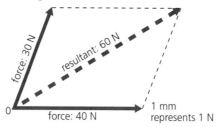

At Extended Level, you will only be required to use this rule when two forces are at right angles.

FORCES AND MOTION

Components of a vector*

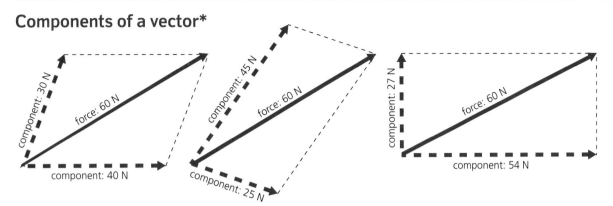

The parallelogram rule also works in reverse: a single vector can be replaced by two vectors having the same effect. Scientifically speaking, a single vector can be **resolved** into two **components**. When using the parallelogram rule in this way, the single vector forms the diagonal.

Above, you can see some of the ways in which a 60 N force can be resolved into two components. There are endless other possibilities.

Components at right angles In working out the effects of a force, it sometimes helps to resolve the force into components at right angles. For example, when a helicopter tilts its main rotor, the force has vertical and horizontal components which lift the helicopter and move it forward:

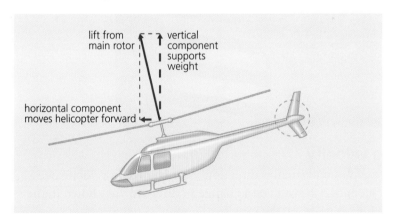

Calculating components

The horizontal and vertical components of a force F can be calculated using trigonometry:

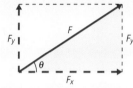

In the tinted triangle above:

$$\cos \theta = \frac{F_x}{F} \quad \text{and} \quad \sin \theta = \frac{F_y}{F}$$

So: $F_x = F \cos \theta$ and $F_y = F \sin \theta$

The horizontal and vertical components of F are therefore as shown below:

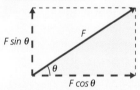

Q

1. How is a vector different from a scalar? Give an example of each.
2. Forces of 12 N and 5 N both act at the same point, but their directions can be varied.
   a. What is their greatest possible resultant?
   b. What is their least possible resultant?
   c. If the two forces are at right angles, find by scale drawing or otherwise the size and direction of their resultant.
3. *On the right, someone is pushing a lawnmower.
   a. By scale drawing or otherwise, find the vertical and horizontal components of the 100 N force.
   b. If the lawnmower weighs 300 N, what is the total downward force on the ground?
   c. If the lawnmower is pulled rather than pushed, how does this affect the total downward force?

Related topics: vectors **2.1**; force **2.7**

2.14 Moving in circles

Objectives: – to know that motion in a circle is a form of acceleration – to explain what centripetal force is and the factors on which it depends.

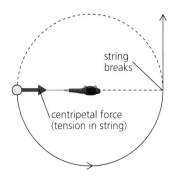

Ⓔ Centripetal force

On the left, someone is whirling a ball around in a horizontal circle at a steady speed. An inward force is needed to make the ball follow a circular path. The tension in the string provides this force. Without it, the ball would travel in a straight line, as predicted by Newton's first law of motion. This is exactly what happens if the string breaks.

This inward force needed to make an object move in a circle is called the **centripetal force**. *More* centripetal force is needed if:

- the *mass* of the object is *increased*
- the *speed* of the object is *increased*
- the *radius* of the circle is *reduced*.

▶ When a motorcycle goes round a corner like this, the sideways friction between the tyres and the road provides the necessary centripetal force.

Centripetal force...
Centripetal force isn't produced by circular motion. It is the force that must be *supplied* to make something move in a circle rather than in a straight line.

...and centrifugal force
When you whirl a ball around on the end of some string, you feel an outward pull on your hand. But there is no such thing as a 'centrifugal force' on the ball itself. If the string breaks, the ball moves off at a tangent. It isn't flung outwards.

Ⓔ Changing velocity

Velocity is speed in a particular direction. So a change in velocity can mean *either* a change in speed *or* a change in direction, as shown in the diagrams below. Diagram B shows what happens during circular motion.

If something has a changing velocity, then it has acceleration – in the same direction as the force. So, with circular motion, the acceleration is towards the centre of the circle. It may be difficult to imagine something accelerating towards a point without getting closer to it, but the object is always moving inwards from the position it would have had if travelling in a straight line.

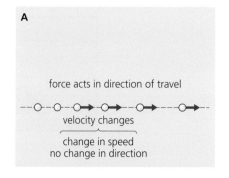

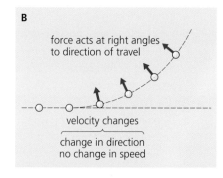

54

FORCES AND MOTION

Orbits*

Satellites around the Earth A satellite travels round the Earth in a curved path called an **orbit**, as shown below. Gravitational pull (in other words, the satellite's weight) provides the centripetal force needed. When a satellite is put into orbit, its speed is carefully chosen so that its path does not take it further out into space or back to Earth. Heavy satellites need the same speed as light ones. If the mass is doubled, twice as much centripetal force is required, but that is supplied by the doubled gravitational pull of the Earth.

Planets around the Sun The Earth and other planets move in approximately circular paths around the Sun. The centripetal force needed is supplied by the Sun's gravitational pull.

Electrons around the nucleus In atoms, negatively charged particles called **electrons** are in orbit around a positively charged **nucleus**. The attraction between opposite charges (sometimes called an **electrostatic force** or **electric force**) provides the centripetal force needed.

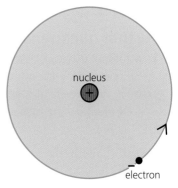

▲ Model of a hydrogen atom: a single electron orbits the nucleus. (According to quantum theory, electron orbits are much more complicated than that shown in this simple model.)

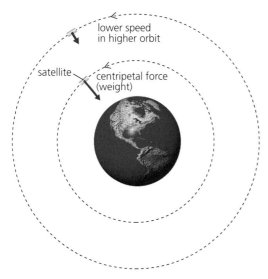

▲ A satellite close to the Earth orbits at a speed of about 29 000 km per hour. The further out the orbit, the lower the gravitational pull, and the less speed is required.

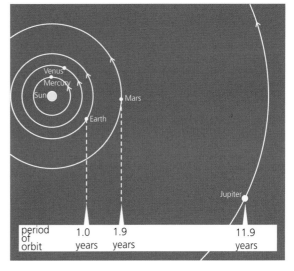

▲ The further a planet is from the Sun, the less speed it has, and the longer it takes to complete one orbit. The time for one orbit is called the **period**.

Q

1 A piece of clay is stuck to the edge of a potter's wheel. Draw a diagram to show the path of the clay if it comes unstuck while the wheel is rotating.

2 A car travels round a bend in the road. What supplies the centripetal force needed?

3 In question 2, how does the centripetal force change if the car
 a has less mass
 b travels at a slower speed
 c travels round a tighter curve?

4 What supplies the centripetal force needed for
 a a planet to orbit the Sun
 b an electron to orbit the nucleus in an atom?

5 A satellite is in a circular orbit around the Earth.
 a Draw a diagram to show any forces on the satellite. Show the direction of the satellite's acceleration.
 b* If the satellite were in a higher orbit, how would this affect its speed?
 c* If the satellite were in a higher orbit, how would this affect the centripetal force required?

Related topics: velocity **2.1**; Newton's 1st law **2.6**; force and acceleration **2.7**; gravity and weight **2.9**; electric charge **8.1**; atoms **11.1**; orbits and satellites **11.4**

Check-up on forces and motion

Further questions

1 a Write down, **in words**, the equation connecting speed, distance and time. [1]
b A car travels at a steady speed of 20 m/s. Calculate the distance travelled in 5 s. [2]

2

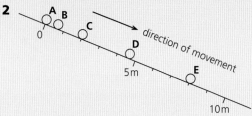

The diagram shows the positions of a ball as it rolled down a track. The ball took 0.5 s to roll from one position to the next. For example, it rolled from **A** to **B** in 0.5 s and from **B** to **C** in 0.5 s and so on.

a Write down:
  i the distance travelled by the ball from **A** to **E**; [1]
  ii the time taken by the ball to reach **E**. [1]
b Calculate the average speed of the ball in rolling from **A** to **E**. Write down the formula that you use and show your working. [3]
c Explain:
  i how you can tell from the diagram that the ball is speeding up; [1]
  ii why the ball speeds up. [1]

E 3

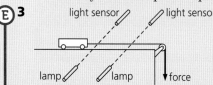

A student measures the acceleration of a trolley using the apparatus above. The light sensors are connected to a computer which is programmed to calculate the acceleration. The results obtained are shown on the acceleration–force graph.

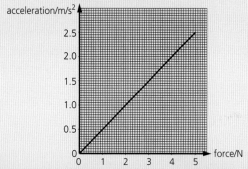

E a i Describe how changing the force affects the acceleration. [2]
  ii Write down, **in words**, the equation connecting force, mass, and acceleration. [1]
  iii Use the data from the graph to calculate the mass of the trolley. [2]
b Sketch the graph and draw the line that would have been obtained for a trolley of larger mass. [1]

4 A car has a mass of 900 kg. It accelerates from rest at a rate of 1.2 m/s^2.
a Calculate the time taken to reach a velocity of 30 m/s. [3]
b Calculate the force required to accelerate the car at a rate of 1.2 m/s^2. [3]
c Even with the engine working at full power, the car's acceleration decreases as the car goes faster. Why is this? [3]

5 The diagram below shows some of the forces acting on a car of mass 800 kg.

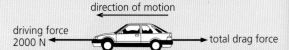

a State the size of the total drag force when the car is travelling at constant speed. [1]
b The driving force is increased to 3200 N.
  i Find the resultant force on the car at this instant. [1]
  ii Write down, **in words**, the equation connecting mass, force and acceleration. [1]
  iii Calculate the initial acceleration of the car. [2]
c Explain why the car will eventually reach a new higher constant speed. [2]

6

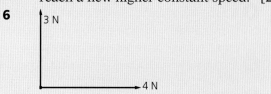

a Using a scale drawing (for example, on graph paper), find the resultant of the forces above. [3]
b Draw diagrams to show how, by changing the direction of one of the forces, it is possible to produce a resultant of **i** 7 N **ii** 1 N. [4]

7 This question is about SPEED and ACCELERATION.
A cycle track is 500 metres long. A cyclist completes 10 laps (that is, he rides completely round the track 10 times).
a How many kilometres has the cyclist travelled? [1]
b On average it took the cyclist 50 seconds to complete one lap (that is, to ride round just once).
   i What was the average speed of the cyclist? [2]
   ii How long in minutes and seconds did it take the cyclist to complete the 10 laps? [2]
c Near the end of the run the cyclist put on a spurt. During this spurt it took the cyclist 2 seconds to increase speed from 8 m/s to 12 m/s. What was the cyclist's acceleration during this spurt? [2]

8 This question is about FORCE and ACCELERATION.
The driver of a car moving at 20 m/s along a straight level road applies the brakes. The car decelerates at a steady rate of 5 m/s^2.
a How long does it take the car to stop? [2]
b What kind of force slows the car down? [1]
c Where is this force applied? [1]
d The mass of the car is 600 kg. What is the size of the force slowing the car down? [2]

9 A girl wearing a parachute jumps from a helicopter. She does not open the parachute straight away. The table shows her speed during the 9 seconds after she jumps.

time in seconds	0	1	2	3	4	5	6	7	8	9
speed in m/s	0	10		30	40	25	17	12	10	10

a Copy and complete the table by writing down the speed at 2 seconds. [3]
b Plot a graph of speed against time. [1]
c How many seconds after she jumped did the girl open her parachute? How do the results show this? [2]
d i What force pulls the girl down? [1]
   ii What force acts upwards? [1]
   iii Which of these forces is larger:
      at 3 seconds?
      at 6 seconds?
      at 9 seconds? [3]

e How will the graph continue after 9 seconds if she is still falling? [1]
f The girl makes a second jump with a larger area parachute. She falls through the air for the same time before opening her new parachute. How will this affect the graph:
   i during the first four seconds? [1]
   ii after this? [1]

10 a Sketch a velocity–time graph for a car moving with uniform acceleration from 5 m/s to 25 m/s in 15 seconds. [3]
b Use the sketch graph to find values for i the acceleration, ii the total distance travelled during acceleration. Show clearly at each stage how you used the graph. [4]

11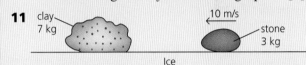

A stone of mass 3 kg is sliding across a frozen pond at a speed of 10 m/s when it collides head on with a lump of clay of mass 7 kg. The stone sticks to the clay and the two slide on together across the ice in the same direction as before. Calculate the following (assume that there is no friction from the ice):
a The momentum of the stone before the collision. [2]
b The total momentum of the stone and clay after the collision. [1]
c The total mass of the stone and clay. [1]
d The speed of the stone and clay after the collision. [2]

12 In the diagram below, someone is swinging a ball round on the end of a piece of string.

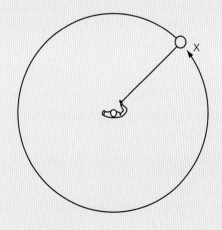

a What name is given to the force needed to make the ball move in a circle? [1]
b Copy and complete the diagram to show where the ball will travel if the string breaks when the ball is at point X. [2]
c Planets move around the Sun in approximately circular orbits. What provides the force necessary for the orbit? [1]

Use the list below when you revise for your IGCSE examination. The spread number, in brackets, tells you where to find more information.

Revision checklist

Core Level
☐ Measuring speed. (2.1)
☐ The difference between speed and velocity. (2.1)
☐ Linking acceleration with changing speed. (2.1)
☐ Representing motion using distance-time and speed-time graphs. (2.2)
☐ Calculating speed from the gradient of a distance-time graph. (2.2)
☐ Recognizing from the shape of a distance-time graph when an object is
 – stationary (at rest)
 – moving at a steady speed
 – moving with changing speed. (2.2)
☐ Recognizing acceleration and deceleration on a speed-time graph. (2.2)
☐ Calculating the distance travelled from a speed-time graph. (2.2)
☐ How the acceleration of free fall, g, is constant. (2.4)
☐ Measuring force. (2.6)
☐ The newton, unit of force. (2.6)
☐ Weight is a gravitational force. (2.6)
☐ Air resistance is a form of friction. (2.6 and 2.8)
☐ How an object moves if the forces on it are balanced. (2.6)
☐ The meaning of resultant force. (2.7)
☐ The resultant of two forces in line. (2.7 and 2.13)
☐ How a force can change the motion of an object. (2.7)
☐ How forces can change shape and volume, as well as motion. (2.7)
☐ The effects of friction. (2.8)
☐ Using the equation weight = mass × g (2.9)

Extended Level
As for Core Level, plus the following:
☐ Calculating acceleration. (2.1)
☐ Deceleration is negative acceleration. (2.1)
☐ Calculating acceleration from the gradient of a speed-time graph. (2.2)
☐ How an object falls in a uniform (constant) gravitational field
 – without air resistance. (2.4)
 – with air resistance. (2.5 and 2.6)
☐ Terminal velocity. (2.6)
☐ Recognizing the difference between uniform (constant) and non-uniform (non-constant) acceleration from the shape of a speed-time graph. (2.5)
☐ Mass as resistance to change in motion (2.7)
☐ The link between force, mass, and acceleration. (2.7)
☐ Defining the newton. (2.7)
☐ The difference between weight and mass. (2.9)
☐ Calculating momentum. (2.11)
☐ The link between force and momentum change. (2.11)
☐ Calculating impulse. (2.11)
☐ The conservation of momentum. (2.12)
☐ The difference between vectors and scalars. (2.13)
☐ Adding vectors using the parallelogram rule. (2.13)
☐ Motion in a circle, and centripetal force. (2.14)

3
Forces and pressure

- TURNING EFFECT OF A FORCE
- CENTRE OF GRAVITY
- BALANCE AND EQUILIBRIUM
- STRETCHING AND COMPRESSING MATERIALS
- PRESSURE
- PRESSURE IN LIQUIDS
- GAS PRESSURE AND VOLUME

Sharks like this are very effective hunters. Their sharp teeth give them a dangerous bite, although because of their long jaws, their biting force is not much more than that of a human. However, when it comes to diving, sharks beat humans easily. Some types can reach depths of over 2000 metres, where the water pressure is far too great for any human diver.

3.1 Forces and turning effects

Objectives: – to explain what is meant by the moment of a force – to understand and use the principle of moments – to give the conditions needed for equilibrium.

Moment of a force

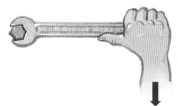

▲ A large force at the end of a long spanner gives a large turning effect.

It is difficult to tighten a nut with your fingers. But with a spanner, you can produce a larger turning effect. The turning effect is even greater if you increase the force or use a longer spanner. The turning effect of a force is called a **moment**. It is calculated like this:

moment of a force = force × perpendicular distance
about a point   from the point

Below, there are some examples of forces and their moments. Moments are described as **clockwise** or **anticlockwise**, depending on their direction. The moment of a force is also called a **torque**.

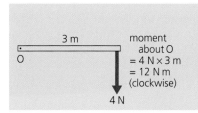
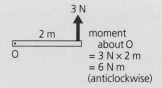
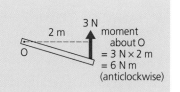

The principle of moments

Unit of force
Force is measured in newtons (N).

In diagram A below, the bar is in a state of balance, or **equilibrium**. Note that the anticlockwise moment about O is equal to the clockwise moment. One turning effect balances the other. In diagram B, there are more forces acting but, once again, the bar is in equilibrium. This time, the *total* clockwise moment about O is equal to the anticlockwise moment.

These examples illustrate the **principle of moments**.

Taking moments
Calculating the moments about a point is called *taking moments* about the point.

If an object is in equilibrium:
the sum of the clockwise moments about any point is equal to the sum of the anticlockwise moments about that point.

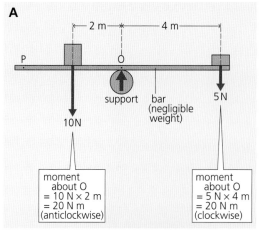
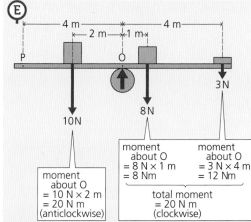

60

FORCES AND PRESSURE

Conditions for equilibrium

If an object is in equilibrium, the forces on it must balance as well as their turning effects. So:

- The sum of the forces in one direction must equal the sum of the forces in the opposite direction.
- The principle of moments must apply.

For example, in diagram A on the opposite page, the upward force from the support must be 15 N, to balance the 10 N + 5 N total downward force. Also, if you take moments about *any* point, for example P, the total clockwise moment must equal the total anticlockwise moment.

When taking moments about P, you need to include the moment of the upward force from the support. This doesn't arise when taking moments about O because the force has no moment about that point.

Clockwise .... or anticlockwise?
In the diagram below, the 500 N force has a *clockwise* moment about A, but an *anticlockwise* moment about B. To decide whether a moment is clockwise or anticlockwise about a point, imagine that the diagram is pinned to the table through the point, then decide which way the force arrow is trying to turn the paper.

Solving a problem

Example Below right, someone of weight 500 N is standing on a plank supported by two trestles. Calculate the upward forces, *X* and *Y*, exerted by the trestles on the plank. (Assume the plank has negligible weight.)

The system is in equilibrium, so the principle of moments applies. You can take moments about any point. But taking moments about A or B gets rid of one of the unknowns, *X* or *Y*.

Taking moments about A:

clockwise moment = 500 N × 2 m = 1000 N m

anticlockwise moment = Y × 5 m

As the moments balance, 5 Y m = 1000 N m

So: Y = 200 N

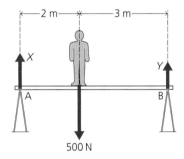

From here, there are two methods of finding *X*. *Either* take moments about B and do a calculation like the one above. *Or* use the fact that *X* + *Y* must equal the 500 N downward force. By either method: *X* = 300 N

1. The moment (turning effect) of a force depends on two factors. What are they?
2. What is the principle of moments? What other rule also applies if an object is in equilibrium?
3. Below, someone is trying to balance a plank with stones. The plank has negligible weight.
   a Calculate the moment of the 4 N force about O.
   b Calculate the moment of the 6 N force about O.

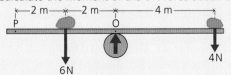

   c Will the plank balance? If not, which way will it tip?
   d What extra force is needed at point P to balance the plank?
   e In which direction must the force at P act?
4. In diagram **B** on the opposite page:
   a What is the upward force from the support?
   b If moments are taken *about point P*, which forces have clockwise moments? What is the total clockwise moment about P?
   c Which force or forces have anticlockwise moments about P? What is the total anticlockwise moment about P?
   d Comparing moments about P, does the principle of moments apply?

Related topics: force, balanced forces 2.6

3.2 Centre of gravity

Objectives: – to explain what an object's centre of gravity is and how it can be found by experiment – to describe the conditions needed for an object to be stable.

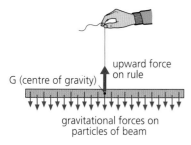

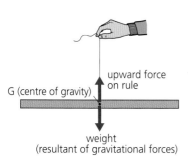

Like other objects, the beam on the left is made up of lots of tiny particles, each with a small gravitational force on it. The beam balances when suspended at one particular point, G, because the gravitational forces have turning effects about G which cancel out.

Together, the small gravitational forces act like a single force at G. In other words, they have a resultant at G. This resultant is the beam's **weight**. G is the **centre of gravity** (or **centre of mass**).

Finding a centre of gravity

In diagram 1 below, the card can swing freely from the pin. When the card is released, the forces on it turn the card until its centre of gravity is vertically under the pin, as in diagram 2. Whichever point the card is suspended from, it will always hang with its centre of gravity vertically under the pin. This fact can be used to find the centre of gravity.

In diagram 3, the centre of gravity lies somewhere along the plumb line, whose position is marked by the line AB. If the card is suspended at a different point, a second line CD can be drawn. The centre of gravity must also lie along this line, so it is at the point where AB crosses CD.

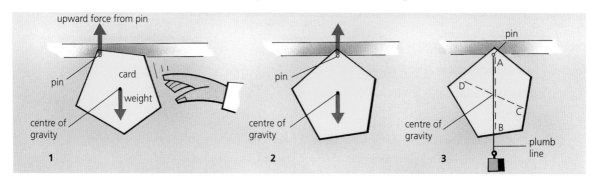

Heavy bar problem

In simple problems, you are often told that a balanced bar has negligible weight. In more complicated problems, you have to include the weight.

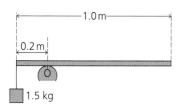

Example If a uniform bar balances, as on the left, with a 1.5 kg mass attached to one end, what is its weight? ($g = 10$ N/kg)

To solve the problem, redraw the diagram to show all the forces and distances, as in the lower diagram. As $g = 10$ N/kg, the 1.5 kg mass has a weight of 15 N. 'Uniform' means that the bar's weight is evenly distributed, so the centre of gravity of the bar (by itself) is at the mid-point, 0.5 m from one end. The bar's weight W acts at this point.

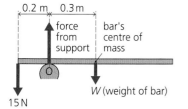

Now take moments about the support, O. The upward force has no moment about this point, but there is an anticlockwise moment of 15 N × 0.2 m and a clockwise moment of W × 0.3 m. As the bar is in equilibrium:

15 N × 0.2 m = W × 0.3 m

So: the bar's weight W is 10 N.

FORCES AND PRESSURE

Stability

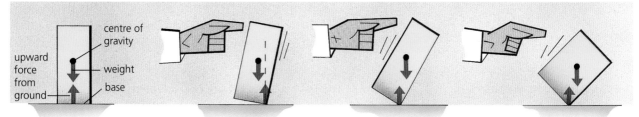

This box is in equilibrium. The forces on it are balanced, and so are their turning effects.

With a small tilt, the forces will turn the box back to its original position.

With a large tilt, the forces will tip the box over.

A box with a wider base and a lower centre of gravity can be tilted further before it falls over.

If the box above is pushed a little and then released, it falls back to its original position. Its position was **stable**. If the box is pushed much further, it topples. It starts to topple as soon as its centre of gravity passes over the edge of its base. From then on, the forces on the box have a turning effect which tips it even further. A box with a wider base and/or a lower centre of gravity is more stable. It can be tilted to a greater angle before it starts to topple.

States of equilibrium
Here are three types of equilibrium:

Stable equilibrium If you tip the cone a little, the centre of gravity stays over the base. So the cone falls back to its original position.

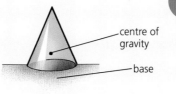

Unstable equilibrium The cone is balanced, but only briefly. Its pointed 'base' is so small that the centre of gravity immediately passes beyond it.

Neutral equilibrium Left alone, the ball stays where it is. When moved, it stays in its new position. Wherever it lies, its centre of gravity is always exactly over the point which is its 'base'.

▲ He will stay balanced – as long as he keeps his centre of gravity over the beam.

1 The stool on the right is about to topple over.
   a Copy the diagram, showing the position of the centre of gravity.
   b Give *two* features which would make the stool more stable.
2 A uniform metre rule has a 4 N weight hanging from one end. The rule balances when suspended from a point 0.1 m from that end.
   a Draw a diagram to show the rule and the forces on it.
   b Calculate the weight of the rule.
3 Draw diagrams to show a drawing pin in positions of stable, unstable, and neutral equilibrium.

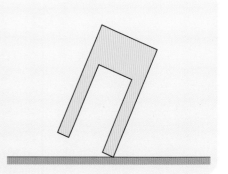

Related topics: resultant force **2.7** and **2.11**; mass, weight, and *g* **2.9**; turning effects, moments, and equilibrium **3.1**

63

3.3 More about moments

Objectives: to do more complicated calculations involving the principle of moments.

Force and moment essentials

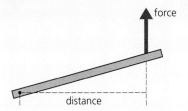

A **moment** is the turning effect of a force:
moment of a force = force × perpendicular distance
about a point   from the point

According to the **principle of moments**:
If a system is in equilibrium (balanced), the sum of the clockwise moments about any point is equal to the sum of the anticlockwise moments about that point.

If an object is in equilibrium, the forces on it must balance and also their turning effects. So:
- The sum of the forces in one direction must equal the sum of the forces in the opposite direction.
- The principle of moments must apply.

Testing the principle of moments

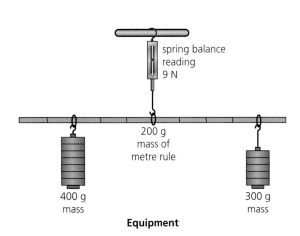

Equipment

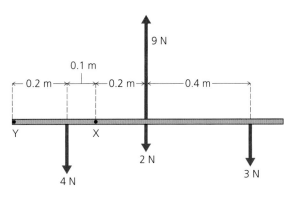

Force diagram

You can test the principle of moments by carrying out an experiment like the one above. Here, a metre rule has been suspended from a spring balance. Weights have been hung from the rule and their positions adjusted so that the system is balanced – it is equilibrium. The second diagram shows all the forces on the rule, including the weight of the rule itself. (Each 100 g of gravity is assumed to weigh 1 N).

E The principle of moments should apply about *any* point. So, for example, choosing point X (and omitting some units for simplicity):

- The 2 N and 3 N forces each have a *clockwise* moment about X. So, sum of clockwise moments = (2 × 0.2) + (3 × 0.6) = 3.2 N m
- The 9 N and 4 N forces each have an *anticlockwise* moment about X. So, sum of anticlockwise moments = (9 × 0.2) + (4 × 0.1) = 3.2 N m

The two sums are equal, as predicted by the principle. You could express this result in another way: calling the clockwise moments positive, and the anticlockwise moments negative, the *net* moment (combined total) is zero.

FORCES AND PRESSURE

Crane problem

Example The diagram on the right shows a model crane. The crane has a counterbalance weighing 400 N, which can be moved further or closer to O to cope with different loads. (With no load or counterbalance, the top section would balance at point O.)
a With the 100 N load shown, how far from O should the counterbalance be placed?
b What is the maximum load the crane can safely lift?

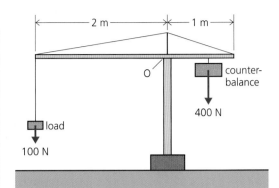

a To prevent the crane falling over, its top section must balance at point O. So the moment of the 400 N force (the counterbalance) must equal the moment of the 100 N force (the load). That follows from the principle of moments.
Let x be the distance of the 400 N force from O.
Taking moments about O:
clockwise moment = anticlockwise moment
$400 \text{ N} \times x = 100 \text{ N} \times 2 \text{ m}$
$x = 0.5 \text{ m}$
So: the counterbalance should be placed 0.5 m from O.

b Let F be the maximum load (in N). With this load on the crane, the counterbalance must produce its maximum moment about O. So it must be the greatest possible distance from O – in other words, 1 m from it. As the crane is in equilibrium, the principle of moments applies:
Taking moments about O:
clockwise moment = anticlockwise moment
$400 \text{ N} \times 1 \text{ m} = F \times 2 \text{ m}$
$F = 200 \text{ N}$
So: the maximum load is 400 N.

> **Centre of gravity essentials**
> Although weight is distributed through an object, it acts as a single, downward force from a point called the **centre of gravity** (or **centre of mass**). For an object to be stable when resting on the ground, its centre of gravity must be over its base. If an object is pushed, and its centre of gravity passes beyond the edge of its base, it will topple over.

1. In the diagram on the right, a plank weighing 120 N is supported by two trestles at points A and B. A man weighing 480 N is standing on the plank.
   a Redraw the diagram, showing all the forces acting on the plank.
   b Calculate the total clockwise moment of the two weights about A.
   c Use the principle of moments to calculate the upward force from the trestle at B.
   d What is the total downward force on the trestles?
   e What is the upward force from the trestle at A?
   f The man now walks past A towards the left-hand end of the plank. What is the upward force from the trestle at B at the instant the plank starts to tip?
   g How far is the man from A as the plank tips?

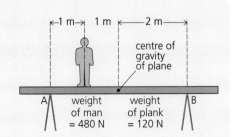

2. In **Testing the principle of moments** on the opposite page, moments were taken about X. Calculate the moments again, only about point Y. Are the sums of the clockwise and anticlockwise moments still equal?

Related topics: balanced forces **2.6**; moments and equilibrium **3.1**; centre of gravity **3.2**

3.4 Stretching and compressing

Objectives: to describe how the load on a spring affects its extension – to interpret load/extension graphs – to explain what is meant by the spring constant.

Elastic and plastic

If you bend a ruler slightly and release it, it springs back to its original shape. Materials that behave like this are **elastic**. However, they stop being elastic if bent or stretched too far. They either break or become permanently deformed (out of shape).

If you stretch or bend Plasticine, it keeps its new shape. Materials that behave like this are **plastic**. (The materials we call 'plastics' were given that name because they are plastic and mouldable when hot.)

Force and weight essentials

Force is measured in newtons (N).

Weight is a force.

On Earth, the weight of an object is 10 N for each kilogram of mass.

Stretching a spring

In the experiment below, a steel spring is stretched by hanging masses from one end. The force applied to the spring is called the **load**. As g is 10 N/kg, the load is 1 N for every 100 g of mass hung from the spring.

As the load is increased, the spring stretches more and more. Its **extension** is the difference between its stretched and unstretched lengths.

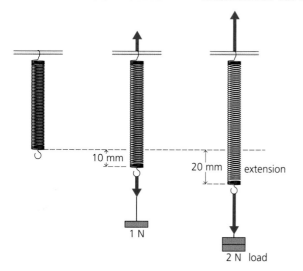

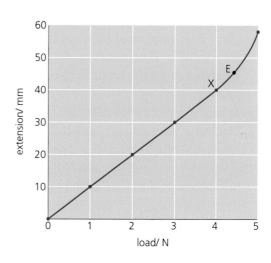

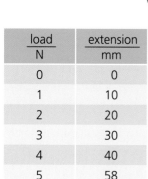

(E) The readings on the left can be plotted as a graph, as above. Up to point X, the graph line has these features:
- The line is straight, and passes through the origin.
- If the load is doubled, the extension is doubled, and so on.
- Extension ÷ load always has the same value (10 mm/N).
- Every 1 N increase in load produces the same extra extension (10 mm).

Mathematically, these can be summed up as follows:

Up to point X, the extension is proportional to the load. X is the **limit of proportionality**.

*Point E marks another change in the spring's behaviour. Up to E, the spring behaves elastically and returns to its original length when the load is removed. E is its **elastic limit**. Beyond E, the spring is left permanently stretched.

FORCES AND PRESSURE

Hooke's law*

In the 1660s, Robert Hooke investigated how springs and wires stretched when loads were applied. He found that, for many materials, the extension and load were in proportion, provided the elastic limit was not exceeded:

> A material obeys Hooke's law if, beneath its elastic limit, the extension is proportional to the load.

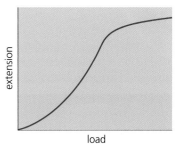

▲ Extension–load graph for rubber

Steel wires do not stretch as much as steel springs, but they obey Hooke's law. Glass and wood also obey the law, but rubber does not.

E Spring constant

For the spring on the opposite page, up to point X on the graph, dividing the load (force) by the extension always gives the same value, 0.1 N/mm. This is called the **spring constant** (symbol k):

> load = spring constant × extension   In symbols: $F = k \times x$

Knowing k, you could use this equation to calculate the extension produced by any load up to the limit of proportionality. For example, for a load of 2.5 N:

$2.5 = 0.1 \times$ extension   (omitting units for simplicity)

Rearranged, this gives: extension $= 2.5/0.1 = 25$ mm

Compressing and bending *

Materials can be compressed as well as stretched. If the compression is elastic, the material will return to its original shape when the forces are removed. When a material is bent, the applied forces produce compression on one side and stretching on the other. If the elastic limit is exceeded, the bending is permanent. This can happen when a metal sheet is dented.

▲ The Oriental Pearl Tower in Shanghai is over half a kilometre high. In high winds, its top can move by a quarter of a metre. But, being elastic, its steel and concrete structure always returns to its original shape.

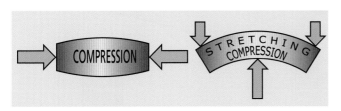

1. What is meant by an *elastic* material?
2. What is meant by the *elastic limit* of a material?
3. *Look at the small graph at the top of the page. Does rubber obey Hooke's law? Explain how you can tell from the graph whether this law is obeyed or not.
4. The table on the right shows the readings taken in a spring-stretching experiment:
   a What is the unstretched length of the spring?
   b Copy and complete the table.
   c Plot a graph of extension against load.

d* Mark the *elastic limit* on your graph.
e Over which section of the graph line is the extension proportional to the load?
f What load would produce a 35 mm extension?
g What load would make the spring stretch to a length of 65 mm?

load/ N	0	1	2	3	4	5	6
length/ mm	40	49	58	67	76	88	110
extension/ mm							

Related topics: forces 2.6; mass, weight, and g 2.9

3.5 Pressure

Objectives: – to know the equation linking force, pressure, and area – to describe situations where the pressure from a solid object needs to be reduced or increased.

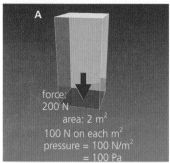

A
force: 200 N
area: 2 m²
100 N on each m²
pressure = 100 N/m²
= 100 Pa

B
force: 200 N
area: 4 m²
50 N on each m²
pressure = 50 N/m²
= 50 Pa

Blocks A and B on the left are resting on soft ground. Both weigh the same and exert the same force on the ground. But the force from block B is spread over a larger area, so the force *on each square metre* is reduced. The **pressure** under block B is less than that under block A.

For a force acting at right angles to a surface, the pressure is calculated like this:

$$\text{pressure} = \frac{\text{force}}{\text{area}} \qquad \text{In symbols: } p = \frac{F}{A}$$

If force is measured in newtons (N) and area in square metres (m²), pressure is measured in newton/square metre (N/m²). 1 N/m² is called 1 **pascal** (Pa):

If a 100 N force is spread over an area of   1 m², the pressure is   100 Pa.
If a 100 N force is spread over an area of   2 m², the pressure is   50 Pa.
If a 100 N force is spread over an area of 0.2 m², the pressure is   500 Pa.
If a 200 N force is spread over an area of 0.2 m², the pressure is   1000 Pa.

For most pressure measurements, the pascal is a very small unit. In practical situations, it is often more convenient to use the **kilopascal** (kPa).
1 kPa = 1000 Pa

Increasing the pressure by *reducing* the area

The studs on a football boot have only a small area of contact with the ground. The pressure under the studs is high enough for them to sink into the ground, which gives extra grip.

The area under the edge of a knife's blade is extremely small. Beneath it, the pressure is high enough for the blade to push easily through the material.

Under the tiny area of the point of a drawing pin, the pressure is far too high for the wood to withstand.

Reducing the pressure by *increasing* the area

Skis have a large area to reduce the pressure on the snow so that they do not sink in too far.

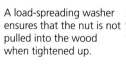

Wall foundations have a large horizontal area. This reduces the pressure underneath so that the walls do not sink further into the ground.

A load-spreading washer ensures that the nut is not pulled into the wood when tightened up.

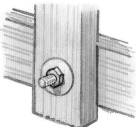

FORCES AND PRESSURE

Typical pressures

20 kPa

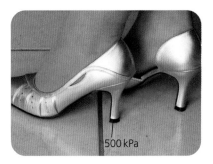
500 kPa

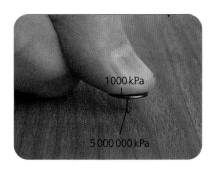
1000 kPa
5 000 000 kPa

Pressure problems

Example 1 The wind pressure on the wall on the right is 100 Pa. If the wall has an area of 6 m², what is the force on it?

To solve this problem, you need to rearrange the pressure equation:

force = pressure × area
      = 100 Pa × 6 m² = 600 N

So the force on the wall is 600 N.

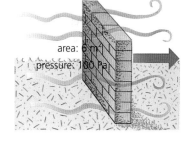

Example 2 A concrete block has a mass of 2600 kg. If the block measures 0.5 m by 1.0 m by 2.0 m, what is the maximum pressure it can exert when resting on the ground? ($g = 10$ N/kg)

As g is 10 N/kg, the 2600 kg block has a weight of 26 000 N, so the force on the ground is also 26 000 N.

To exert *maximum* pressure, the block must be resting on the side with the *smallest* area. This is the side measuring 1.0 m × 0.5 m, as shown on the right. Its area = 1.0 m × 0.5 m = 0.5 m². So:

$$\text{pressure} = \frac{\text{force}}{\text{area}} = \frac{26\,000\text{ N}}{0.5\text{ m}^2} = 52\,000\text{ Pa}$$

So the maximum pressure is 52 000 Pa, or 52 kPa.

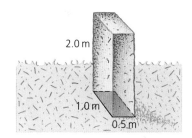

Assume that $g = 10$ N/kg, and that all forces are acting at right angles to any area mentioned.

1 A force of 200 N acts on an area of 4 m².
   a What pressure is produced?
   b What would the pressure be if the same force acted on half the area?

2 What force is produced if:
   a A pressure of 1000 Pa acts on an area of 0.2 m²?
   b A pressure of 2 kPa acts on an area of 0.2 m²?

3 Explain why a tractor's big tyres stop it sinking too far into soft soil.

4 A rectangular block of mass 30 kg measures 0.1 m by 0.4 m by 1.5 m.
   a Calculate the weight of the block.
   b Draw diagrams to show how the block must rest on the ground to exert **i** *maximum* pressure **ii** *minimum* pressure.
   c Calculate the maximum and minimum pressures in part **b**.

Related topics: force **2.6**; mass, weight, and *g* **2.9**

3.6 Pressure in liquids

Objectives: to know the factors affecting the pressure in a liquid — to know how to calculate the pressure in a liquid.

Pressure acts in all directions.

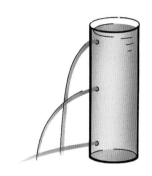
Pressure increases with depth.

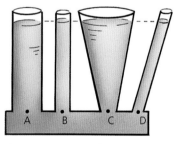
The pressure at points A, B, C, and D is the same.

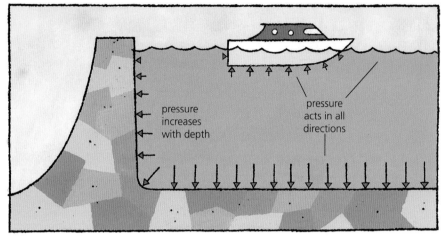

A liquid is held in its container by its weight. This causes pressure on the container, and pressure on any object in the liquid.

The following properties apply to any stationary liquid in an open container. The experiments on the left demonstrate three of them.

Pressure acts in all directions The liquid pushes on every surface in contact with it, no matter which way the surface is facing. For example, the deep-sea vessel below has to withstand the crushing effect of sea water pushing in on it from all sides, not just downwards.

Pressure increases with depth The deeper into a liquid you go, the greater the weight of liquid above and the higher the pressure. Dams are made thicker at the bottom to withstand the higher pressure there.

Pressure depends on the density of the liquid The more dense the liquid, the higher the pressure at any particular depth.

Pressure doesn't depend on the shape of the container Whatever the shape or width, the pressure at any particular depth is the same.

▶ Deep-sea diving vessels are built to withstand the crushing effect of sea water whose pressure pushes inwards from all directions.

FORCES AND PRESSURE

Useful connections

For calculations like those below, you need to know the connections between these:

volume (in m³)   density (in kg/m³)   mass (in kg)   weight (in N)   g (10 N/kg)

For example, you might know the volume and density of a liquid, but need to find its weight. For this, the equations required are:

$$\text{density} = \frac{\text{mass}}{\text{volume}} \qquad \text{weight} = \text{mass} \times g$$

From these equations, it follows that:

$$\text{mass} = \text{density} \times \text{volume} \qquad \text{weight} = \text{density} \times \text{volume} \times g$$

Pressure and weight essentials

For a force acting at right angles to a surface:

$$\text{pressure} = \frac{\text{force}}{\text{area}}$$

If force is in newtons (N) and area in square metres (m²), then pressure is in pascals (Pa).

E Calculating the pressure in a liquid

The container on the right has a base area A. It is filled to a depth h with a liquid of density ρ (Greek letter 'rho'). To calculate the pressure on the base due to the liquid, you first need to know the weight of the liquid on it:

volume of liquid = base area × depth = Ah

mass of liquid = density × volume = ρAh

weight of liquid = mass × g = $\rho Ah g$ (g = 10 N/kg)

So: force on base = $\rho g A h$

This force is acting on an area A, so: $\text{pressure} = \frac{\text{force}}{\text{area}} = \frac{\rho g A h}{A} = \rho g h$

At a depth h in a liquid of density ρ:

$$\text{pressure} = \rho g h$$

or, for a change in depth Δh, the change in pressure $\Delta p = \rho g \Delta h$

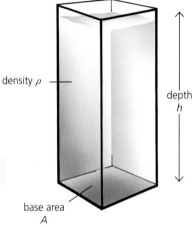

Example If the density of water is 1000 kg/m³, what is the pressure due to the water at the bottom of a swimming pool 2 m deep?

pressure = $\rho g h$ = 1000 kg/m³ × 10 N/kg × 2 m = 20 000 Pa

Q

g = 10 N/kg; density of water = 1000 kg/m³; density of paraffin = 800 kg/m³

1 In the diagram on the right:
  a How does the pressure at A compare with the pressure at B?
  b How does the pressure at B compare with the pressure at D?
  c How does the pressure at A compare with the pressure at C?
  d If the water in the system were replaced with paraffin, how would this affect the pressure at B?

2 A rectangular storage tank 4 m long by 3 m wide is filled with paraffin to a depth of 2 m. Calculate:
  a the volume of paraffin
  b the mass of paraffin
  c the weight of paraffin
  d the pressure at the bottom of the tank due to the paraffin

3 In the diagram on the right, calculate the pressure at B due to the water.

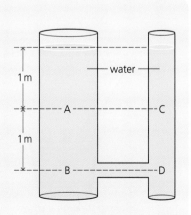

Related topics: density 1.4; mass, weight, and g 2.9; pressure 3.5

3.7 Pressure from the air*

Objectives: to describe the characteristics of atmospheric pressure – to know about the instruments used to measure atmospheric pressure and gas pressure.

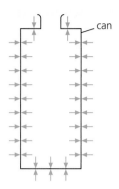

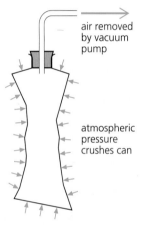

▲ **Demonstrating atmospheric pressure** When the air is removed from the can, there is nothing to resist the outside pressure, and the can is crushed.

The atmosphere is a deep ocean of air which surrounds the Earth. In some ways, it is like a liquid:
- Its pressure acts in all directions.
- Its pressure becomes less as you rise up through it (because there is less and less weight above).

Unlike a liquid, air can be compressed (squashed). This makes the atmosphere denser at lower levels. The atmosphere stretches hundreds of kilometres into space, yet the bulk of the air lies within about 10 kilometres of the Earth's surface.

Atmospheric pressure

At sea level, atmospheric pressure is about 100 kPa (100 000 newtons per square metre) – equivalent to the weight of ten cars pressing on every square metre. But you aren't crushed by this huge pressure because it is matched by the pressure in your lungs and blood system.

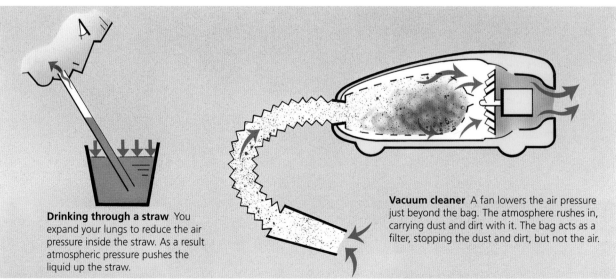

Drinking through a straw You expand your lungs to reduce the air pressure inside the straw. As a result atmospheric pressure pushes the liquid up the straw.

Vacuum cleaner A fan lowers the air pressure just beyond the bag. The atmosphere rushes in, carrying dust and dirt with it. The bag acts as a filter, stopping the dust and dirt, but not the air.

FORCES AND PRESSURE

The mercury barometer

Instruments that measure atmospheric pressure are called **barometers**. The barometer on the right contains the liquid metal mercury. Atmospheric pressure has pushed mercury up the tube because the space at the top of the tube has no air in it. It is a **vacuum**. At sea level, atmospheric pressure will support a column of mercury 760 mm high, on average. For convenience, scientists sometimes describe this as a pressure of 760 'millimetres of mercury'. However, it is easily converted into pascals and other units, as you can see below.

The actual value of atmospheric pressure varies slightly depending on the weather. Rain clouds form in large areas of lower pressure, so a fall in the barometer reading may mean that rain is on the way. Atmospheric pressure also decreases with height above sea level. This idea is used in the **altimeter**, an instrument fitted in aircraft to measure altitude.

Barometer

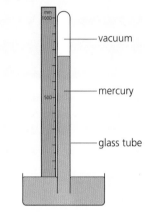

Standard atmospheric pressure

The pressure that will support a column of mercury 760.0 mm high is known as **standard atmospheric pressure**, or 1 **atmosphere** (1 atm). Its value in pascals can be found by calculating the pressure due to such a column.

At a depth h in a liquid of density ρ, the pressure $= \rho g h$, where g is 9.807 N/kg (or 10 N/kg if less accuracy is needed). As the density of mercury is 13 590 kg/m^3, and the height of the column is 0.760 0 m:

$$1 \text{ atm} = \rho g h = 13\,590 \text{ kg/m}^3 \times 9.807 \text{ N/kg} \times 0.760\,0 \text{ m} = 101\,300 \text{ Pa}$$

In calculations, for simplicity, you can assume that 1 atm = 100 000 Pa. In weather forecasting, the **millibar (mb)** is often used as a pressure unit. 1 mb = 100 Pa, so standard atmospheric pressure is approximately 1000 millibars.

Manometer

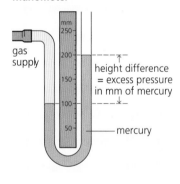

The manometer

A **manometer** measures pressure *difference*. The one in the diagram on the right is filled with mercury. The height difference shows the *extra* pressure that the gas supply has in addition to atmospheric pressure. This extra pressure is called the **excess pressure**. To find the actual pressure of the gas supply, you add atmospheric pressure to this excess pressure.

1. Write down *two* ways in which the pressure in the atmosphere is like the pressure in a liquid.
2. Explain why, when you 'suck' on a straw, the liquid travels up it.
3. If a mercury barometer were carried up a mountain, how would you expect the height of the mercury column to change?
4. Look at the diagram of the manometer on this page. If atmospheric pressure is 760 mm of mercury:
   a What is the excess pressure of the gas supply (in mm of mercury)?
   b What is the actual pressure of the gas supply (in mm of mercury)?
   c What is the actual pressure of the gas supply (in Pa)?
5.* If, on a particular day, atmospheric pressure is 730 mm of mercury, what is this **a** in pascals **b** in atmospheres **c** in millibars?
6. The density of mercury is 13 590 kg/m^3, the density of water is 1000 kg/m^3, and g is 9.81 N/kg.
   a What is the pressure (in Pa) at the bottom of a column of water 1 metre long?
   b If a barometer is made using water instead of mercury, and a very long tube, how high is the water column when atmospheric pressure is 1 atm (760 mm of mercury)?

Related topics: density 1.4; pressure 3.5; calculating the pressure in a liquid 3.6

3.8 Gas pressure and volume

Objectives: – to know how the pressure of a gas depends on its volume – to know and use the equation linking pressure and volume.

▲ When this balloon rises, the pressure, volume, and temperature can all change.

When dealing with a fixed mass of gas, there are always three factors to consider: *pressure, volume, and temperature*. A change in one of these factors always produces a change in at least one of the others. Often all three change at once. This happens, for example, when a balloon rises through the atmosphere, or gases expand in the cylinders of a car engine.

This spread deals with a simpler case: how the pressure of a gas depends on its volume if the temperature is kept constant. The link between the pressure and the volume can be found from the following experiment.

Linking pressure and volume (at constant temperature)

The equipment for the experiment is shown in the diagram below left, where the gas being studied is a *fixed mass* of *dry air*. The air is trapped in a glass tube. Its volume is reduced in stages by pumping air into the reservoir so that oil is pushed further up the tube. Each time the volume is reduced, the pressure of the trapped air is measured on the gauge.

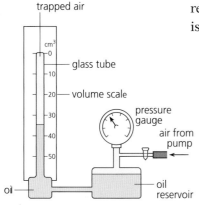

pressure kPa	volume cm³
200	50
250	40
400	25
500	20
1000	10

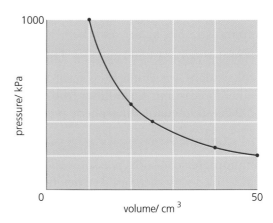

Squashing the air warms it up slightly. So before taking each reading, you have to wait a few moments for the air to return to its original temperature. The gauge actually measures the pressure in the reservoir, but this is the same as in the tube because the oil transmits the pressure.

Above, you can see some typical readings and the graph they produce. Results like this show that the relationship between the pressure and volume is an **inverse proportion**. It has these features:

1. If the volume *halves*, the pressure *doubles*, and so on.
2. *Pressure* × *volume* always has the same value (10 000 in this case).
3. If *pressure* is plotted against $\frac{1}{volume}$, the graph is a straight line through the origin, as shown on the left.

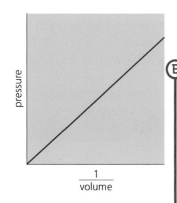

The findings can be expressed as a law:

> For a fixed mass of gas at constant temperature, the pressure is inversely proportional to the volume.

This is known as **Boyle's law**.

FORCES AND PRESSURE

(E) Here is another way of writing Boyle's law. If the pressure of a gas changes from p_1 to p_2 when the volume is changed from V_1 to V_2:

$$p_1 \times V_1 = p_2 \times V_2 \quad \text{(at constant temperature)}$$

Example An air bubble has a volume of 2 cm³ when released at a depth of 20 m in water. What will its volume be when it reaches the surface? Assume that the temperature does not change and that atmospheric pressure is equivalent to the pressure from a column of water 10 m deep.

In this case: p_1 = atmospheric pressure + pressure due to 20 m of water
= 1 atm + 2 atm = 3 atm

Also: $p_2 = 1$ atm, $V_1 = 2$ cm³, and V_2 is to be found.

As the temperature does not change, Boyle's law applies. So:

$p_1 \times V_1 = p_2 \times V_2$ (at constant temperature)

So: $3 \times 2 = 1 \times V_2$ (omitting units for simplicity)

This gives $V_2 = 6$, so on the surface, the volume of the bubble is 6 cm³.

Pressure essentials

pressure = $\frac{\text{force}}{\text{area}}$

If force is measured in newtons (N) and area in square metres (m²), pressure is measured in pascals (Pa): 1 Pa = 1 N/m². Standard atmospheric pressure, called 1 atmosphere (atm), is approximately 100 000 Pa.

Explaining Boyle's law*

The **kinetic theory**, summarized on the right, explains Boyle's law like this. In a gas, the molecules are constantly striking and bouncing off the walls of the container. The force of these impacts causes the pressure. If the volume of the gas is halved, as shown below, there are twice as many molecules *in each cubic metre*. So, every second, there are twice as many impacts with each square metre of the container walls. So the pressure is doubled.

A gas that exactly obeys Boyle's law is called an **ideal gas**. Real gases come close to this provided they have a low density, a temperature well above their liquefying point, and are not full of water vapour. Unless these conditions are met, attractions between molecules affect their behaviour. An ideal gas has no attractions between its molecules.

The kinetic theory

According to this theory, a gas is made up of tiny, moving particles (usually molecules). These are spaced out with almost no attractions between them, and move about freely at high speed. The higher the temperature, then on average, the faster they move.

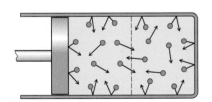

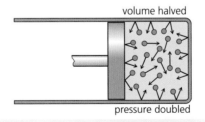
volume halved
pressure doubled

Q

1 If you squash a balloon, the pressure inside it rises. How does the kinetic theory explain this?

2 A balloon contains 6 m³ of helium at a pressure of 100 kPa. As the balloon rises through the atmosphere, the pressure falls and the balloon expands. Assuming that the temperature does not change, what is the volume of the balloon when the pressure has fallen to
a 50 kPa **b** 40 kPa?

3 The readings below are for a fixed mass of gas at constant temperature:

pressure/ atm	5.0	4.0	2.0	1.0	0.5	0.4
volume/ cm³	4	5	10	20	40	50

a How can you tell that the gas obeys Boyle's law?
b Use a calculator to work out values for 1/volume. Plot a graph of pressure against 1/volume and describe its shape.

Related topics: pressure 3.5; air pressure 3.7; kinetic theory 5.1; temperature 5.2; water vapour 5.8

Check-up on forces and pressure

Further questions

1 The diagram shows a pair of nutcrackers. Forces F are applied to the handles of the nutcrackers.

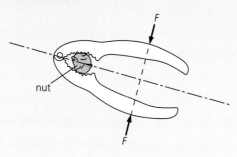

a The forces on the nut are bigger than F. Explain this. [1]

b The nut does not crack. State **two** changes that could be made to crack the nut. [2]

2 The diagram below shows a uniform metre rule, weight W, pivoted at the 75 cm mark and balanced by a force of 2 N acting at the 95 cm mark.

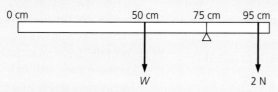

a Calculate the moment of the 2 N force about the pivot. [2]

b Use the principle of moments to calculate the value of W in N. [2]

3 The diving bell below contains trapped air at the same pressure as the water outside. At the surface, air pressure is 100 kPa. As the bell descends, the pressure on it increases by 100 kPa for every 10 m of depth.

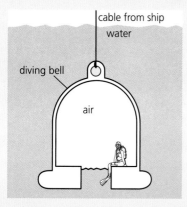

a What is the pressure on the diver at depths of 0 m, 10 m, 20 m, and 30 m? [2]

b At the surface, the bell holds 6 m³ of air. If the bell is lowered to a depth of 20 m, and no more air is pumped into it, what will be the volume of the trapped air? (Assume no change in temperature.) [3]

4 The figure shows an empty wheelbarrow which weighs 80 N.

The operator pulls upwards on the handles with a force of 20 N to keep the handles horizontal.

The point marked M is the centre of gravity of the wheelbarrow.

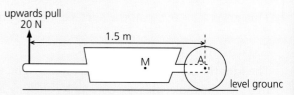

a Copy the figure and draw arrows to show the other two vertical forces that act on the wheelbarrow. [2]

b Determine
  i the moment of the 20 N force about the centre of the wheel A,
  ii the distance between points A and M. [3]

5 The following results were obtained when a spring was stretched.

load /N	1.0	3.0	4.5	6.0	7.5
length of spring /cm	12.0	15.5	19.0	22.0	25.0

a Use the results to plot a graph of length of spring against load. [1]

b Use the graph to find the
  i unloaded length of the spring, [1]
  ii extension produced by a 7.0 N load, [1]
  iii load required to increase the length of the spring by 5.0 cm. [1]

6 a A glass window pane covers an area of 0.6 m². The force exerted by air pressure on the outside of the glass window pane is 60 000 N. Calculate the pressure of the air. Write down the formula that you use and show your working. [3]

b Explain why the window does not break under this force. [1]

7 A fitness enthusiast is trying to strengthen her calf muscles.
She uses the exercise machine below. Her heels apply a force to the padded bar. This lifts the heavy weights.

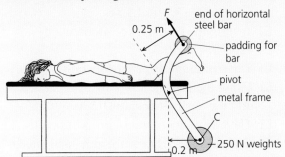

a The centre of gravity of the weights is at C. Draw a diagram to show **where** and in which **direction** the force of gravity acts on the weights. Label this force W. [2]

b The narrow steel bar is padded. Why does this feel more comfortable when lifting the weights? [2]

E c The heels press against the pad with a force F and cause a turning effect about the pivot. Calculate the value of F when the weights are in the position shown in the diagram. Show your working. [3]

d Why does it become harder to lift the weights when they move to the right? [2]

8 Three concrete blocks can be stacked in two different ways as shown below.

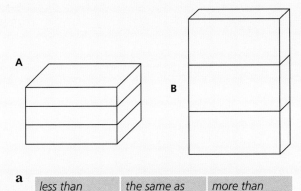

a
| less than | the same as | more than |

Copy and complete the paragraph below using a phrase from the boxes above. Each phrase may be used once, more than once or not at all.
The force of stack **A** on the ground is _____ the force of stack **B**.

The pressure on the ground from stack **B** is _____ the pressure from stack **A**, because the area in contact with the ground for **B** is _____ for **A**. [3]

b Write down, **in words**, the equation connecting pressure, force, and area. [1]

c If the weight of stack **A** is 500 N and the area in contact with the ground is 200 cm², calculate the pressure on the ground in N/cm². [2]

9 The figure shows a tyre used on a large earth-moving vehicle.

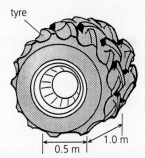

a When the vehicle is loaded, the area of **each** tyre in contact with the ground is a rectangle of sides 1.0 m and 0.5 m.
  i Calculate the area in m² of contact of **one** tyre with the ground.
  ii The vehicle has four of these tyres. Calculate the total area in m² of contact with the ground. [4]

b When the vehicle is loaded, it weighs 100 000 N.
Calculate the pressure in N/m² exerted on the ground by the tyres. [3]

10 A rectangular storage tank has a base measuring 3 m by 2 m. The tank is filled with water to a depth of 2 m. The density of the water is 1000 kg/m³, and g is 10 N/kg. Draw a diagram of the tank with the water in it, and mark all the dimensions on your drawing. Then calculate the following:
a The volume of water in the tank. [2]
b The mass of water in the tank. [2]
c The weight of water in the tank (in N). [2]
d The pressure at the bottom of the tank. [2]

FORCES AND PRESSURE

Use the list below when you revise for your IGCSE examination. The spread number, in brackets, tells you where to find more information.

Revision checklist

Core Level
- ☐ Factors affecting the moment (turning effect) of a force. (3.1)
- ☐ How to calculate the moment of a force. (3.1)
- ☐ Applying the principle of moments to a balanced beam. (3.1)
- ☐ The conditions applying when an object is equilibrium. (3.1)
- ☐ The meaning of centre of gravity. (3.2)
- ☐ Finding the centre of gravity of flat sheet by experiment. (3.2)
- ☐ How the position of the centre of gravity affects stability. (3.2)
- ☐ How forces can produce a change in size and shape. (3.4)
- ☐ How the extension changes with load when a spring is stretched. (3.4)
- ☐ How to obtain extension-load graphs by experiment. (3.4)
- ☐ How to interpret extension-load graphs. (3.4)
- ☐ How pressure depends on force and area. (3.5)
- ☐ Using the equation linking pressure, force, and area. (3.5)
- ☐ How the pressure beneath the surface of a liquid changes with depth and density of the liquid. (3.6)
- ☐ Describing, using ideas about particles (molecules), how the pressure of a gas changes with volume when the temperature is kept constant. (3.8)

Extended Level
As for Core Level, plus the following:
- ☐ Solving problems using the principle of moments. (3.1 and 3.3)
- ☐ Testing the principle of moments by experiment. (3.3)
- ☐ The meaning of spring constant. (3.4)
- ☐ Using the equation linking extension, load, and the spring constant. (3.4)
- ☐ The meaning of limit of proportionality. (3.4)
- ☐ Calculating the pressure at a particular depth in a liquid: the equation linking pressure, depth, and g (3.6)
- ☐ How, if a gas is at constant temperature and obeys Boyle's law, pV is constant. (3.9)
- ☐ Using the equation $p_1V_1 = p_2V_2$ for a gas at constant temperature. (3.9)

4
Forces and energy

- WORK
- ENERGY
- CONSERVATION OF ENERGY
- POTENTIAL AND KINETIC ENERGY
- EFFICIENCY
- POWER
- POWER STATIONS
- ENERGY RESOURCES
- RENEWABLE AND NON-RENEWABLE ENERGY
- ENERGY FROM THE SUN

The Niagara Falls, on the USA–Canada border. The photograph shows the highest section of the falls, where the water tumbles over 50 metres to the river below. Nearly three million litres of water flow over the falls every second. Most of the energy is wasted, but some is harnessed by a hydroelectric power station which generates electricity for the surrounding area.

4.1 Work and energy

Objectives: – to define work done – to know the unit of work and energy – to know that energy can be held in different energy stores.

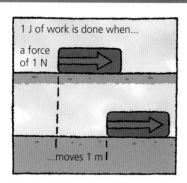

1 J of work is done when...
a force of 1 N
...moves 1 m

Work

In everyday language, work might be writing an essay or digging the garden. But to scientists and engineers, work has a precise meaning: work is done whenever a force makes something move. The greater the force and the greater the distance moved, the more work is done.

The SI unit of work is the **joule (J)**:

> 1 joule of work is done when a force of 1 newton (N) moves an object 1 metre in the direction of the force.

Work is calculated using this equation:

> work done = force × distance moved in the direction of the force

In symbols: $W = F \times d$

For example, if a 4 N force moves an object 3 m, the work done is 12 J.

1 kilojoule (kJ) = 1000 J (10^3 J)
1 megajoule (MJ) = 1 000 000 J (10^6 J)

Energy

A compressed spring has energy; so does a tankful of petrol. Both can be used to do work. Like work, energy is measured in joules (J). Although people talk about energy being stored or released, energy isn't a 'thing'. If, for example, a compressed spring stores 100 joules of energy, this is just a measurement of how much work could be done by the spring if the energy were released.

Energy can be stored in different ways. These are described on the opposite page. To understand them, you need to know the following:

- Moving objects store energy. For example, a moving ball can do work by knocking something over.
- Materials are made up of particles (atoms and molecules) that are constantly in motion. For example in a solid, the particles are vibrating. If the solid is heated, its temperature rises, and the particles move faster.

▲ Particles vibrating in a solid. The particles have energy because of their motion.

▶ A fully flexed bow stores about 300 joules of energy.

80

FORCES AND ENERGY

Energy stores

It sometimes helps to add a label to the word energy to describe how it is being stored. Here are the names used to describe the different **energy stores**:

Kinetic energy This is energy stored by an object because of its motion. 'Kinetic' means 'moving'.

Gravitational potential energy This is energy stored by objects lifted upwards against the force of gravity. It is released when they fall downwards.

Elastic (strain) energy A stretched rubber band stores energy, so does a compressed spring.

Chemical energy This describes energy stored in the chemical bonds between atoms. Fuels store energy in this way, so do batteries and foods. It can be released by chemical reactions – a fuel burning for example.

Electrostatic energy If electric charges attract each other but are held apart, they store energy. For more on the forces between charges, see spread 8.1.

Nuclear energy An atom has a nucleus at its centre. This is made up of particles bound together by strong forces. In some atoms, if the particles become rearranged, or the nucleus splits, stored energy is released.

Thermal energy In all materials, the particles are moving (see opposite page), so they store energy. When a hot object cools, its particles slow down, so some of the stored energy is released – people commonly called it heat. Thermal energy is related to **internal energy**. For more on this, see spread 5.1.

Magnetic energy* If two magnets attract each other but are held apart, they store energy. For more on the forces between magnetic poles, see spread 9.1.

In many of the above, objects or particles store energy because of their position. The general name for this is **potential energy**.

Energy pathways

Everywhere around us, things are rising, falling, speeding up, slowing down, heating, cooling, charging, discharging, burning... In other words, energy is being **transferred** (shifted) from one store to another. Here are the four **pathways** it can take:

- Mechanically, by a force moving something.
- Electrically, by a current.
- By heating because of a temperature difference.
- By radiation such as light waves and sound waves.

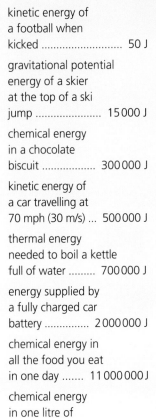

Typical energy values

kinetic energy of a football when kicked .... 50 J

gravitational potential energy of a skier at the top of a ski jump .... 15 000 J

chemical energy in a chocolate biscuit .... 300 000 J

kinetic energy of a car travelling at 70 mph (30 m/s) ... 500 000 J

thermal energy needed to boil a kettle full of water .... 700 000 J

energy supplied by a fully charged car battery .... 2 000 000 J

chemical energy in all the food you eat in one day .... 11 000 000 J

chemical energy in one litre of petrol .... 35 000 000 J

Electrical energy?

An electric current transfers energy. It doesn't store it. That's why electrical energy doesn't appear in the list of energy stores. However the name is still commonly used when calculating the energy supplied by battery or generator (see spread 8.12).

Q

1. How much work is done if a force of 12 N moves an object a distance of 5 m?
2. If you use a 40 N force to lift a bag, and do 20 J of work, how far do you lift it?
3. Express the following amounts of energy in joules:
   a 10 kJ   b 35 MJ   c 0.5 MJ   d 0.2 kJ
4. Using information in the chart of energy values on this page, estimate how many fully charged car batteries are needed to store the same amount of energy as one litre of petrol.
5. a Write down three ways in which the falling apple on the right stores energy.
   b Using the energy chart on this page as a guide, decide which of the apple's energy stores has the most energy.

Related topics: scientific notation **1.1**; SI units **1.2**; force **2.6**; particles **5.1**; electrons in circuits **8.4**

4.2 Energy transfers

Objectives: – to know the law of conservation of energy, and what happens to wasted energy – to know the link between work done and energy transferred.

Conservation of energy

Money doesn't disappear when you spend it. It goes somewhere else! Energy is similar: it never disappears. People talk about 'using energy' but what really happens is that the energy is moved from one store to another store:

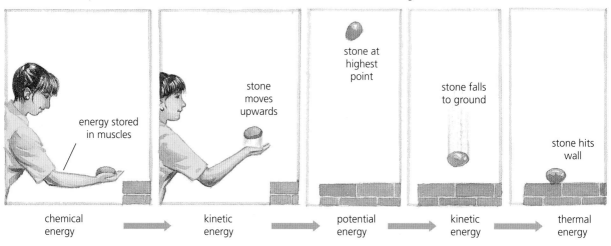

The diagram above shows a sequence of energy transfers. In the last transfer, energy is moved from a kinetic store to a thermal store. When the stone hits the ground, it makes the particles (atoms and molecules) in the ground move faster, so the materials warm up a little.

During each transfer, the total quantity of energy stays the same. This is an example of the **law of conservation of energy**:

> Energy can be stored or transferred, but it cannot be created or destroyed.

Wasting energy

Work and energy essentials

Work is done whenever a force makes something move.

work done
= force × distance moved

Things have energy if they can be used to do work.

Work and energy are both measured in joules (J).

The diagram above shows the energy transfers as a simple chain. In reality, at different stages, some energy is transferred elsewhere. For example, muscles action transfers less than 20% of the energy stored in your food to a kinetic store. The rest is wasted as thermal energy – which is why you heat up when you exercise. And when an object moves through air, some energy is transferred to a thermal store because of friction (air resistance). Even energy transferred by sound waves ends up being stored thermally. The stone, ground, and surroundings become a little warmer than before.

E The diagram at the top of the next page shows how *all* of the energy from the thrower's chemical store is eventually transferred to a thermal store – although most of it is far too spread out to detect. Despite the energy *dissipated* (wasted), the law of conservation of energy still applies. The *total quantity* of energy is unchanged.

FORCES AND ENERGY

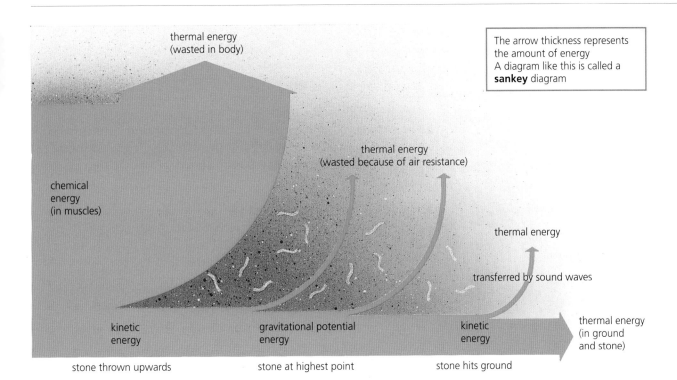

The arrow thickness represents the amount of energy
A diagram like this is called a **sankey** diagram

Work done and energy transferred

In the diagram on the right, a brick is dropped. As it falls to the ground, it loses height and speeds up. Before it falls, it has 20 J of energy in its gravitational potential energy store. When it about to strike the ground, all 20 J has been transferred to its kinetic store (assuming no air resistance). Looked at in another way, 20 J of work has been done by the force of gravity.

Whenever work is done:

work done = energy transferred

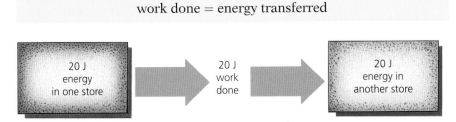

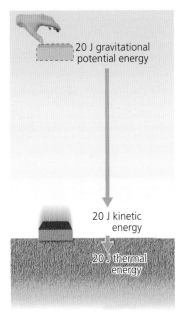

1. 50 J of work must be done to lift a vase from the ground up on to a shelf.
   a When the vase is on the shelf, what is its gravitational potential energy?
   b If the vase falls from the shelf, how much kinetic energy does it have just before it hits the ground? (Assume that air resistance is negligible.)
   c What happens to this energy after the vase has hit the ground?
2. What is the law of conservation of energy?
3. On the right, you can see someone's idea for an electric fan that costs nothing to run. The electric motor which turns the fan also turns a generator. This produces electricity for the motor, so no battery or mains supply is needed! Explain why this idea will not work.

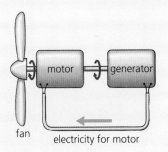

Related topics: work and energy stores **4.1**

4.3 Calculating PE and KE

Objectives: – to know how to calculate kinetic energy and gravitational potential energy – to solve problems involving both.

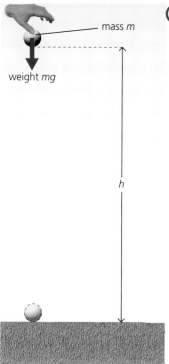

The ball on the left has potential energy because of the Earth's gravitational pull on it and its position above the ground. This is called **gravitational potential energy** (**PE**). If the ball falls, it gains **kinetic energy** (**KE**). Both PE and KE can be calculated.

Calculating PE

The gravitational potential energy of the ball on the left is equal to the work which would be done if the ball were to fall to the ground. Assuming no air resistance, it is also equal to the work done in lifting the ball a distance h up from the ground in the first place:

$$\text{downward force on ball} = \text{weight of ball} = mg$$

So: $\quad$ upward force needed to lift ball $= mg$

So: $\quad$ work done in lifting ball $=$ force $\times$ distance moved
$$= mgh$$

For an object of mass m at a vertical height h above the ground:

$$\text{gravitational potential energy} = mgh$$

For example, if a 2 kg mass is 3 m above the ground, and g is 10 N/kg:
gravitational PE $= 2 \text{ kg} \times 3 \text{m} \times 10 \text{ N/kg} = 60 \text{ J}$

Calculating KE

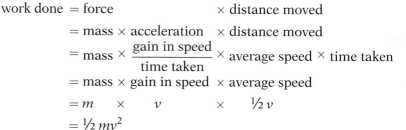

The kinetic energy of the ball above is equal to the work which the ball could do by losing all of its speed. Assuming no air resistance, it is also equal to the work done on the ball in increasing its speed from zero to v in the first place:

$$\begin{aligned}
\text{work done} &= \text{force} \times \text{distance moved} \\
&= \text{mass} \times \text{acceleration} \times \text{distance moved} \\
&= \text{mass} \times \frac{\text{gain in speed}}{\text{time taken}} \times \text{average speed} \times \text{time taken} \\
&= \text{mass} \times \text{gain in speed} \times \text{average speed} \\
&= m \times v \times \tfrac{1}{2}v \\
&= \tfrac{1}{2}mv^2
\end{aligned}$$

For an object of mass m and speed v:

$$\text{kinetic energy} = \tfrac{1}{2}mv^2$$

For example, if a 2 kg mass has a speed of 3 m/s:
kinetic energy $= \tfrac{1}{2} \times 2\text{kg} \times (3 \text{ m/s})^2 = \tfrac{1}{2} \times 2 \times 3^2 \text{ J} = 9 \text{ J}$

Units
Mass is measured in kilograms (kg).
Force is measured in newtons (N).
Weight is a force.
Work is measured in joules (J).
Energy is measured in joules (J).

Useful equations
average speed $= \dfrac{\text{distance moved}}{\text{time taken}}$
acceleration $= \dfrac{\text{gain in speed}}{\text{time taken}}$
force $=$ mass $\times$ acceleration
weight $=$ mass $\times g$
$\quad (g = 10 \text{ N/kg})$
work done $=$ force $\times$ distance moved
work done $=$ energy transformed

FORCES AND ENERGY

E Scalar energy

Energy is a **scalar** quantity: it has magnitude (size) but no direction. So you do not have to allow for direction when doing energy calculations.

On the right, objects A and B have the same mass and are at the same height above the ground. B was lifted vertically but A was moved up a smooth slope. Although A had to be moved further, less force was needed to move it, and the work done was the same as for B. As a result, both objects have the same PE. The PE (mgh) depends on the vertical gain in height h and not on the particular path taken to gain that height.

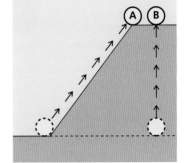

KE and PE problems

Example If the stone on the right is dropped, what is its kinetic energy when it has fallen half-way to the ground? ($g = 10$ N/kg)

In problems like this, you don't necessarily have to use KE = $\frac{1}{2}mv^2$. When the stone falls, its *gain in KE* is equal to its *loss in PE*, so you can calculate that instead:

height lost by stone = 2 m

So: gravitational PE lost by stone = mgh = 4 kg × 10 N/kg × 2 m = 80 J

So: KE gained by stone = 80 J

As the stone started with no KE, this is the stone's KE half-way down.

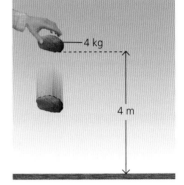

Example The stone on the right slides down a smooth slope. What is its speed when it reaches the bottom? ($g = 10$ N/kg)

This problem can also be solved by considering energy changes. At the top of the slope, the stone has extra gravitational PE. When it reaches the bottom, all of this PE has been transformed into KE.

gravitational PE at top of slope = mgh = 4 kg × 10 N/kg × 5 m = 200 J

So: kinetic energy at bottom of slope = 200 J

So: $\frac{1}{2}mv^2$ = 200 J

So: $\frac{1}{2} \times 4$ kg $\times v^2$ = 200 J

This gives: $v = 10$ m/s

So the stone's speed at the bottom of the slope is 10 m/s.

Note: if the stone fell vertically, it would start with the same gravitational PE and end up with the same KE, so its final speed would still be 10 m/s.

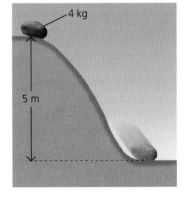

Assume that g is 10 N/kg and that air resistance and other frictional forces are negligible.

1. An object has a mass of 6 kg. What is its gravitational potential energy
   a 4 m above the ground   b 6 m above the ground?
2. An object of mass 6 kg has a speed of 5 m/s.
   a What is its kinetic energy?
   b What is its kinetic energy if its speed is doubled?
3. A ball of mass 0.5 kg has 100 J of kinetic energy. What is the speed of the ball?
4. A ball has a mass of 0.5 kg. Dropped from a cliff top, the ball hits the sea below at a speed of 10 m/s.
   a What is the kinetic energy of the ball as it is about to hit the sea?
   b What was the ball's gravitational potential energy before it was dropped?
   c From what height was the ball dropped?
   d A stone of mass 1 kg also hits the sea at 10 m/s. Repeat stages **a**, **b**, and **c** above to find the height from which the stone was dropped.

Related topics: speed and acceleration **2.1**; force, mass, and acceleration **2.7**; mass and weight **2.9**; work and energy **4.1–4.2**

4.4 Efficiency and power

Objectives: – to explain what efficiency is – to know, and be able to use, the equation for calculating power.

Force, work, and energy essentials
Work is measured in joules (J).

Energy is measured in joules (J).

work done = energy transferred

Force is measured in newtons (N).

work done = force × distance moved

Inputs and outputs
In any system, the *total* energy output must equal the *total* energy input. That follows from the law of conservation of energy. Therefore, the equations on the right could also be written with 'total energy output' replacing 'total energy input'.

Typical power outputs
washing machine motor	250 W
athlete	400 W
small car engine	35 000 W
large car engine	150 000 W
large jet engine	75 000 000 W

1 kilowatt (kW) = 1000 W

Engines and motors do work by making things move. Petrol and diesel engines need the energy stored in their fuel. Electric motors rely on the energy transferred from a battery or generator. The human body is also a form of engine. It needs the energy stored in food.

Efficiency

An engine does useful work with some of the energy supplied to it, but the rest is wasted as thermal energy (heat). The **efficiency** of an engine can be calculated as follows:

$$\text{efficiency} = \frac{\text{useful work done}}{\text{total energy input}} \quad \text{or} \quad \text{efficiency} = \frac{\text{useful energy output}}{\text{total energy input}}$$

For example, if a petrol engine does 25 J of useful work for every 100 J of energy supplied to it, then its efficiency is ¼, or 25%. In other words, its useful energy output is ¼ of its total energy input.

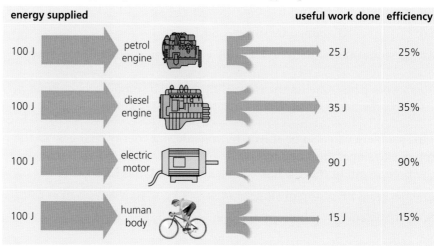

The chart above shows the efficiencies of some typical engines and motors. The low efficiency of fuel-burning engines is not due to poor design. When a fuel burns, it is impossible to transfer its thermal energy to kinetic (motion) energy without wasting much of it.

Power

A small engine can do just as much work as a big engine, but it takes longer to do it. The big engine can do work at a faster rate. The rate at which work is done is called the **power**.

The SI unit of power is the **watt (W)**. A power of 1 watt means that work is being done (or energy transferred) at the rate of 1 joule per second. Power can be calculated as follows:

$$\text{power} = \frac{\text{work done}}{\text{time taken}} \quad \text{or} \quad \text{power} = \frac{\text{energy transferred}}{\text{time taken}}$$

For example, if an engine does 1000 joules of useful work in 2 seconds, its power output is 500 watts (500 joules per second).

FORCES AND ENERGY

As energy and power are related, there is another way of calculating the efficiency of an engine:

$$\text{efficiency} = \frac{\text{useful power output}}{\text{total power input}}$$

The **horsepower (hp)** is a power unit which dates back to the days of the early steam engines: 1 hp = 746 W (about ¾ kilowatt)

Power problems

Example 1 The crane on the right lifts a 100 kg block of concrete through a vertical height of 16 m in 20 s. If the power input to the motor is 1000 W, what is the efficiency of the motor?

On Earth, g is 10 N/kg, so a 100 kg block has a weight of 1000 N. Therefore, a force of 1000 N is needed to lift the block. When the block is lifted:

$$\text{work done} = \text{force} \times \text{distance} = 1000 \text{ N} \times 16 \text{ m} = 16\,000 \text{ J}$$

$$\text{useful power output} = \frac{\text{useful work done}}{\text{time taken}} = \frac{16\,000 \text{ J}}{20 \text{ s}} = 800 \text{ W}$$

$$\text{efficiency} = \frac{\text{useful power output}}{\text{total power input}} = \frac{800 \text{ W}}{1000 \text{ W}} = 0.8$$

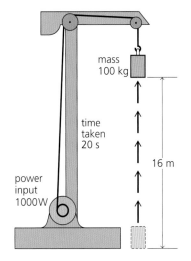

So the motor has an efficiency of 80%.

Example 2* The car on the right has a steady speed of 30 m/s. If the total frictional force on the car is 700 N, what useful power output does the engine deliver to the driving wheels?

As the speed is steady, the engine must provide a forward force of 700 N to balance the total frictional force. In 1 second, the 700 N force moves 30 m, so: work done = force × distance = 700 N × 30 m = 21 000 J.

As the engine does 21 000 J of useful work in 1 second, its useful power output must be 21 000 W, or 21 kW.

Problems of this type can also be solved with this equation:

$$\text{useful power output} = \text{force} \times \text{speed}$$

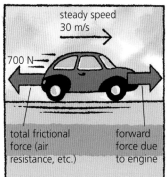

g = 10 N/kg

1 An engine does 1500 J of useful work with each 5000 J of energy supplied to it.
   a What is its efficiency?
   b What happens to the rest of the energy supplied?
2 If an engine does 1500 J of work in 3 seconds, what is its useful power output?
3 A motor has a useful power output of 3 kW.
   a What is its useful power output in watts?
   b How much useful work does it do in 1 s?
   c How much useful work does it do in 20 s?
   d If the power input to the motor is 4 kW, what is the efficiency?
4 Someone hauls a load weighing 600 N through a vertical height of 10 m in 20 s.
   a How much useful work does she do?
   b How much useful work does she do in 1 s?
   c What is her useful power output?
5 A crane lifts a 600 kg mass through a vertical height of 12 m in 18 s.
   a What weight (in N) is the crane lifting?
   b What is the crane's useful power output?
6* With frictional forces acting, a forward force of 2500 N is needed to keep a lorry travelling at a steady speed of 20 m/s along a level road. What useful power is being delivered to the driving wheels?

Related topics: SI units **1.2**; force, mass, weight, and g **2.9**; law of conservation of energy **4.2**; work and energy **4.2–4.3**

87

4.5 Energy for electricity (1)

Objectives: – to describe the main features of a thermal power station – to explain why some energy is always wasted when electricity is generated.

▶ Part of a thermal power station. The large, round towers with clouds of steam coming from them are cooling towers.

Industrial societies spend huge amounts of energy. Much of it is supplied by electricity which comes from **generators** in **power stations**.

Thermal power stations

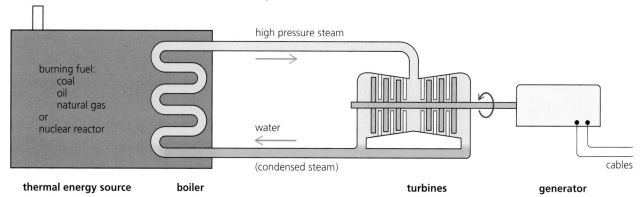

In most power stations, the generators are turned by **turbines**, blown round by high pressure steam. To produce the steam, water is heated in a boiler. The thermal energy comes from burning fuel (coal, oil, or natural gas) or from a **nuclear reactor**. Nuclear fuel does not burn. Its energy is released by nuclear reactions which split uranium atoms. The process is called **nuclear fission**. Future reactors may use **nuclear fusion:** see 10.7.

Once steam has passed through the turbines, it is cooled and condensed (turned back into a liquid) so that it can be fed back to the boiler. Some power stations have huge cooling towers, with draughts of air up through them. Others use the cooling effect of nearby sea or river water.

▲ A turbine

▶ Block diagram of what happens in a thermal power station

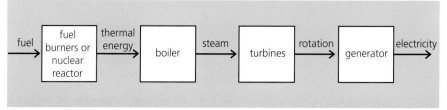

FORCES AND ENERGY

E Energy spreading

Thermal power stations waste more energy than they deliver. Most is lost as thermal energy in the cooling water and waste gases. For example, the efficiency of a typical coal-burning power station is only about 35% – in other words, only about 35% of the energy in its fuel is transformed into electrical energy. The diagram below shows what happens to the rest:

efficiency
$$= \frac{\text{useful energy output}}{\text{energy input}}$$
$$= \frac{\text{useful power output}}{\text{power input}}$$

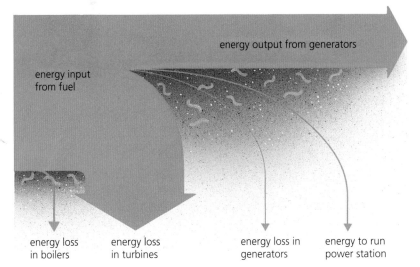

energy input from fuel

energy output from generators

energy loss in boilers | energy loss in turbines | energy loss in generators | energy to run power station

The power output of power stations is usually measured in megawatts (MW) or in gigawatts (GW):

1 MW = 1 000 000 W
          (1 million watts)

1 GW = 1000 MW
          (1 billion watts)

◀ Typical energy-flow chart for a thermal power station. A chart like this is called a **Sankey diagram**. The thickness of each arrow represents the amount of energy.

E Engineers try to make power stations as efficient as possible. But once energy is in thermal form, it cannot all be used to drive the generators. Thermal energy is the energy of randomly moving particles (such as atoms and molecules). It has a natural tendency to spread out. As it spreads, it becomes less and less useful. For example, the concentrated energy in a hot flame could be used to make steam for a turbine. But if the same amount of thermal energy were spread through a huge tankful of water, it would only warm the water by a few degrees. This warm water could not be used as an energy source for a turbine.

District heating* The unused thermal energy from a power station does not have to be wasted. Using long water pipes, it can heat homes, offices, and factories in the local area. This works best if the power station is run at a slightly lower efficiency so that hotter water is produced.

Combined cycle gas turbine power stations
These are smaller units which can be brought up to speed or shut off very quickly, as the demand for electricity varies. In them, natural gas is used as the fuel for a jet engine. The shaft of the engine turns one generator. The hot gases from the jet are used to make steam to drive another generator.

1. Write down *four* different types of fuel used in thermal power stations.
2. In a thermal power station:
   a What is the steam used for?
   b What do the cooling towers do?
3. The table on the right gives data about the power input and losses in two power stations, X and Y.
   a Where is most energy wasted?
   b In what form is this wasted energy lost?
   c What is the electrical power output of each station? (You can assume that the table shows all the power losses in each station.)
   d What is the efficiency of each power station?

	power station X coal	power station Y nuclear
power input from fuel in MW	5600	5600
power losses in MW:		
– in reactors/boilers	600	200
– in turbines	2900	3800
– in generators	40	40
power to run station in MW	60	60
electrical power output in MW	?	?

Related topics: energy 4.1–4.2; efficiency and power 4.4; generators 9.9; electricity supply 9.12; nuclear energy 10.6–10.7

Energy for electricity (2)

Objectives: – to describe the alternatives to thermal power stations, and give the pros and cons of each type.

Energy units

The electricity supply industry uses the **kilowatt-hour (kWh)** as its unit of energy measurement:

1 kWh is the energy supplied by a 1 kW power source in 1 hour.

As 1 watt = 1 joule per second (J/s), a 1 kW power source supplies energy at the rate of 1000 joules per second. So in 1 hour, or 3600 seconds, it supplies 3600 × 1000 joules (J). Therefore:

1 kWh = 3 600 000 J

Reactions for energy*

When fuels burn, they combine with oxygen in the air. With most fuels, including fossil fuels, the energy is released by this chemical reaction:

$$\underbrace{\text{fuel} + \text{oxygen}}_{\text{these are used up}} \xrightarrow{\text{burning}} \underbrace{\text{carbon dioxide} + \text{water}}_{\text{these waste gases are made}} + \textit{thermal energy}$$

There may be other waste gases as well. For example, burning coal produces some sulfur dioxide. Natural gas, which is mainly methane, is the 'cleanest' (least polluting) of the fuels burned in power stations.

In a nuclear power station, the nuclear reactions produce no waste gases like those above. However, they do produce radioactive waste.

Pollution problems

Thermal power stations can cause pollution in a variety of ways:
- Fuel-burning power stations put extra carbon dioxide gas into the atmosphere. This traps the Sun's energy and is adding to global warming. Coal-burning power stations emit almost twice the amount of carbon dioxide per kJ output compared with those burning natural gas.
- Unless low-sulfur coal is used, or desulfurization (FGD) units are fitted, coal-burning power stations emit sulfur dioxide, which is harmful to health and causes acid rain.
- Transporting fuels can cause pollution. For example, there may be a leak from an oil tanker.
- The radioactive waste from nuclear power stations is highly dangerous. It must be carried away and stored safely in sealed containers for many years – in some cases, thousands of years.
- Nuclear accidents are rare. But when they do occur, radioactive gas and dust can be carried thousands of kilometres by winds.

▲ One effect of acid rain

Power from water and wind

Some generators are turned by the force of moving water or wind. There are three examples on the next page. Power schemes like this have no fuel costs, and give off no polluting gases. However, they can be expensive to build, and need large areas of land. Compared with fossil fuels, moving water and wind are much less concentrated sources of energy:

1 kWh of electrical energy can be supplied using…

…0.5 litres of oil (burning)

…5000 litres of fast-flowing water (20 m/s)

FORCES AND ENERGY

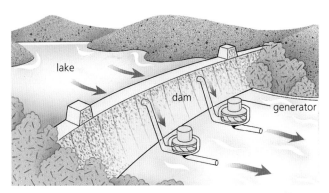

Hydroelectric power scheme River and rain water fill up a lake behind a dam. As water rushes down through the dam, it turns turbines which turn generators.

Pumped storage scheme This is a form of hydroelectric scheme. At night, when power stations have spare capacity, power is used to pump water from a lower reservoir to a higher one. During the day, when extra electricity is needed, the water runs down again to turn generators.

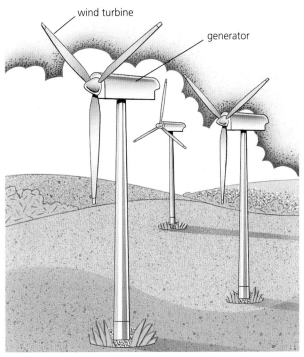

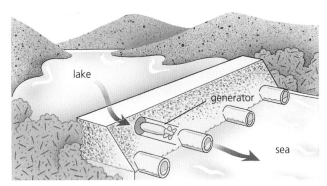

Tidal power scheme A dam is built across a river where it meets the sea. The lake behind the dam fills when the tide comes in and empties when the tide goes out. The flow of water turns the generators.

Wind farm This is a collection of **aerogenerators** – generators driven by giant wind turbines ('windmills').

power station (1 MW = 1 000 000 W)	A coal (non-FGD)	B combined cycle gas	C nuclear	D wind farm	E large tidal scheme
power output in MW	1800	600	1200	20	6000
efficiency (fuel energy → electrical energy)	35%	45%	25%	–	–
The following are on a scale 0–5					
build cost per MW output	2	1	5	3	4
fuel cost per kWh output	5	4	2	0	0
atmospheric pollution per kWh output	5	3	<1	0	0

1 What is the source of energy in a hydroelectric power station?

2 The table above gives data about five different power stations, A–E.

  a C has an efficiency of 25%. What does this mean?

  b Which power station has the highest efficiency? What are the other advantages of this type of power station?

  c Which power station cost most to build?

  d Which power station has the highest fuel cost per kWh output?

  e Which power station produces most atmospheric pollution per kWh output?

  f Why do two of the power stations have a zero rating for fuel costs and atmospheric pollution?

Related topics: efficiency and power **4.4**; energy resources **4.7–4.8**; calculating energy in kWh **8.12**

4.7 Energy resources

Objectives: – to describe the various renewable and non-renewable energy resources available, and give the pros and cons of using each type.

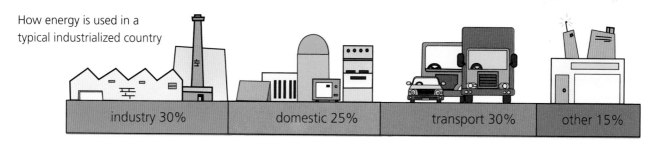

How energy is used in a typical industrialized country

industry 30% | domestic 25% | transport 30% | other 15%

Most of the energy that we use comes from fuels that are burned in power stations, factories, homes, and vehicles. Nearly all of this energy originally came from the Sun. To find out how, see the next spread, 4.8.

The Sun is 75% hydrogen. It releases energy by a process called **nuclear fusion** (see spread 10.7). One day, it may be possible to harness this process on Earth, but until this can be done, we will have to manage with other resources.

The energy resources we use on Earth can be **renewable** or **non-renewable**. For example, wood is a renewable fuel. Once used, more can be grown to replace it. Oil, on the other hand, is non-renewable. It took millions of years to form in the ground, and cannot be replaced.

Non-renewable energy resources

Fossil fuels Coal, oil, and natural gas are called fossil fuels: they formed from the remains of plants and tiny sea creatures that lived millions of years ago. They are a very concentrated source of energy. Petrol, diesel, and jet fuel are all extracted from crude oil (oil as it occurs natural in the ground) oil and it is the raw material for making most plastics.

Natural gas is the 'cleanest' of the fossil fuels (see spread 4.6). At present, it is mostly taken from the same underground rock formations that contain oil – the gas formed with the oil and became trapped above it. However, over the next decades, more and more gas may be extracted from a rock called shale (see left).

Problems When fossil fuels burn, their waste gases pollute the atmosphere. Globally, the most serious concern is the amount of carbon dioxide being produced. There is around 30% more in the atmosphere today than there was 50 years ago, and there is little doubt that this is adding to global warming (see panel on the next page).

Nuclear fuels Most contain uranium. 1 kg of nuclear fuel stores as much energy as 55 tonnes of coal. In nuclear power stations, the energy is released by fission, a process in which the nuclei of uranium atoms are split.

Problems High safety standards are needed. The waste from nuclear fuel is very dangerous and stays radioactive for thousands of years. Nuclear power stations are expensive to build, and expensive to decommission (close down and dismantle at the end of their working life).

Shale gas and fracking

Shale gas (see below right) is extracted from shale by a process called **fracking** (hydraulic fracturing). High-pressure water is pumped into the rock, fracturing it, and opening up cracks so that the trapped gas can flow out. Some see shale gas as a major source of energy for the future. Others have deep concerns about the environmental impact of extracting it.

Where to find out more

For more detailed information on… see spread…

hydroelectric energy	4.6
tidal energy	4.6
wind energy	4.6
solar panel	5.8
energy and mass	10.6
nuclear fission	10.6
nuclear fusion	10.7

Renewable energy resources

Hydroelectric energy A river fills a lake behind a dam. Water flowing down from the lake turns generators.

Problems Expensive to build. Few areas of the world are suitable. Flooding land and building a dam causes environmental damage.

Tidal energy Similar to hydroelectric energy, but a lake fills when the tide comes in and empties when it goes out.

Problems As for hydroelectric energy.

Wind energy Generators are driven by wind turbines ('windmills').

Problems Large, remote, windy sites are needed. Winds are variable. The wind turbines are noisy and can spoil the landscape.

Wave energy Generators are driven by the up-and-down motion of waves at sea.

Problems Difficult to build – few devices have been successful.

Geothermal energy 'Geothermal' means heat from the Earth. Water is pumped down to hot rocks deep underground and rises as steam. In areas of volcanic activity, the steam comes naturally from hot springs.

Problems Deep drilling is difficult and expensive.

Solar energy (energy radiated from the Sun) Solar panels absorb this energy and use it to heat water. Solar cells are made from materials that can deliver an electric current when they absorb the energy in light.

Problems Variable amounts of sunshine in some countries. Solar cells are expensive, and must be large to deliver useful amounts of power. A cell area of around 10 m^2 is needed to power an electric kettle.

Biofuels These are fuels made from plant or animal matter. They include wood, alcohol from sugar cane, and methane gas from rotting waste.

Problems Huge areas of land are needed to grow plants.

Saving energy

Burning fossil fuels causes pollution. But the alternatives have their own environmental problems. That is one reason why we need to be less wasteful with energy by using vehicles less, and more efficiently, and recycling more waste materials. Also, better insulation in buildings would mean less need for heating in cold countries and for air conditioning in hot ones.

A global warning

By burning huge amounts of fossil fuels, industrial societies are putting more carbon dioxide into the atmosphere than is being removed by natural processes. The extra carbon dioxide is trapping more of the Sun's energy, and there is little doubt that this is causing global warming. The result: more extreme weather events, glacial melting, and coastal flooding. Most scientists believe that we have little time left to tackle the problem.

▲ In Brazil, many cars use alcohol as a fuel instead of petrol. The alcohol is made from sugar cane, which is grown as a crop.

To answer these questions, you may need information from the illustration on the next spread, 4.8.

1 Some energy resources are *non-renewable*. What does this mean? Give *two* examples.
2 Give *two* ways of generating electricity in which no fuel is burned and the energy is renewable.
3 The energy in petrol originally came from the Sun. Explain how it got into the petrol.
4 Describe *two* problems caused by using fossil fuels.
5 Describe *two* problems caused by the use of nuclear energy.
6 What is *geothermal* energy? How can it be used?
7 What is *solar* energy? Give *two* ways in which it can be used.
8 Three of the energy resources described in this spread make use of moving water. What are they?
9 Give *four* practical methods of saving energy so that we use less of the Earth's energy resources.

Related topics: power stations **4.5–4.6**; energy from the Sun **4.8**; solar panels **4.8**; nuclear reactors **10.6–10.7**

4.8 How the world gets its energy

Objective: – to understand that most of the world's energy originally came from the Sun.

Solar panels
These absorb energy radiated from the Sun and use it to heat water.

Solar cells
These use the energy in sunlight to produce small amounts of electricity.

Energy in food
We get energy from the food we eat. The food may be from plants, or from animals which fed on plants.

Biofuels from plants
Wood is an important fuel in many countries. When wood is burned, it releases energy that the tree once took in from the Sun. In some countries, sugar cane is grown and fermented to make alcohol. This can be used as a fuel instead of petrol.

Biofuels from waste
Rotting animal and plant waste can give off methane gas. This is similar to natural gas and can be used as a fuel. Marshes, rubbish tips, and sewage treatment works are all sources of methane. Some waste can also be used directly as fuel by burning it.

Batteries
Some batteries (e.g. car batteries) have to be given energy by charging them with electricity. Others are manufactured from chemicals which already store energy. But energy is needed to produce the chemicals in the first place.

E The Sun
The Sun radiates energy because of nuclear fusion reactions deep inside it (see spread 10.7). Its output is equivalent to that from 3×10^{26} electric hotplates. Just a tiny fraction of this reaches the Earth.

Energy in plants
Plants take in energy from sunlight falling on their leaves. They use it to turn water and carbon dioxide from the air into new growth. The process is called photosynthesis. Animals eat plants to get the energy stored in them.

Fossil fuels
Fossil fuels (coal, oil, and natural gas) were formed from the remains of plants and tiny sea creatures which lived many millions of years ago. Industrial societies rely on fossil fuels for most of their energy. Many power stations burn fossil fuels.

Fuels from oil
Many fuels can be extracted from oil (crude). These include: petrol, diesel fuel, jet fuel, paraffin, central heating oil, bottled gas.

The tides
The gravitational pull of the Moon (and to a lesser extent, the Sun) creates gentle bulges in the Earth's oceans. As the Earth rotates, different places have high and low tides as they pass in and out of the bulges. The motion of the tides carries energy with it.

Tidal energy
In a tidal energy scheme, an estuary is dammed to form an artificial lake. Incoming tides fill the lake; outgoing tides empty it. The flow of water in and out of the lake turns generators.

Nucleus of the atom
Radioactive materials have atoms with unstable nuclei (centres) which break up and release energy. The material gives off the energy slowly as thermal energy. Energy can be released more quickly by splitting heavy nuclei (fission). Energy can also be released by joining light nuclei (fusion), as happens in the Sun.

Weather systems
These are driven by energy radiated from the Sun. Heated air rising above the equator causes belts of wind around the Earth. Winds carry water vapour from the oceans and bring rain and snow.

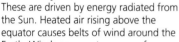

Geothermal energy
Deep underground, the rocks are hotter than they are on the surface. The thermal energy comes from radioactive materials naturally present in the rocks. It can make steam for heating buildings or driving generators.

Nuclear energy
In a reactor, nuclear fission reactions release energy from the nuclei of uranium atoms. This heats water to make steam for driving generators.

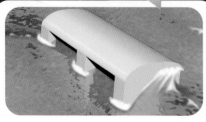

Wave energy
Waves are caused by the wind (and partly by tides). Waves cause a rapid up-and-down movement on the surface of the sea. This movement can be used to drive generators.

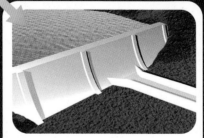

Hydroelectric energy
An artificial lake forms behind a dam. Water rushing down from this lake is used to turn generators. The lake is kept full by river water which once fell as rain or snow.

Wind energy
For centuries, people have been using the power of the wind to move ships, pump water, and grind corn. Today, huge wind turbines are used to turn generators.

Check-up on forces and energy

Further questions

1

wound up spring	batteries connected to motors
rotating flywheel	stretched rubber bands

   a State which of the above **change shape** when their stored energy is transferred. [2]

   b* Describe how the energy from a rotating flywheel can be transferred to moving parts of a child's toy. [2]

2 The diagram below shows a pendulum which was released from position **A**.

   a How is the energy of the pendulum stored at
     i A,   **ii B**,   **iii C**? [3]

   b Eventually the pendulum would stop moving.
    Explain what has happened to the initial energy of the pendulum. [2]

3

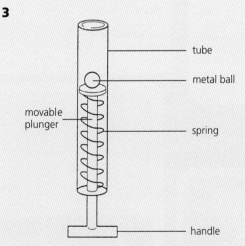

A type of toy catapult consists of a movable plunger which has a spring attached as shown above. The handle was pulled down to fully compress the spring and on release the metal ball of mass 0.1 kg (weight 1 N) was projected 0.75 m vertically.

   a **i** What name is used for the energy stored in a compressed spring? [1]
     ii What happens to this stored energy when the handle of the plunger is released? [2]

(E) **b** Calculate the maximum gravitational potential energy acquired by the metal ball from the catapult. Write down the formula that you use and show your working. Take the acceleration due to gravity to be 10 m/s^2. [3]

   c Explain why the maximum gravitational potential energy gained by the metal ball is less than the original stored energy of the spring. [3]

4 **a** Name **four** renewable energy sources that are used to generate electricity. [4]

   b Most renewable sources have no fuel costs. Give **two other advantages** of using renewable energy rather than fossil fuels. [2]

   c There can be problems with using renewable energy sources. Give **one** example. [1]

   d* If most renewable energy sources have no fuel costs, why isn't the electricity they supply free? [2]

(E) **5** A drop hammer is used to drive a hollow steel post into the ground. The hammer is placed inside the post by a crane. The crane lifts the hammer and then drops it so that it falls onto the baseplate of the post.

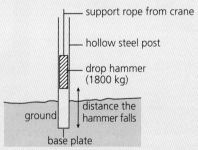

The hammer has a mass of 1800 kg. Its velocity is 5 m/s just before it hits the post.

   a Calculate the kinetic energy of the hammer just before it hits the post. [3]

   b How much gravitational potential energy is transferred from the hammer as it falls? Assume that it falls freely. [1]

   c Calculate the distance the hammer has fallen. (Assume $g = 10$ N/kg) [3]

6 A crate of mass 300 kg is raised by an electric motor through a height of 60 m in 45 s. Calculate:
   a The weight of the crane ($g = 10$ N/kg) [2]
   b The useful work done. [2]
   c The useful power of the motor. [2]
   d The efficiency of the motor, if it takes a power of 5000 W from its electricity supply. [2]

7

electrical appliance	power rating/ kW	power rating/ W
television	0.1	100
electric kettle		2000
food mixer	0.6	

The table above shows the power rating of three electrical appliances.
   a Copy the table and fill in the blank spaces. [2]
   b State which appliance transfers the least amount of energy per second. [1]
   c State which appliance transfers some of the energy it receives to kinetic energy. [1]
   d What happens to the rest of the energy? [1]

8 a Explain what you understand by the phrase *non-renewable energy resources*. [2]
   b Explain why most non-renewable energy resources are burned. [1]
   c Name a non-renewable energy resource which is not burned. [1]

9 a The chemical energy stored in a fossil fuel is transferred to thermal energy when the fuel is burned. Describe how this energy is then used to produce electricity at a power station. [2]
   b What is the environmental impact of generating electricity using fossil fuels? [2]

10 a

fuel/ energy resource	renewable fuel	must be burned to release energy	found in the Earth's crust
coal	no	yes	yes
wood		yes	
uranium			yes

The table above shows that coal is **not** a renewable fuel. It releases energy when burned and is found in the Earth's crust. Copy and complete the table for the other fuels/energy resources named. [2]
   b i Explain how fossil fuels were produced. [1]
      ii State **two** reasons why we should use less fossil fuels. [2]

11 a Most of the energy available on Earth comes, or has come, from the Sun. Some energy resources on Earth store the Sun's energy from millions of years ago. Name one of these resources. [1]
   b Copy and complete the sentence below to say what a fuel does. [2]
      A fuel is a material which supplies _____ when it _____.
   c Explain the difference between renewable and non-renewable fuels. [1]
   d Copy and complete the following table to give examples of some fuels and their uses. The first one has been done for you. [4]

description	example	use
a gaseous fuel	hydrogen	rocket fuel
a liquid fuel		
a solid fuel		
a renewable fuel		
a non-renewable fuel		

12 Copy and complete the following sentences about household electrical devices. Use words from the list below. Each word may be used once, more than once or not at all.

chemical current thermal kinetic light sound

   a In an iron, energy delivered by a _____ is transferred to _____ energy.
   b The energy supplied to a vacuum cleaner is transferred to _____ energy and unwanted _____ energy.
   c In a torch, _____ stored in the battery is carried by a _____ and some is then carried away by _____ waves.
   d Some of the energy supplied to a radio is carried away by _____ waves and the rest is transferred to _____ energy. [9]

13 The list below contains the names of some energy stores.

chemical elastic gravitational nuclear

Copy and complete the table below by naming the energy stored in each one. Use words from the list. Each word may be used once, more than once or not at all. [3]

a bow about to fire an arrow	
water at the top of a waterfall	
a birthday cake	

FORCES AND ENERGY

Use the list below when you revise for your IGCSE examination. The spread number, in brackets, tells you where to find more information.

Revision checklist

Core Level
- ☐ How work done depends on force and distance moved. (4.1)
- ☐ The equation linking work done, force, and distance moved. (4.1)
- ☐ The joule, unit of work and energy. (4.1)
- ☐ The different energy stores. (4.1)
- ☐ The different ways in which energy can be transferred from one store to another. (4.1)
- ☐ The law of conservation of energy. (4.2)
- ☐ How energy is transferred whenever work is done. (4.2)
- ☐ How power depends on work done and time taken (or energy transferred and time taken). (4.4)
- ☐ The watt, unit of power. (4.4)
- ☐ How thermal power stations (fuel-burning power stations and nuclear power stations) produce electricity. (4.5)
- ☐ The alternatives to thermal power stations. (4.6–4.8)
- ☐ The difference between renewable and non-renewable energy resources. (4.7)
- ☐ Non-renewable energy resources:
  – fossil fuels
  – nuclear fuels
  The advantages and disadvantages of each type, including environmental impact. (4.6–4.8)
- ☐ Renewable energy resources:
  – hydroelectric energy
  – tidal energy
  – wind energy
  – wave energy
  – geothermal energy
  – solar energy (solar cells and solar panels)
  The advantages and disadvantages of each type, including environmental impact. (4.6–4.8)
- ☐ How high efficiency means less energy wasted. (4.7)

Extended Level
As for Core Level, plus the following:
- ☐ How the law of conservation of energy applies in a series of energy changes. (4.2)
- ☐ Calculating gravitational potential energy (PE) (4.3)
- ☐ Calculating kinetic energy (KE). (4.3)
- ☐ Solving problems on PE and KE. (4.3)
- ☐ Using the equation linking power, energy transformed (or work done) and time taken. (4.4)
- ☐ Calculating efficiency knowing the energy input and energy output. (4.4)
- ☐ Calculating efficiency knowing the power input and power output. (4.4)
- ☐ How to interpret energy flow diagrams (Sankey diagrams). (4.5)
- ☐ How, in a series of energy changes, energy tends to spread out and become less useful. (4.5)
- ☐ How the Sun is the source of energy for most of our energy resources on Earth. (4.7 and 4.8)
- ☐ How energy is released by nuclear fusion in the Sun. (4.7, 4.8, and 10.7)
- ☐ How research is being carried out to develop nuclear fusion reactors for power stations. (10.7)

5
Thermal effects

- PARTICLES IN SOLIDS, LIQUIDS, AND GASES
- TEMPERATURE
- EXPANSION OF SOLIDS AND LIQUIDS
- HEATING GASES
- THERMAL CONDUCTION
- CONVECTION
- THERMAL RADIATION
- LIQUIDS AND VAPOURS
- SPECIFIC HEAT CAPACITY

Typhoon aircraft takes off. The glow comes from hot gases in its jet engines, where the temeperature can reach more than 1500 °C. At high altitudes, jet aircraft like this leave 'vapour trails' across the sky. However, the trails are not really vapour, but millions of tiny droplets, formed when water vapour from the engines condenses in the cold atmosphere.

5.1 Moving particles

Objectives: – to describe the motion of particles in solids, liquids, and gas – to give experimental evidence for this motion – to explain what internal energy is.

▶ Water can exist in three forms: solid, liquid, and gas. (The gas is called water vapour, and it is present in the air.) Like all materials, water is made up of tiny particles. Which form it takes depends on how firmly its particles stick together.

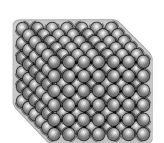

Solid Particles vibrate about fixed positions.

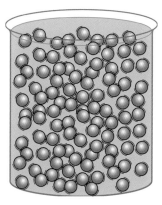

Liquid Particles vibrate, but can change positions.

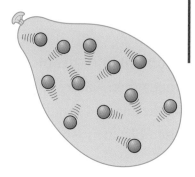

Gas Particles move about freely.

Solids, liquids, and gases

Every material is a solid, a liquid, or a gas. Scientists have developed a model (description) called the **kinetic theory** to explain how solids, liquids, and gases behave. According to this theory, matter is made up of tiny particles which are constantly in motion. The particles attract each other strongly when close, but the attractions weaken if they move further apart.

Solid A solid, such as iron, has a fixed shape and volume. Its particles are held closely together by strong forces of attraction called **bonds**. They vibrate backwards and forwards but cannot change positions.

Liquid A liquid, such as water, has a fixed volume but can flow to fill any shape. The particles are close together and attract each other. But they vibrate so vigorously that the attractions cannot hold them in fixed positions, and they can move past each other.

Gas A gas, such as hydrogen, has no fixed shape or volume and quickly fills any space available. Its particles are well spaced out, and virtually free of any attractions. They move about at high speed, colliding with each other and the walls of their container.

Ⓔ What are the particles?

Everything is made from about 100 simple substances called **elements**. An **atom** is the smallest possible amount of an element. In some materials, the 'moving particles' of the kinetic theory are atoms. However, in most materials, they are groups of atoms called **molecules**. Below, each atom is shown as a coloured sphere. This is a simplified model (description) of an atom. Atoms have no colour or precise shape.

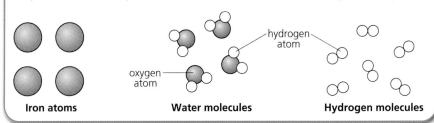

Iron atoms   Water molecules   Hydrogen molecules

THERMAL EFFECTS

Brownian motion: evidence for moving particles

Smoke is made up of millions of tiny bits of ash or oil droplets. If you look at smoke through a microscope, as on the right, you can see the bits of smoke glinting in the light. As they drift through the air, they wobble about in zig-zag paths. This effect is called **Brownian motion**, after the scientist Robert Brown who first noticed the wobbling, wandering motion of pollen grains in water, in 1827.

(E) The kinetic theory explains Brownian motion as follows. The bits of smoke are just big enough to be seen, but have so little mass that they are jostled about as thousands of particles (gas molecules) in the surrounding air bump into them at random.

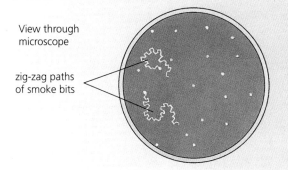
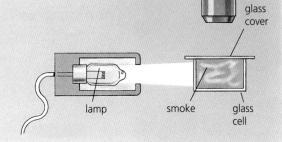

(E) Energy of particles

The particles (atoms or molecules) in solids, liquids, and gases have kinetic energy because they are moving. They also have potential energy because their motion keeps them separated and opposes the bonds trying to pull them together. The particles in gases have the most potential energy because they are furthest apart.

The total kinetic and potential energies of all the atoms or molecules in a material is called its **internal energy**. The hotter a material is, the faster its particles move, and the more internal energy it has.

If a hot material is in contact with a cold one, the hot one cools down and loses internal energy, while the cold one heats up and gains internal energy. The energy transferred is known as **heat**.

The term **thermal energy** is often used for both internal energy and heat.

Kinetic energy
Energy because of motion.

Potential energy
Energy stored because of position.

1 Say whether each of the following describes a *solid*, a *liquid*, or a *gas*:
   a Particles move about freely at high speed.
   b Particles vibrate and cannot change positions.
   c Fixed shape and volume.
   d Particles vibrate but can change positions.
   e No fixed shape or volume.
   f Fixed volume but no fixed shape.
   g Virtually no attractions between particles.

2 Smoke is made up of millions of tiny bits of ash or oil droplets.
   a What do you see when you use a microscope to study illuminated smoke floating in air?
   b What is the effect called?
   c How does the kinetic theory explain the effect?

3 If a gas is heated up, how does this affect the motion of its particles?

4 What is meant by the *internal energy* of an object?

Related topics: energy **4.1**; changing state **5.10**; atoms and elements **10.1**

5.2 Temperature

Objectives: – to know the fixed points are on the Celsius scale – to know what the Kelvin scale is, and how to convert from 0C to K.

The Celsius scale

A **temperature scale** is a range of numbers for measuring the level of hotness. Everyday temperatures are normally measured on the **Celsius** scale (sometimes called the 'centigrade' scale). Its unit of temperature is the **degree Celsius** (**°C**). The numbers on the scale were specially chosen so that pure ice melts at 0 °C and pure water boils at 100 °C (under standard atmospheric pressure of 101 325 pascals). These are its two **fixed points**. Temperatures below 0 °C have negative (−) values.

Sun's centre	15 000 000 °C
Sun's surface	6000 °C
bulb filament	2500 °C
bunsen flame	1500 °C
boiling water	100 °C
human body	37 °C
warm room	20 °C
melting ice	0 °C
food in freezer	−18 °C
liquid oxygen	−180 °C
absolute zero	−273 °C

Thermometers*

Temperature is measured using a **thermometer**. One simple type is shown below. The glass bulb contains a liquid – either mercury or coloured alcohol – which expands when the temperature rises and pushes a 'thread' of liquid further along the scale.

Every thermometer depends on some *property* (characteristic) of a material that varies with temperature. For example, the thermometer above contains a liquid whose volume increases with temperature. The two thermometers below use materials whose electrical properties vary with temperature.

All thermometers agree at the fixed points. However, at other temperatures, they may not agree exactly because their chosen properties may not vary with temperature in quite the same way.

▼ **Clinical thermometers** like the one below measure the temperature of the human body very accurately. Their range is only a few degrees either side of the average body temperature of 37 °C. When removed from the body, they keep their reading until reset.

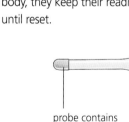

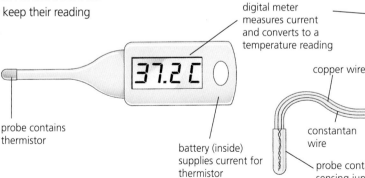

Thermistor thermometer The thermistor is a device which becomes a much better electrical conductor when its temperature rises. This means that a higher current flows from the battery, causing a higher reading on the meter.

Thermocouple thermometer Two different metals are joined to form two junctions. A temperature difference between the junctions causes a tiny voltage which makes a current flow. The greater the temperature difference, the greater the current.

THERMAL EFFECTS

What is temperature?

In any object, the particles (atoms or molecules) are moving, so they have kinetic energy. They move at varying speeds, but the higher the temperature, then – on average – the faster they move.

If a hot object is placed in contact with a cold one, as on the right, there is a transfer of thermal energy from one to the other. As the hot object cools down, its particles lose kinetic energy. As the cold object heats up, it particles gain kinetic energy. When both objects reach the same temperature, the transfer of energy stops because the average kinetic energy per particle is the same in both:

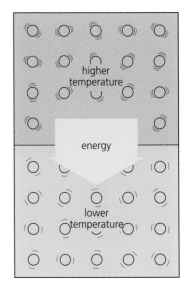

> Objects at the *same temperature* have the *same average kinetic energy per particle*. The higher the temperature, the greater the average kinetic energy per particle.

Temperature is not the same as heat. For example, a spoonful of boiling water has exactly the same temperature (100 °C) as a saucepanful of boiling water, but would store far less thermal energy (heat).

Absolute zero and the Kelvin scale

As the temperature falls, the particles in a material lose kinetic energy and move more and more slowly. At −273 °C, they can go no slower. This is the lowest temperature there is, and it is called **absolute zero**. The rules of atomic physics do not allow particles to have zero energy, but at absolute zero they would have the minimum energy possible.

In scientific work, temperatures are often measured using the **Kelvin scale**. Its temperature unit, the **kelvin (K)**, is the same size as the degree Celsius, but the scale uses absolute zero as its zero (0 K). You convert from one scale to the other like this:

The Kelvin scale is a **thermodynamic** scale. It is based on the average kinetic energy of particles, rather than on a property of a particular substance.

The **constant volume hydrogen thermometer** contains trapped hydrogen gas whose pressure increases with temperature. It gives the closest match to the thermodynamic scale and is used as a standard against which other thermometers are calibrated (marked).

Kelvin temperature/K = Celsius temperature/°C + 273

	absolute zero	melting ice	boiling water
Celsius scale	−273 °C	0 °C	100 °C
Kelvin scale	0 K	273 K	373 K

Where greater accuracy is required, absolute zero is taken to be −273.15 °C.

1 −273   0   100   273   373
   Say which of the above is the temperature of
   a boiling water in °C
   b boiling water in K
   c absolute zero in °C
   d absolute zero in K
   e melting ice in °C
   f melting ice in K.

2* Every thermometer depends on some property of a material that varies with temperature. What property is used in each of the following?
   a A mercury-in-glass thermometer.
   b A thermistor thermometer.

3 Blocks A and B above are identical apart from their temperature.
   a How does the motion of the particles in A compare with that in B?
   b In what direction is thermal energy transferred?
   c When does the transfer of thermal energy cease?

Related topics: kinetic energy **4.1** and **4.3**; motion of particles in solids, liquids, and gases **5.1**; expansion of liquids **5.3**; motion of particles in a gas **5.4**; heating gases **5.4**; thermistors **8.6**

5.3 Expanding solids and liquids

Objectives: – to know that most solids expand by a small amount when heated – to describe some of the uses of thermal expansion, and problems associated with it.

Kinetic theory essentials
According to the kinetic theory, solids and liquids are made up of tiny, vibrating particles (atoms or molecules) which attract each other. The higher the temperature, then on average, the faster the particles vibrate.

If a concrete or steel bar is heated, its volume will increase slightly. The effect is called **thermal expansion**. It is usually too small to notice, but unless space is left for it, it can produce enough force to crack the concrete or buckle the steel. Most solids expand when heated. So do most liquids – and by more than solids. If a liquid is stored in a sealed container, a space must be left at the top to allow for expansion.

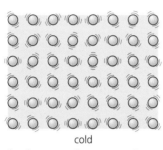
cold

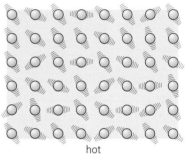
hot

The kinetic theory explains thermal expansion as follows. When, say, a steel bar is heated, its particles speed up. Their vibrations take up more space, so the bar expands slightly in all directions. If the temperature falls, the reverse happens and the material contracts (gets smaller).

Comparing expansions

invar (metal)	0.1 mm
Pyrex glass	0.3 mm
platinum alloy	0.9 mm
glass	0.9 mm
concrete	1 mm
steel	1 mm
brass	2 mm
aluminium	3 mm

increase in length of a 1 m bar for a 100°C rise in temperature

The chart on the left shows how much 1 metre lengths of different materials expand when their temperature goes up by 100 °C. For greater lengths and higher temperature increases, the expansion is more.

When choosing materials for particular jobs, it can be important to know how much they will expand. Here are two examples:

Steel rods can be used to reinforce concrete because both materials expand equally. If the expansions were different, the steel might crack the concrete on a hot day.

If an ordinary glass dish is put straight into a hot oven, the outside of the glass expands before the inside and the strain cracks the glass. Pyrex expands much less than ordinary glass, so should not crack.

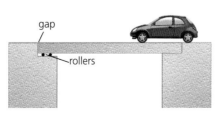

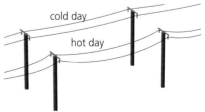

Allowing for expansion...
Gaps are left at the ends of bridges to allow for expansion. One end of the bridge is often supported on rollers so that movement can take place.

... and contraction
When overhead cables are suspended from poles or pylons, they are left slack, partly to allow for the contraction that would happen on a very cold day.

THERMAL EFFECTS

Using expansion

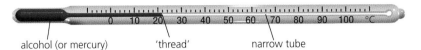

alcohol (or mercury)   'thread'   narrow tube

In the thermometer above, the liquid in the bulb expands when the temperature rises. The tube is made narrow so that a small increase in volume of the liquid produces a large movement along the tube, as explained in the previous spread, 5.3.

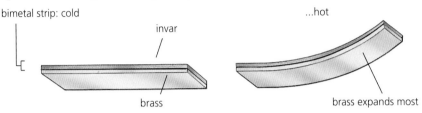

bimetal strip: cold — invar — brass ...hot — brass expands most

In the **bimetal strip** above, thin strips of two different metals are bonded together. When heated, one metal expands more than the other, which makes the bimetal strip bend. Bimetal strips are used in some **thermostats** – devices for keeping a steady temperature. The thermostat shown on the right is controlling an electric heater. More modern designs often use an electronic circuit containing a thermistor, rather than a bimetal strip.

Water and ice*

When hot water cools, it contracts. However, when water freezes it *expands* as it turns into ice. The force of the expansion can burst water pipes and split rocks with rainwater trapped in them.

Water expands on freezing for the following reason. In liquid water, the particles (water molecules) are close together. But in ice, the molecules link up in a very open structure that actually takes up *more* space than in the liquid – as shown in the diagram on the right.

Ice has a lower *density* than liquid water – in other words, each kilogram has a greater volume. Because of its lower density, ice floats on water. When liquid water is cooled, the molecules start forming into an open structure at 4 °C, just before freezing point is reached. As a result, water expands very slightly as it is cooled from 4 °C to 0 °C. It takes up least space, and therefore has its maximum density, at 4 °C.

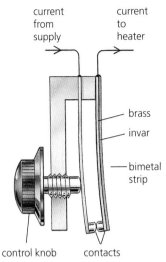

▲ **Bimetal thermostat** When the temperature rises, the bimetal strip bends, the contacts separate, and the current to the heater is cut off. When the temperature falls, the bimetal strip straightens, and the current is switched on again. In this way, an approximately steady temperature is maintained.

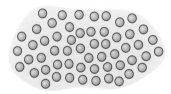

molecules in liquid water

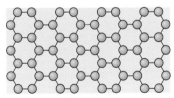

molecules in ice

1 Explain the following:
  a A metal bar expands when heated.
  b Overhead cables are hung with plenty of slack in them.
  c It would *not* be a good idea to reinforce concrete with aluminium rods.
  d A bimetal strip bends when heated.
  e* Water expands when it freezes.

2 This question is about the thermostat in the diagram at the top of the page.
  a Why does the power to the heater get cut off if the temperature rises too much?
  b To maintain a *higher* temperature, which way would you move the control knob? – to the *right* so that it moves towards the contacts, or to the *left*? Explain your answer.

Related topics: density 1.4; kinetic theory and particles 5.1; thermometers 5.2; thermistors 8.6

5.4 Heating gases

Objectives: – to describe how a temperature change can affect the pressure or the volume of a gas – to know how the expansions of solids, liquids, and gases compare.

Kinetic theory essentials
According to the kinetic theory, a gas is made up of tiny, moving particles (usually molecules). These move about freely at high speed and bounce off the walls of their container. The higher the temperature, then on average, the faster they move.

Unlike a solid or liquid, a gas does not necessarily expand when heated. That is because its volume depends on the container it is in. When dealing with a fixed mass of gas, there are always three factors to consider: pressure, volume, and temperature. Depending on the circumstances, a change in temperature can produce a change in pressure, or volume, or both.

This spread deals with the effects of a change in temperature. To find out more about the link between the pressure of a gas and its volume when the temperature doesn't change, see spread 3.8.

How pressure changes with temperature (at constant volume)

In the experiment on the left, air is trapped in a flask of fixed volume. The temperature of the air is changed in stages by heating the water – or putting a hotter or colder material (melting ice, for example) in the container. At each stage, the pressure is measured on the gauge.

The table shows some typical readings:

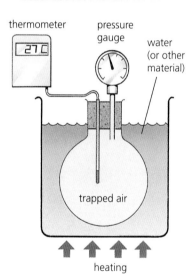

temperature / °C	20	80	140	200
pressure / kPa	102	123	144	165

As the temperature of the air rises, so does the pressure. This is because the molecules move faster. There is a greater change in momentum

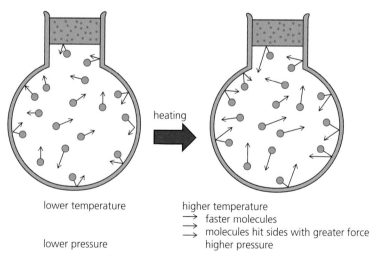

when they hit the sides of the flask, so a greater force:

The cylinders used for storing gas are strong enough to withstand any extra pressure due to normal rises in temperature. It is dangerous to throw aerosol cans on bonfires because they might burst. However, that is mainly because more of the liquid propellant in the can turns to gas.

THERMAL EFFECTS

How volume changes with temperature (at constant pressure)

In the experiment on the right, trapped gas (air) is heated at constant pressure. This is atmospheric pressure because only the short length of liquid separates the air from the atmosphere outside. As the temperature rises, the volume of the gas increases – the gas expands.

Here is an experiment to show the opposite effect. Take an empty plastic bottle (of the type used for bottled water). Screw the top on tightly. Put the bottle in a freezer for about 5 minutes, then see if you notice any difference.

E Comparing expansions of solids, liquids, and gases

At constant pressure, gases expand much more than liquids which, in turn, expand more than solids. For example, for the same volume of material and the same rise in temperature (starting at room temperature):

Water expands 7 times as much as steel

Air (at constant pressure) expands 16 times as much as water.

It is the strength of the attractions between the particles (molecules, for example) that makes the difference. In a solid, the attractions are very strong. If the temperature rises and the particles move faster, this has very little effect on their separation because they are so tightly held together. In a liquid, the attractions are weaker, so the expansion is greater. In a gas, the attractions are extremely weak, so the expansion is much more.

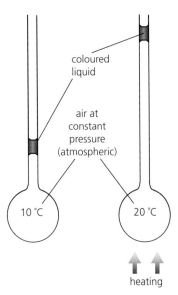

▲ In this experiment, the pressure of the gas stays constant. As the temperature increases, so does the volume.

◄ Before its flight, this balloon is filled with cold air using a motorized fan. Then the gas burner raises the temperature of the air to 100 °C or more. There is no change in pressure (it stays at atmospheric), but a large increase in volume.

Q

1. How does the kinetic theory explain the following?
   a. A gas exerts a pressure on its container walls.
   b. The pressure increase with temperature (assuming that the volume does not change).
2. If a gas is heated at constant pressure, what happens to its volume?
3. Comparing a *solid* with a *liquid*, which would you expect to expand the most when heated? Use the kinetic theory to explain your answer.
4. Comparing a *liquid* with a *gas*, which would you expect to expand the most when heated? Use the kinetic theory to explain your answer.

Related topics: momentum 2.11; gas pressure and volume 3.8; kinetic theory 5.1; temperature 5.2; expansion of solids and liquids 5.4

5.5 Thermal conduction

Objectives: – to give examples of materials that are good thermal conductors – to explain why metals are the best conductors – to describe some uses of insulators.

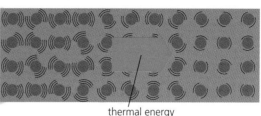

high temperature — thermal energy transferred by conduction — lower temperature

Very good conductors
metals e.g.   copper
              aluminium
              iron
silicon
graphite

All materials are made up of tiny, moving particles (atoms or molecules). The higher the temperature, the faster the particles move.

If one end of a metal bar is heated as above, the other end eventually becomes too hot to touch. Thermal energy (heat) is transferred from the hot end to the cold end as the faster particles pass on their extra motion to particles all along the bar. The process is called **conduction**.

More thermal energy is transferred every second if:
- the temperature difference across the ends of the bar is *increased*
- the cross-sectional ('end-on') area of the bar is *increased*
- the length of the bar is *reduced*.

Thermal conductors and insulators

Very poor conductors (insulators)
glass
water
plastics
rubber
wood
wool

materials containing trapped air
{ wool
  glass wool (fibreglass)
  plastic foam
  expanded polystyrene }

The materials above are arranged in order of conducting ability starting with the best.

Some materials are much better **conductors** than others. Very poor conductors are called **insulators**. Many materials are in between the two.

Metals are the best thermal conductors. Non-metal solids tend to be poor conductors; so do most liquids. Gases are the worst of all. Many materials are insulators because they contain tiny pockets of trapped air. You use this idea when you put on lots of layers of clothes to keep you warm. There are some more examples at the top of the next page.

You can sometimes tell how well something conducts just by touching it. A metal door handle feels cold because it quickly conducts thermal energy away from your hand, which is warmer. A polystyrene tile feels warm because it insulates your hand and stops it losing thermal energy.

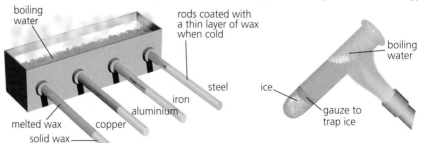

▲ Comparing four good thermal conductors. Ten minutes or so after the boiling water has been tipped into the tank, the length of melted wax shows which material is the best conductor.

▲ This experiment shows that water is a poor thermal conductor. The water at the top of the tube can be boiled without the ice melting.

THERMAL EFFECTS

Using insulating materials

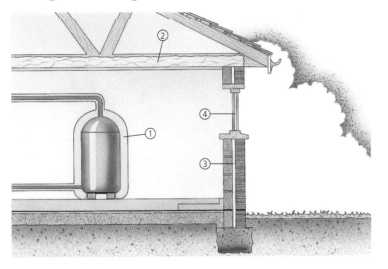

▲ Feathers give good thermal insulation, especially when fluffed up to trap more air.

In countries where buildings need to be heated, good insulation means lower fuel bills. Above are some of the ways in which insulating materials are used to reduce heat losses from a house:
1. Plastic foam lagging round the hot water storage tank.
2. Glasswool or mineral wool insulation in the loft.
3. Wall cavity filled with plastic foam, beads, or mineral wool.
4. Double-glazed windows: two sheets of glass with air between them.

How materials conduct

When a material is heated, the particles move faster, push on neighbouring particles, and speed those up too. All materials conduct like this but, in metals, energy is also transferred by another, much quicker method.

In atoms, there are tiny particles called **electrons**. Most are firmly attached, but in metals, some are 'loose' and free to drift between the atoms. When a metal is heated, these **free electrons** speed up. As they move randomly within the metal, they collide with atoms and make them vibrate faster. In this way, thermal energy is rapidly transferred to all parts.

An electric current is a flow of electrons – so metals are good electrical conductors as well as good thermal conductors.

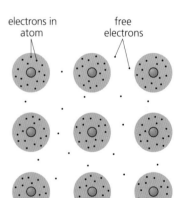

▲ Atoms in a metal

1. Explain each of the following:
   a. A saucepan might have a copper bottom but a plastic handle.
   b. Wool and feathers are good insulators.
   c. A Glass wool aluminium window frame feels colder than a wooden window frame when you touch it.
   d. It is much safer picking up hot dishes with a dry cloth than a wet one.
2. Give *three* ways in which insulating materials are used to reduce thermal energy losses from a house.
3. A hot water tank loses thermal energy even when lagged. How could the energy loss be reduced?
4. Look at the experiment shown on the opposite page, comparing four thermal conductors.
   a. Which of the metals is the best conductor?
   b. In experiments like this, it is important to make sure that the test is fair. Write down *three* features of this experiment which make it a fair test.
5. Why are metals much better thermal conductors than most other materials?

Related topics: energy 4.1; particles of matter 5.1; temperature 5.2; electrical conductors 8.1

5.6 Convection

Objectives: – to explain what convection is – to describe situations where thermal energy is transferred by convection.

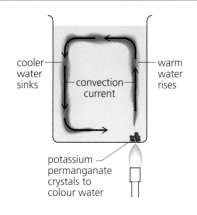

cooler water sinks

warm water rises

convection current

potassium permanganate crystals to colour water

Liquids and gases are poor thermal conductors, but if they are free to circulate, they can carry thermal energy from one place to another very quickly.

Convection in a liquid

In the experiment on the left, the bottom of the beaker is being gently heated in one place only. As the water above the flame becomes warmer, it expands and becomes less dense. It rises upwards as cooler, denser water sinks and displaces it (pushes it out of the way). The result is a circulating stream, called a **convection current**. Where the water is heated, its particles (water molecules) gain energy and vibrate more rapidly. As the particles circulate, they transfer energy to other parts of the beaker.

Convection does not occur if the water is heated at the top rather than at the bottom. The warmer, less dense water stays at the top.

Convection in air

Convection can occur in gases as well as liquids. For example, warm air rises when it is displaced by cooler, denser air sinking around it.

Heated by the Sun, warm air rises above the equator as it is displaced by cooler, denser air sinking to the north and south. The result is huge convection currents in the Earth's atmosphere. These cause winds across all oceans and continents. Convection also causes the onshore and offshore breezes which sometimes blow at the coast during the summer:

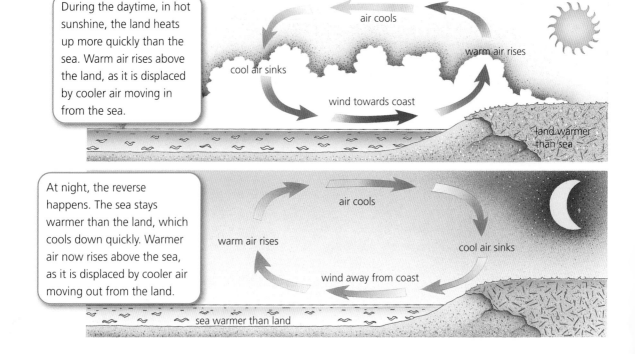

During the daytime, in hot sunshine, the land heats up more quickly than the sea. Warm air rises above the land, as it is displaced by cooler air moving in from the sea.

At night, the reverse happens. The sea stays warmer than the land, which cools down quickly. Warmer air now rises above the sea, as it is displaced by cooler air moving out from the land.

THERMAL EFFECTS

Using convection in the home

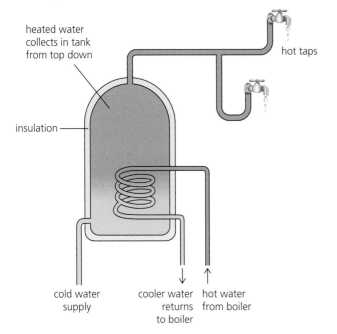

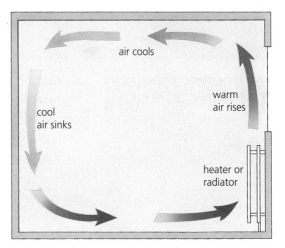

Room heating Warm air rising above a convector heater or radiator carries thermal energy all around the room – though unfortunately, the coolest air is always around your feet.

Hot water system In the system above, hot water for the taps comes from a large storage tank. The water is heated by a coil of copper pipe: hot water from a boiler flows through this and is recirculated by a pump. In the tank, the heated water rises to the top by convection. In this way, a supply of hot water collects from the top down. The tank is insulated to reduce thermal energy losses by conduction and convection.

Practical systems are more complicated than the one shown. There is additional pipework to allow the water to expand safely when heated. Also, there may be an extra circuit for radiators.

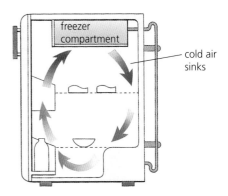

Refrigerator Cold air sinks below the freezer compartment. This sets up a circulating current of air which cools all the food in the refrigerator.

1 Explain the following:
   a A radiator quickly warms all the air in a room, even though air is a poor thermal conductor.
   b The smoke from a bonfire rises upwards.
   c Anyone standing near a bonfire feels a draught.
   d The freezer compartment in a refrigerator is placed at the top.
   e A refrigerator does not cool the food inside it properly if the food is too tightly packed.
2 On a hot summer's day, coastal winds often blow in from the sea.
   a What causes these winds?
   b Why do the winds change direction at night?

3 Some hot water systems have an immersion heater – an electrical heating element in the storage tank. In the tank below, should the heating element be placed at A or at B? Explain your answer.

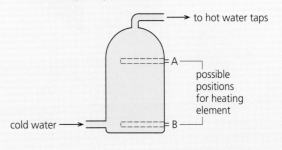

Related topics: work and forms of energy **4.01**

111

5.7 Thermal radiation

Objectives: – to know which surfaces are the best emitters and best absorbers – to explain what happens when incoming and outgoing energy transfers aren't equal.

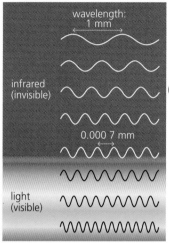

▲ Thermal radiation is mainly infrared waves, but very hot objects also give out light waves.

On Earth, we are warmed by the Sun. Its energy travels to us in the form of **electromagnetic waves**. These include invisible **infrared waves** as well as light, and they can travel through a vacuum (empty space). They heat up things that absorb them, so are often called **thermal radiation**.

(E) All objects give out some thermal radiation. The higher their surface temperature and the greater their surface area, the more energy they radiate per second. Thermal radiation is a mixture of different **wavelengths**, as shown on the left. Warm objects radiate infrared. But if they become hotter, they also emit shorter wavelengths which may include light. That is why a radiant heater or grill starts to glow 'red hot' when it heats up.

Emitters and absorbers

Some surfaces are better at emitting (sending out) thermal radiation than others. For example, a black saucepan cools down more quickly than a similar white one because it emits energy at a faster rate.

Good emitters of thermal radiation are also good absorbers, as shown in the chart below. White or silvery surfaces are poor absorbers because they reflect most of the thermal radiation away. That is why, in hot, sunny countries, houses are often painted white to keep them cool inside.

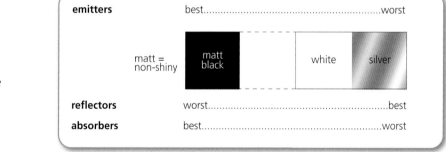

▶ This chart shows how some surfaces compare as emitters, reflectors, and absorbers of thermal radiation.

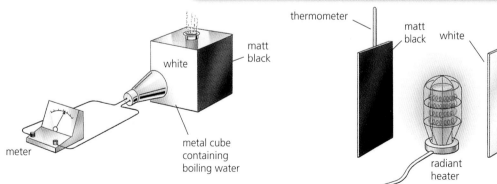

(E) **Comparing emitters** The metal cube is filled with boiling water which heats the surfaces to the same temperature. The thermal radiation detector is placed in turn at the same distance from each surface and the meter readings compared.

(E) **Comparing absorbers** The metal plates are placed at the same distance from a radiant heater. To find out which surface absorbs thermal radiation most rapidly, the rises in temperature are compared.

THERMAL EFFECTS

The Earth in balance?

If an object absorbs thermal radiation faster than it radiates, it heats up. If it radiates faster than it absorbs, it cools down. The temperature changes until a balance is reached. This is true of the Earth, which is absorbing incoming radiation from the Sun, but also radiating energy into space.

Without an atmosphere, the Earth's average surface temperature would be −18 °C. However, carbon dioxide and water vapour in the atmosphere trap some of the incoming energy. As a result, the average surface temperature is closer to 15 °C. Over the last 60 years, the amount of carbon dioxide in the atmosphere has gone up by more than 30%, mainly due to the increased burning of fossil fuels. As a result, the flows of thermal radiation are out of balance, and the Earth is very slowly warming up. We have global warming.

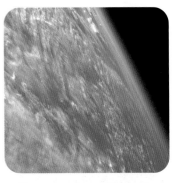

▲ The atmosphere affects the rate at which the Earth absorbs and emits thermal radiation.

More transfers

In many situations, the transfer of energy takes place by more than one process. Here are two examples:

▲ Stand in front of this wood-burning stove and you will feel the warming effect of the thermal radiation coming from it. However, the room is mostly warmed by air that is heated by the stove, then rises and circulates by convection.

▲ A car has a 'radiator' to stop its engine overheating. Unwanted thermal energy is carried by liquid coolant to an array of pipes, conducted to metal fins, then carried away by air flowing across the fins. Only a little of the energy is transferred by radiation.

1 white   silvery   matt black
Which of the above surfaces is the best at
  a absorbing thermal radiation
  b emitting thermal radiation
  c reflecting thermal radiation?

2 When a warm object is heated up, the thermal radiation it emits changes. Give *two* ways in which the thermal radiation changes.

3 Where in the radiator of a car engine is energy transferred by conduction?

4 In experiments like those on the opposite page, it is important to make sure that each test is fair.
  a Write down *three* features of the *Comparing emitters* experiment that make it a fair test.
  b Repeat for the *Comparing absorbers* experiment.

5 A hot metal sphere is absorbing 50 J of energy every second from a nearby radiant source. Explain what will happen to the temperature of the sphere if
  a the sphere is radiating 40 J of energy per second
  b the sphere is radiating 60 J of energy per second.

6 In the solar panel above, why does the panel have
  a a blackened layer at the back
  b a network of water pipes?

Related topics: energy **4.1**; global warming **4.7**; solar energy **4.7–4.8**; thermal energy **4.1** and **5.1**; conduction **5.5**; convection **5.6**; electromagnetic waves **7.10–7.11**

5.8 Liquids and vapours

Objectives: – to explain what is meant by evaporation, boiling, condensation, and solidification – to know that evaporation has a cooling effect.

Kinetic theory essentials

According to the kinetic theory, every material is made up of tiny, moving particles (usually molecules). These move at varying speeds. But the higher the temperature, then on average, the faster they move.

In a liquid, attractions keep the particles together. In a gas, the particles have enough energy to overcome the attractions, stay spaced out, and move around freely.

Gas and vapour

A gas is called a vapour if it can be turned back into a liquid by compressing it.

▼ When a liquid **evaporates**, faster particles escape from its surface to form a gas. However, unless the gas is removed, some of the particles will return to the liquid.

Evaporation

Even on a cool day, rain puddles can vanish and wet clothes dry out. The water becomes an invisible gas (called water vapour) which drifts away in the air. When a liquid below its boiling point changes into a gas, this is called **evaporation**. It happens because some particles in the liquid move faster than others. The faster ones near the surface have enough energy to escape and form a gas.

There are several ways of making a liquid evaporate more quickly:

Increase the temperature Wet clothes dry faster on a warm day because more of the particles (water molecules) have enough energy to escape.

Increase the surface area Water in a puddle dries out more quickly than water in a cup because more of its molecules are close to the surface.

Reduce the humidity* If air is very *humid*, this means that it already has a high water vapour content. In humid air, wet washing dries slowly because molecules in the vapour return to the liquid almost as fast as those in the liquid escape. In less humid air, wet washing dries more quickly.

Blow air across the surface Wet clothes dry faster on a windy day because the moving air carries escaping water molecules away before many of them can return to the liquid.

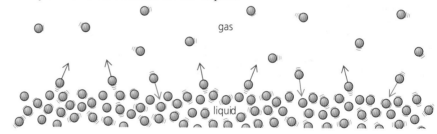

Boiling

Boiling is a very rapid form of evaporation. When water boils, as in the photograph on the left, vapour bubbles form deep in the liquid. They expand, rise, burst, and release large amounts of vapour.

Even cold water has tiny vapour bubbles in it, but these are squashed by the pressure of the atmosphere. At 100 °C, the vapour pressure in the bubbles is strong enough to overcome atmospheric pressure, so the bubbles start to expand and boiling occurs. At the top of Mount Everest, where atmospheric pressure is less, water would boil at only 70 °C.

The cooling effect of evaporation

Evaporation has a cooling effect. For example, if you wet your hands, the water on them starts to evaporate. As it evaporates, it takes thermal energy away from your skin. So your hands feel cold.

THERMAL EFFECTS

E) The kinetic theory explains the cooling effect like this. If faster particles escape from the liquid, slower ones are left behind, so the temperature of the liquid is less than before.

Refrigerators use the cooling effect of evaporation. In the refrigerator on the right, the process works like this:
1. In the pipes in the freezer compartment, a liquid called a **refrigerant** evaporates and takes thermal energy from the food and air.
2. The vapour is drawn away by the pump, which compresses it and turns it into a liquid. This releases thermal energy, so the liquid heats up.
3. The hot liquid is cooled as it passes through the pipes at the back, and the thermal energy is carried away by the air.

Overall, thermal energy is transferred from the things inside the fridge to the air outside.

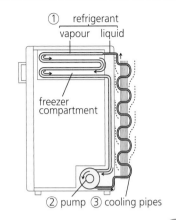

Sweating also uses the cooling effect of evaporation. You start to sweat if your body temperature rises more than about 0.5 °C above normal. The sweat, which is mainly water, comes out of tiny pores in your skin. As it evaporates, it takes thermal energy from your body and cools you down.

On a humid ('close') day, sweat cannot evaporate so easily, so it is more difficult to stay cool and comfortable.

Condensation

When a gas changes back into a liquid, this is called **condensation**. For example, cold air can hold less water vapour than warm air, so if humid air is suddenly cooled, some of the water vapour may condense. It may become billions of tiny water droplets in the air – we see these as clouds, mist, or fog. Or it may become condensation on windows or other surfaces. If condensation freezes, the result is frost.

Changes of state
A change from liquid to gas (or gas to liquid) is called a change of state. A change from liquid to solid is another **change of state**
It is called **solidification**, and water freezing is an example. When it happens, the particles continue to vibrate but do not have enough energy to overcome their bonds and change positions.

Condensation can be seen ...on mirrors

...as clouds in the sky

...and as clouds of 'steam' from a kettle (the vapour itself is invisible)

1. A puddle and a small bowl are next to each other. There is the same amount of water in each.
   a. Explain why the puddle dries out more rapidly than the water in the bowl.
   b. Give *two* changes that would make the puddle dry out even more rapidly.
2. If you are wearing wet clothes, and the water evaporates, it cools you down. How does the kinetic theory explain the cooling effect?
3. Give *two* practical uses of the cooling effect of evaporation.
4.* Explain why, on a humid day
   a. you may feel hot and uncomfortable
   b. you do not feel so uncomfortable if there is a breeze blowing.
5. What is the difference between evaporation and boiling?
6. Why does condensation form on cold windows?

Related topics: atmospheric pressure **3.7**; solids, liquids, gases, and kinetic theory **5.1**; change of state and latent heat **5.10**

5.9 Specific heat capacity

Objectives: – to know what is meant by the specific heat capacity of a material – to do calculations involving it, and describe how it can be measured.

Internal energy
If a material absorbs thermal energy, its internal energy increases. For more about internal energy, see spread 5.1.

If a material takes in thermal energy, then unless it is melting or boiling, its temperature rises. However, some materials have a greater capacity for absorbing thermal energy than others. For example, if you heat a kilogram each of water and aluminium, the water must be supplied with nearly five times as much energy as the aluminium for the same rise in temperature:

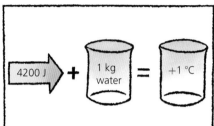
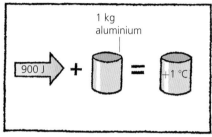

▲ 4200 joules of energy are needed to raise the temperature of 1 kg of water by 1 °C.

▲ 900 joules of energy are needed to raise the temperature of 1 kg of aluminium by 1 °C.

Scientifically speaking, water has a **specific heat capacity** of 4200 J/(kg °C). Aluminium has a specific heat capacity of only 900 J/(kg °C). Other specific heat capacities are shown in the table below left.

Units
Energy is measured in joules (J). Temperature is measured in °C or in kelvin (K). Both scales have the same size 'degree', so a 1 °C *change* in temperature is the same as a 1 K change.

The energy that must be transferred to an object to increase its temperature can be calculated using this equation:

energy transferred = mass × specific heat capacity × temperature change

In symbols: energy transferred = $mc\Delta\theta$

where m is the mass in kg, c is the specific heat capacity in J/(kg °C), and $\Delta\theta$ represents the temperature *change* in °C (or in K).

The same equation can also be used to calculate the energy transferred

	specific heat capacity
	J/(kg °C)
water	4200
alcohol	2500
ice	2100
aluminium	900
concrete	800
glass	700
steel	500
copper	400

Example If 2 kg of water cools from 70 °C to 20 °C, how much thermal energy does it lose?

In this case, the temperature change is 50 °C.

So: energy transferred = $mc\Delta\theta$ = 2 × 4200 × 50 J
$\qquad\qquad\qquad\qquad\qquad$ = 420 000 J

Thermal capacity*

The quantity *mass × specific heat capacity* is called the **thermal capacity** (or **heat capacity**). For example, if there is 2 kg of water in a kettle:
thermal capacity of the water = 2 kg × 4200 J/(kg °C) = 8400 J/ °C

This means that, for each 1 °C rise in temperature, 8400 joules of energy must be supplied to the water in the kettle. A greater mass of water would have a higher thermal capacity.

THERMAL EFFECTS

Linking energy and power

$$\text{power} = \frac{\text{energy}}{\text{time}}$$

So: energy = power × time

Energy is measured in joules (J).
Power is measured in watts (W).
Time is measured in seconds (s).

(E) Measuring specific heat capacity

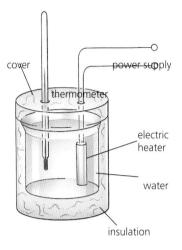

Water A typical experiment is shown on the right. Here, the beaker contains 0.5 kg of water. When the 100 watt electric heater is switched on for 230 seconds, the temperature of the water rises by 10 °C. From these figures, a value for the specific heat capacity of water can be calculated:

(Omitting some of the units for simplicity)
energy transferred to water = $mc\Delta\theta$ = 0.5 × c × 10
energy supplied by heater = power × time = 100 × 230 = 23 000 J
so: 0.5 × c × 10 = 23 000
Rearranged and simplified, this gives c = 4600
so the specific heat capacity of water is 4600 J/(kg °C).

This method makes no allowance for any thermal energy lost to the beaker or the surroundings, so the value of c is only approximate.

Aluminium (or other metal) The method is as above, except that a block of aluminium is used instead of water. The block has holes drilled in it for the heater and thermometer. As before, c is calculated from this equation:

    power × time = $mc\Delta T$ (assuming no thermal energy losses)

Storing thermal energy

Because of its high specific heat capacity, water is a very useful substance for storing and carrying thermal energy. For example, in central heating systems, water carries thermal energy from the boiler to the radiators around the house. In car cooling systems, water carries unwanted thermal energy from the engine to the radiator.

Night storage heaters use concrete blocks to store thermal energy. Although concrete has a lower specific heat capacity than water, it is more dense, so the same mass takes up less space. Electric heating elements heat up the blocks overnight, using cheap, 'off-peak' electricity supplied through a special meter. The hot blocks release thermal energy through the day as they cool down.

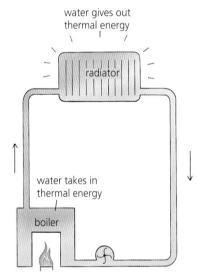

▲ In most central heating systems, water is used to carry the thermal energy.

The specific heat capacities of copper and water are given in the table on the opposite page.

1 Water has a very high specific heat capacity. Give *two* practical uses of this.

2 a How much thermal energy is needed to raise the temperature of 1 kg of copper by 1 °C?
  b If a 10 kg block of copper cools from 100 °C to 50 °C, how much thermal energy does it give out?
  c If, in part **b**, the copper were replaced by water, how much thermal energy would this give out?

3 A 210 W heater is placed in 2 kg of water and switched on for 200 seconds.
  a How much energy is needed to raise the temperature of 2 kg of water by 1 °C?
  b How much energy does the heater supply?
  c Assuming that no thermal energy is wasted, what is the temperature rise of the water?

Related topics: density 1.4; thermal energy 4.1 and 5.1; internal energy 5.1; temperature 5.2; electrical power 8.12

5.10 Latent heat

Objectives: – to know what is meant by latent heat – to know how to measure the specific latent heats of ice and of liquid water.

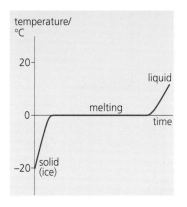

Water can be a solid (ice), a liquid, or a gas called water vapour (or steam). These are its three **phases**, or **states**.

Solid to liquid

If ice from a cold freezer is put in a warm room, it absorbs thermal energy. The graph on the left shows what happens to its temperature. While melting, the ice goes on absorbing energy, but its temperature does not change: it stays at 0 °C, the melting point. The energy absorbed is called the **latent heat of fusion**. It is needed to separate the particles so that they can form the liquid. If the liquid changes back to a solid, the energy is released again.

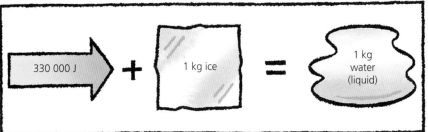

*Ice has a **specific latent heat of fusion** of 330 000 J/kg. This means that 330 000 joules of energy must be transferred to change each kilogram of ice into liquid water at the same temperature (0 °C). For any known mass, the energy transferred can be calculated using this equation:

energy transferred = mass × specific latent heat

In symbols: energy transferred = mL

For example, if 2 kg of ice is melted (at 0 °C):
energy transferred = mL = 2 kg × 330 000 J/kg = 660 000 J

Kinetic theory essentials

According to the kinetic theory, materials are made up of tiny, moving particles (usually molecules). In solids, the particles are held together by strong attractions. In liquids, they have more energy and are less strongly held. In gases, they have enough energy to overcome the attractions, stay spaced out, and move around freely.

***Measuring the specific latent heat of fusion of ice**
In the experiment on the left, a 100 watt heater is switched on for 300 seconds. By weighing the water collected in the beaker, it is found that 0.10 kg of ice has melted. From these figures, a value for L can be calculated:

(Omitting some of the units for simplicity)
energy transferred when ice melts = mL = 0.10 L
energy supplied by heater = power × time = 100 W × 300 s = 30 000 J
So: 0.10 L = 30 000, which gives L = 300 000
So the specific latent heat of fusion of ice is 300 000 J/kg.

This method makes no allowance for any thermal energy received from the funnel or surroundings, so the value of L is only approximate.

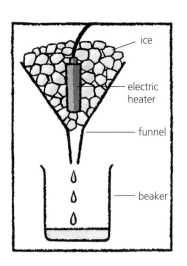

Linking energy and power

power = $\dfrac{\text{energy}}{\text{time}}$

So: energy = power × time

Energy is measured in joules (J).
Power is measured in watts (W).
Time is measured in seconds (s).

Liquid to gas

If you heat water in a kettle, the temperature rises until the water is boiling at 100 °C, then stops rising. If the kettle is left switched on, the water absorbs more and more thermal energy, but this just turns more and more of the boiling water into steam, still at 100 °C. The energy absorbed is called **latent heat of vaporization**. Most is needed to separate the particles so that they can form a gas, but some is required to push back the atmosphere as the gas forms.

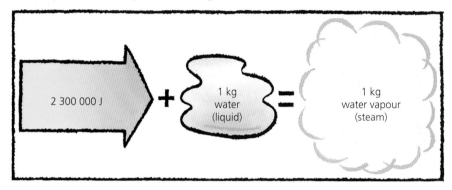

▲ A jet of steam releases latent heat when it condenses (turns liquid). This idea can be used to heat drinks quickly.

*Water has a **specific latent heat of vaporization** of 2 300 000 J/kg. This means that 2 300 000 joules of energy must be transferred to change each kilogram of liquid water into steam at the same temperature (100 °C).

To calculate the energy transferred when any known mass of liquid changes into a gas at the same temperature, you use the equation on the opposite page. However, L is now the specific latent heat of *vaporization*.

*Measuring the specific latent heat of vaporization of water
In the experiment on the right, the can contains boiling water. When the 100 watt heater has been switched on for 500 seconds, the change in the mass balance's reading shows that 0.020 kg of water has boiled away. From these figures, a value for L can be calculated:

(Omitting some of the units for simplicity)
energy transferred when water is vaporized = mL = 0.020 L
energy supplied by heater = power × time = 100 W × 500 s = 50 000 J
So: 0.020 L = 50 000, which gives L = 2 500 000

So the specific latent heat of vaporization of water is 2 500 000 J/kg.

This method makes no allowance for any thermal energy lost to the surroundings, so the value of L is only approximate.

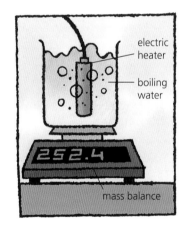

Specific latent heat of fusion of ice = 330 000 J/kg; specific latent heat of vaporization of water = 2 300 000 J/kg

1 Some crystals were melted to form a hot liquid, which was then left to cool. As it cooled, the readings in the table below were taken.
  a What was happening to the liquid between 10 and 20 minutes after it started to cool?
  b What is the melting point of the crystals in °C?

Time/minutes	0	5	10	15	20	25	30
Temperature/°C	90	75	68	68	68	62	58

2 Energy is needed to turn water into water vapour (steam). How does the kinetic theory explain this?

3* How much energy is needed to change
  a 10 kg of ice into water at the same temperature
  b 10 kg of water into water vapour at the same temperature?

4* A 460 watt water heater is used to boil water. Assuming no thermal energy losses, what mass of steam will it produce in 10 minutes?

Related topics: kinetic theory and thermal energy **5.1**; evaporation, boiling, and condensation **5.8**;. electrical power **8.12**

Check-up on thermal effects

Further questions

1. Explain in terms of molecules:
   a. the process of evaporation [3]
   b. why the pressure of the air inside a car tyre increases when the car is driven at high speed. [2]

2. Which of the following describes particles in a solid at room temperature?
   A Close together and stationary.
   B Close together and vibrating.
   C Close together and moving around at random.
   D Far apart and moving at random. [1]

3. In sunny countries, some houses have a solar heater on the roof. It warms up water for the house. The diagram below shows a typical arrangement.

   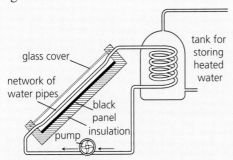

   a. Why is the panel in the solar heater black? [1]
   b. Why is there an insulating layer behind the panel? [1]
   c. How does the water in the tank get heated? [2]
   d. On average, each square metre of the solar panel above receives 1000 joules of energy from the Sun every second. Use this figure to calculate the power input (in kW) of the panel if its surface area is 2 m². [2]
   e. The solar heater in the diagram has an efficiency of 60% (it wastes 40% of the solar energy it receives). What area of panel would be needed to deliver thermal energy at the same rate, on average, as a 3 kW electric immersion heater? [2]
   f. i What are the advantages of using a solar heater instead of an immersion heater? [2]
      ii What are the disadvantages? [2]

4. The Earth absorbs thermal radiation from the Sun. At the same time, it emits thermal radiation into space.
   a. If the average temperature of the Earth's surface were to stay steady, what would this tell you about the two radiation flows? [2]
   b. What effect does the atmosphere have on the thermal radiation emitted by the Earth? [1]
   c. The evidence suggests that the Earth's average surface temperature is slowly increasing. What does this tell you about the two radiation flows? [2]

5. a. The table gives the melting and boiling points for lead and oxygen.

	melting point in °C	boiling point in °C
lead	327	1744
oxygen	−219	−183

   i. At 450 °C will the lead be a solid, a liquid or a gas? [1]
   ii. At −200 °C will the oxygen be a solid, liquid or a gas? [1]

   b. The graph shows how the temperature of a pure substance changes as it is heated.

   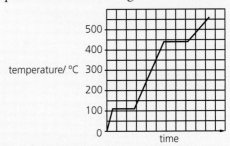

   i. At what temperature does the substance boil? [1]
   ii. Sketch the graph and mark with an **X** any point where the substance exists as both a liquid and gas at the same time. [1]

   c. i. All substances consist of particles. What happens to the average kinetic energy of these particles as the substance changes from a liquid to a gas? [1]
      ii. Explain, in terms of particles, why energy must be given to a liquid if it is to change to a gas. [2]

6. The diagram on the next page shows a refrigerator. In and around a refrigerator, heat is transferred by **conduction**, by **convection**, and

by **evaporation**. Decide which process is mainly responsible for the heat transfer in each of the examples listed at the top of the next page.

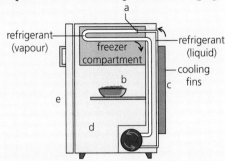

a Thermal energy is absorbed as liquid refrigerant changes to vapour in the pipework. [1]
b Cool air sinking from the freezer compartment transfers thermal energy from the food. [1]
c Thermal energy is lost to the outside air through the cooling fins at the back. [1]
d Some thermal energy from the kitchen enters the refrigerator through its outer panels. [1]
e Some thermal energy enters the refrigerator every time the door is opened. [1]

7 The diagram below shows a hot water storage tank. The water is heated by an electric immersion heater at the bottom.

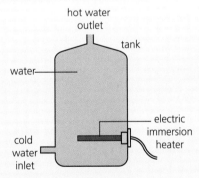

a How could thermal energy loss from the tank be reduced? What materials would be suitable for the job? [2]
b Why is the heater placed at the bottom of the tank rather than the top? [2]
c The heater has a power output of 3 kW.
 i What does the 'k' stand for in 'kW'? [1]
 ii How much energy (in joules) does the heater deliver in one second? [1]
 iii How much energy (in joules) does the heater deliver in 7 minutes? [2]
d The tank holds 100 kg of water. The specific heat capacity of water is 4200 J/(kg °C).
 i How much energy (in joules) is needed to raise the temperature of 1 kg of water by 1 °C? [1]
 ii How much energy (in joules) is needed to raise the average temperature of all the water in the tank by 1 °C? [2]
 iii If the heater is switched on for 7 minutes, what is the average rise in temperature of the water in the tank (assuming that no heat is lost)? [2]

8 The diagram below shows a type of heater used in some schools.

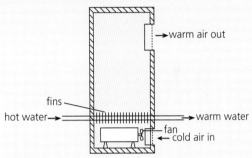

Hot water is pumped from the boiler into pipes inside the heater. Fins are attached to those pipes. Cold air is drawn into the base of the heater by an electric fan.
a Why are fins attached to the pipes inside the heater? [2]
b 600 kg of water pass through the heater every hour. The temperature of the water falls by 5 °C as it passes through the heater. Calculate the amount of heat energy transferred from the water every hour. The specific heat capacity of water is 4200 J/(kg °C). [3]

9 The graph below shows how the temperature of some liquid in a beaker changed as it was heated until it was boiling.

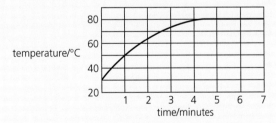

a What was the boiling point of the liquid? [1]
b State and explain what difference, if any, there would be in the final temperature if the liquid was heated more strongly. [2]
c State **one** difference between boiling and evaporation. [1]

MEASUREMENTS AND UNITS

Use the list below when you revise for your IGCSE. The spread number, in brackets, tells you where to find more information.

Revision checklist

Core Level
- ☐ The kinetic theory of matter. (5.1)
- ☐ Solids, liquids, and gases and the motion of their particles (e.g. molecules). (5.1)
- ☐ Brownian motion. (5.1)
- ☐ How an increase in temperature increases the internal energy of an object. (5.1 and 5.9)
- ☐ The melting and boiling temperatures of water. (5.2)
- ☐ The link between temperature and the motion of particles. (5.2)
- ☐ Absolute zero, the lowest possible temperature. (5.2)
- ☐ Converting temperatures between °C and kelvin. (5.2)
- ☐ How most solids and liquids expand when heated. (5.3)
- ☐ Problems caused by thermal expansion, and the uses of thermal expansion. (5.3)
- ☐ How the pressure of a gas is caused by the motion of its particles (molecules). (5.4)
- ☐ Why pressure increases with temperature for a gas at constant volume. (5.4)
- ☐ Why volume increases with temperature for a gas at constant pressure. (5.4)
- ☐ Good and poor thermal conductors. (5.5)
- ☐ Convection currents and why they occur. (5.6)
- ☐ The nature of thermal radiation. (5.7)
- ☐ How different surfaces compare as emitters, reflectors, and absorbers of thermal radiation. (5.7)
- ☐ Everyday uses and effects of thermal conduction, convection, and radiation. (5.5-5.7)
- ☐ Evaporation: the cause and cooling effect. (5.8)
- ☐ What happens during condensation. (5.9)

Extended Level
As for Core Level, plus the following:
- ☐ How the properties of solids, liquids, and gases depend on the motion and arrangement of their particles (e.g. molecules) and the forces between them. (5.1)
- ☐ Temperature and the average kinetic energy of particles. (5.1)
- ☐ Why Brownian motion occurs. (5.1)
- ☐ Why thermal expansion occurs. (5.3)
- ☐ How the expansions of solids, liquids and gases compare. (5.3 and 5.4)
- ☐ How gas pressure is caused by particles colliding with surfaces. (5.4)
- ☐ Explaining why, when heated (at constant pressure), gases expand much more than liquids, and liquids more than solids. (5.4)
- ☐ Why some materials are better thermal conductors than others. (5.5)
- ☐ How the temperature of an object (including the Earth) depends on the rates at which it is absorbing and emitting energy carried by thermal radiation. (5.7)
- ☐ The difference between evaporation and boiling. (5.8)
- ☐ Factors affecting the rate at which a liquid evaporates. (5.8)
- ☐ Why evaporation has a cooling effect. (5.8)
- ☐ Specific heat capacity and its measurement. (5.9)
- ☐ Using the equation energy transferred = $mcD\theta$ (5.9)

6
Waves and sounds

- TRANSVERSE AND LONGITUDINAL WAVES
- DIFFRACTION AND OTHER WAVE EFFECTS
- SOUND WAVES
- SPEED, FREQUENCY, AND WAVELENGTH
- ECHOES
- FREQUENCY AND PITCH
- AMPLITUDE AND LOUDNESS
- ULTRASOUND

This tree frog from Asia uses the large, inflatable sac under its throat to amplify the sound of its voice. Only the males can do this, and their calls can travel ten times further than sounds from other frogs. The sound itself is generated when air from the sac is blown past two stretched membranes in the bottom of the frog's mouth, making them vibrate.

chapter 6   **123**

6.1 Transverse and longitudinal waves

Objectives: – to know the difference between transverse and longitudinal waves – to know the equation linking the speed, frequency and wavelength of waves.

If you drop a stone into a pond, ripples spread across the surface. The tiny waves carry energy, as you can tell from the movements they cause at the water's edge. But there is no flow of water across the pond and no matter is transferred. The wave effect is just the result of up-and-down motions in the water.

Waves are not only found on water. Sound travels as waves, so does light. Waves can also travel along stretched springs like those in the experiments below. These show that there are two main types of waves.

Drawing waves

Transverse waves can be drawn as above.

wavefronts

Waves can also be drawn using lines called **wavefronts**. Think of each wavefront as the 'peak' of a transverse wave or the compression of a longitudinal wave.

Examples of...
transverse waves
electromagnetic waves: radio waves, microwaves, infrared, light, ultraviolet, X-rays, gamma rays

longitudinal waves
sound waves

Seismic waves...
are earthquakes waves: P (primary) waves can travel huge distances through the Earth. They are longitudinal. S (secondary) are slower. They are transverse. Surface waves* are the most destructive. They shake the ground from side to side.

Transverse waves

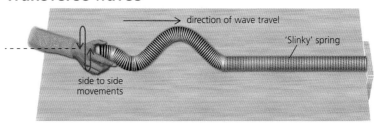

When the end coil of the spring is moved sideways, it pulls the next coil sideways a fraction of a second later... and so on along the spring. In this way, the sideways motion (and its energy) is passed from coil to coil, and a travelling wave effect is produced.

The to-and-fro movements of the coils are called **oscillations**. When the oscillations are up and down or from side to side like those above, the waves are called **transverse waves**. In transverse waves, the oscillations are at right angles to the direction of travel.

Light waves are transverse waves, although it is electric and magnetic fields which oscillate, rather than any material.

Longitudinal waves

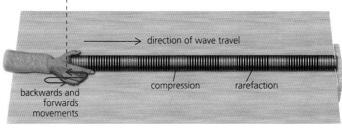

E Moving the end coil of the spring backwards and forwards also produces a travelling wave effect. However, the waves are bunched-up sections of coils with stretched-out sections in between. These sections are known as **compressions** and **rarefactions**.

When the oscillations are backwards-and-forwards like those above, the waves are called **longitudinal waves**. In longitudinal waves, the oscillations are in the direction of travel.

Sound waves are longitudinal waves. When you speak, compressions and rarefactions travel out through the air.

WAVES AND SOUNDS

Describing waves

On the right, transverse waves are being sent along a rope. Here are some of the terms used to describe these and other waves:

Speed The speed of the waves is measured in metres per second (m/s).

Frequency This is the number of waves passing any point per second. The SI unit of frequency is the **hertz** (**Hz**). For example, if the hand on the right makes four oscillations per second, then four waves pass any point per second, and the frequency is 4 Hz. The time for one oscillation is called the **period**. It is equal to 1/frequency. If the frequency is 4 Hz, the period is 1/4 s (0.25 s).

Wavelength This is the distance between any point on a wave and the equivalent point on the next.

Amplitude This is the maximum distance a point moves from its rest position when a wave passes.

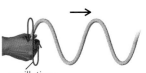
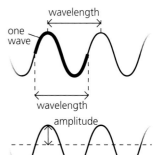

The wave equation

The speed, frequency, and wavelength of any set of waves are linked by this equation:

$$\text{speed} = \text{frequency} \times \text{wavelength}$$

In symbols: $v = f\lambda$ (λ = Greek letter *lambda*)

where speed is in m/s, frequency in Hz, and wavelength in m.

The following example shows why the equation works:

> The waves on the right are travelling across water.
> Each wave is 2 m long, so the wavelength is 2 m.
>
> One second later...
>
> 3 waves have passed the flag, so the frequency is 3 Hz.
> The waves have moved 3 wavelengths (3 × 2 m) to the right so their speed is 6 m/s.
>
> Therefore: 6 m/s = 3 Hz × 2 m
>   (speed)   (frequency)   (wavelength)

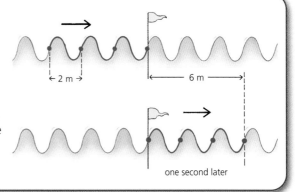

! Frequency (in Hz) is the number of oscillations per second.

Period (in seconds) is the time for one oscillation.

$$\text{frequency} = \frac{1}{\text{period}}$$

Q

1 The waves in A below are travelling across water.
  a Are the waves *transverse* or *longitudinal*?
  b What is the wavelength of the waves?
  c What is the amplitude of the waves?
  d If two waves pass the flag every second, what is
   i the frequency   ii* the period?
  e Use the wave equation to calculate the speed of the waves in A.
  f What is the wavelength of the waves in diagram B below?
  g If the waves in B have the same speed as those in A, what is their frequency?

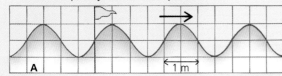
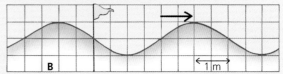

Related topics: SI units **1.2**; period **1.3**; speed **2.1**; speed of sound **6.4**; frequency **6.5**

6.2 Wave effects

Objectives: – to describe the results of wave experiments in a ripple tank – to know that waves can be reflected, refracted, and diffracted.

The properties of waves can be studied using a **ripple tank** like the one below. Ripples (tiny waves) are sent across the surface of water. Obstacles are put in their path to see what effects are produced.

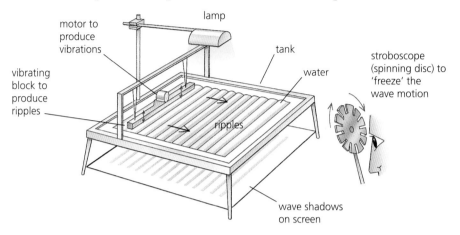

Reflection

A vertical surface is put in the path of the waves. The waves are reflected from the surface at the same angle as they strike it.

Refraction

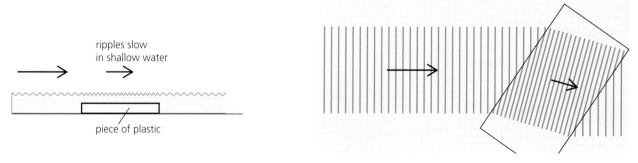

A flat piece of plastic makes the water more shallow, which slows the waves down. When the waves slow, they change direction. The effect is called **refraction**.

Refraction can be explained as follows.

The waves keep oscillating up and down at the same rate (frequency), so when they slow, the wavefronts close up on each other. That follows from the wave equation on the right. As the frequency is unchanged, a decrease in speed must cause a decrease in wavelength. From the last diagram on the opposite page, you can see that if the wavefronts close up on each other, their direction of travel must change, unless they are travelling at right angles to the boundary.

The wave equation
speed = frequency × wavelength

distance per second (m/s) | number of oscillations per second (Hz) | distance between wavefronts (m)

Diffraction

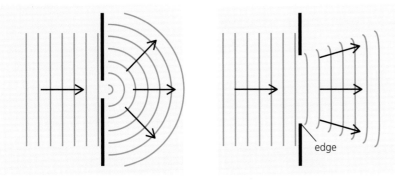

◀ Diffraction of waves passing through a gap. The size of the gap affects how much diffraction occurs.

The waves bend round the sides of an obstacle, or spread out as they pass through a gap. The effect is called **diffraction**.

 Diffraction is only significant if the size of the gap is about the same as the wavelength. Wider gaps produce less diffraction.

Diffraction at an edge
Diffraction mainly occurs at each edge. Longer wavelengths would produce more diffraction.

Wave evidence

Sound, light, and radio signals all undergo reflection, refraction, and diffraction. This suggests that they travel as waves. For example:

a Light reflects from mirrors; sound reflects from hard surfaces.
b Light bends when it passes from air into glass or water.
c Sound bends around obstacles such as walls and buildings, which is why you can hear around corners.
d Light spreads when it passes through tiny holes and slits. This suggests that light waves must have much shorter wavelengths than sound.
e Some radio signals can bend round very large obstacles such as hills. This suggests that radio waves must have long wavelengths.

Q
1 Say whether each of the effects **b** to **e** above is an example of *reflection*, *refraction*, or *diffraction*.
2 On the right, waves are moving towards a harbour.
   a What will happen to waves striking the harbour wall at A?
   b What will happen to waves slowed by the submerged sandbank at B?
   c What will happen to waves passing through the harbour entrance at C?
   d If the harbour entrance were wider, what difference would this make?

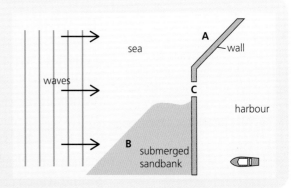

Related topics: waves and the wave equation **6.1**; reflection of sound **6.4**; light waves **7.1**; reflection of light **7.1–7.3**; refraction of light **7.4**; and **7.6**; radio waves **7.11** and **7.12**

6.3 Sound waves

Objectives: – to know that sound waves are longitudinal – to know that they need a medium (material) to travel through, and to know their other properties.

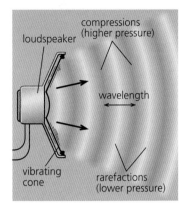

When a loudspeaker cone vibrates, it moves forwards and backwards very fast. This squashes and stretches the air in front. As a result, a series of compressions ('squashes') and rarefactions ('stretches') travel out through the air. These are **sound waves**. When they reach your ears, they make your ear-drums vibrate and you hear a sound.

The nature of sound waves

Sound waves are caused by vibrations Any vibrating object can be a source of sound waves. As well as loudspeaker cones, examples include vibrating guitar strings, the vibrating air inside a trumpet, and the vibrating prongs of a tuning fork. Also, when hard objects (such as cymbals and steel drums) are struck, they vibrate and produce sound waves.

Wavefront essentials
For convenience, waves are often drawn using lines called **wavefronts**. In the case of sound waves, you can think of each wavefront as a compression.

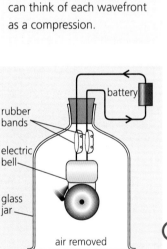

▲ Sound cannot travel through a vacuum. When the air is removed from this jar, the bell goes quiet, even though the hammer is still striking the metal. (The rubber bands reduce the sound transmitted by the connecting wires.)

Sound waves are longitudinal waves The air oscillates backwards and forwards as the compressions and rarefactions pass through it. When a compression passes, the air pressure rises. When a rarefaction passes, the pressure falls. The distance from one compression to the next is the **wavelength**.

Sound waves need a material to travel through This material is called a **medium**. Without it, there is nothing to pass on any oscillations. Sound cannot travel through a vacuum (completely empty space).

Sound waves can travel through solids, liquids, and gases Most sound waves reaching your ear have travelled through air. But you can also hear when swimming underwater, and walls, windows, doors, and ceilings can all transmit (pass on) sound.

Sound waves can be reflected and refracted (see the next spread, 6.4)

WAVES AND SOUNDS

Sounds waves can be diffracted You can hear someone through an open window even if you cannot see them. That is because sound waves are diffracted by everyday objects: they spread through gaps or bend round obstacles of similar size to their wavelength (typically from a few centimetres to a few metres).

Displaying sounds

Sound waves can be displayed graphically using a microphone and an **oscilloscope** as on the right. When sound waves enter the microphone, they make a crystal or a metal plate inside it vibrate. The vibrations are changed into electrical oscillations, and the oscilloscope uses these to make a spot oscillate up and down on the screen. It moves the spot steadily sideways at the same time, producing a wave shape called a **waveform**. The waveform is really a graph showing how the air pressure at the microphone varies with time. It is *not* a picture of the sound waves themselves: sound waves are *not* transverse (up-and-down).

Reducing sounds

Hard surfaces reflect sounds and can cause **echoes** (see spread 6.4). In large rooms and halls, the soft materials in curtains, carpets, and padded furniture help reduce the problem by absorbing the energy in sound waves.

The bricks, wood, and steel used in buildings are all good transmitters of sound waves. To stop unwanted sounds getting in or passing from one room to the next, panels backed with foam or fibrewool can be used to cut down sound transmission.

▲ If you live near an airport, double (or even triple) glazed windows are essential in situations like this. Glass is a good transmitter of sound waves, but glass sheets with an air layer sandwiched between let much less sound through.

▲ Looking like giant mushrooms, these acoustic diffusers hang from the ceiling of the Albert Hall in London. Made of fibreglass, their job is to scatter reflected sounds so that echoes don't spoil the music being performed below.

1 Give an example which demonstrates each of the following:
   a Sound can travel through a gas.
   b Sound can travel through a liquid.
   c Sound can travel through a solid.
2 Explain each of the following:
   a Sound cannot travel though a vacuum.
   b It is possible to hear round corners.
3 a Sound waves are *longitudinal* waves. Explain what this means.
   b If sound waves are longitudinal, why are transverse (up-and-down) 'waves' seen on the screen of the oscilloscope above when someone whistles into the microphone?
4 What happens to sound waves if they strike a hard surface, such as a wall?

Related topics: air pressure **3.7**; longitudinal waves **6.1**; diffraction **6.2**; loudspeaker **9.5**

6.4 Speed of sound and echoes

Objectives: – to know the speed of sound waves in air, and describe how it can be measured – to know how it compares with the speed in solids and in liquids.

▶ Sound is much slower than light, so you hear lightning after you see it. Sound takes about 3 seconds to travel one kilometre. Light does it in almost an instant, so a 3 second gap between the flash and the crash means that the lightning is about a kilometre away.

Sound wave essentials
Sound waves are a series of compressions ('squashes') and rarefactions ('stretches') that travel through the air or other material.

The speed of sound
In air, the speed of sound is about 330 metres per second (m/s), or 760 mph. That is slower than Concorde but about four times faster than a racing car.

The speed of sound depends on the temperature of the air*
Sound waves travel faster through hot air than through cold air.

The speed of sound does not depend on the pressure of the air*
If atmospheric pressure changes, the speed of sound waves stays the same.

The speed of sound is different through different materials
Sound waves travel faster through liquids than through gases, and fastest of all through solids. There are some examples on the left.

Speed of sound through...

air (dry) at 0 °C	330 m/s
air (dry) at 30 °C	350 m/s
water (pure) at 0 °C	1400 m/s
concrete	5000 m/s

Measuring the speed of sound
The speed of sound in air can be measured as shown below. A sound is made by hitting a metal block or plate with a hammer. When the control unit receives a pulse of sound from microphone A, it starts the clock. When it receives a pulse from microphone B, it stops it.

If B is 1.00 metre further away from the source of sound than A, and the clock records a time of 3.0 milliseconds (0.003 s):

$$\text{speed of sound} = \frac{\text{distance travelled}}{\text{time taken}} = \frac{1.00 \text{ m}}{0.003 \text{ s}} = 330 \text{ m/s}$$

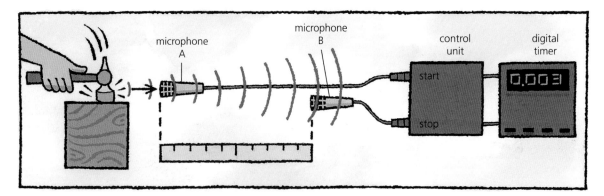

Refraction of sound*

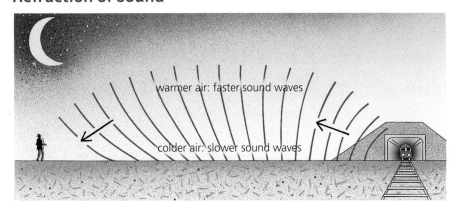

Distant trains and traffic often sound louder (and closer) at night. The reason is this. During the night time, when the ground cools quickly, air layers near the ground become colder than those above. Sound waves travel more slowly through this colder air. As a result, waves leaving the ground tend to bend back towards it, instead of spreading upwards. A bending effect like this, caused by a change in speed, is called **refraction**.

Echoes

Hard surfaces such as walls reflect sound waves. When you hear an **echo**, you are hearing a reflected sound a short time after the original sound. In the diagram on the right, the sound has to travel to the wall *and back again*. The time it takes is the **echo time**. So:

$$\text{speed of sound} = \frac{\text{distance travelled}}{\text{time taken}} = \frac{2 \times \text{distance to wall}}{\text{echo time}}$$

If the speed of sound is known, and the echo time is measured accurately, the distance to the wall can be calculated from the above equation. The principle is used in several devices, including the following:

- **Echo-sounder** This measures the depth of water under a boat. It sends pulses of sound waves towards the sea-bed and measures the echo time. The longer the time, the deeper the water (see spread 6.6).
- **Radar*** This uses the echo-sounding principle, but with microwaves instead of sound waves. It detects the positions of aircraft or ships by measuring the 'echo times' of microwave pulses reflected from them.
- **Parking sensors*** These set off warning bleeps when a car is getting too close to an obstacle. Some work like radar; some use sound pulses.

Assume that the speed of sound in air is 330 m/s.

1 a Why do you hear lightning after you see it?
  b If lightning strikes, and you hear it 4 seconds after you see it, how far away is it?
2 Does sound travel faster through
  a a *solid* or a *gas*?   **b*** *cold air* or *warm air*?
3 When sound waves change direction because their speed changes, what is this effect called?

4 A ship is 220 metres from a large cliff when it sounds its foghorn.
  a When the echo is heard on the ship, how far has the sound travelled?
  b What time delay is there before the echo is heard?
  c The ship changes its distance from the cliff. When the echo time is 0.5 seconds, how far is the ship from the cliff?

Related topics: refraction **6.2**; sound waves **6.3**; echo-sounding **6.6**; speed of light **7.10** and **11.5**; microwaves **7.11** and **7.12**

6.5 Characteristics of sound waves

Objectives: – to explain how frequency and pitch are related, and how amplitude and loudness are related – to know how to interpret the waveforms.

Sound wave essentials

Sound waves are a series of compressions ('squashes') and rarefactions ('stretches') that travel through the air or other material.

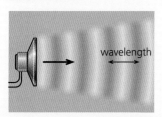

Frequency and pitch

Sound waves are caused by vibrations – for example, the rapid, backwards-and-forwards oscillations of a loudspeaker cone.

The number of oscillations per second is called the **frequency**. It is measured in **hertz (Hz)**. If a loudspeaker cone has a frequency of 100 Hz, it is oscillating 100 times per second and giving out 100 sound waves per second.

Different frequencies sound different to the ear. You hear *high* frequencies as *high* notes: musicians say that they have a **high pitch**. You hear *low* frequencies as *low* notes: they have a **low pitch**.

The human ear can detect frequencies ranging from about 20 Hz up to 20 000 Hz, although the ability to hear high frequencies decreases with age.

pitch		frequency
high	upper limit of hearing	20 000 Hz
	whistle	10 000 Hz
	high note (soprano)	1000 Hz
	low note (bass)	100 Hz
low	drum note	20 Hz

1000 Hz = 1 kilohertz (kHz)

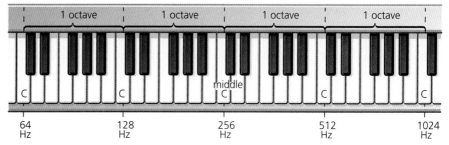

Octaves* Musical scales are based on these. If the pitch of a note increases by one octave, the frequency doubles, as shown on the keyboard above. This keyboard is tuned to **scientific pitch**. Bands and orchestras normally use frequencies that differ slightly from those shown.

The diagrams below show what happens if two steady notes, an octave apart, are picked up by a microphone and displayed on the screen of an oscilloscope. As the higher note has double the frequency of the lower note, the peaks occur twice as often and are only half as far apart.

▶ The **waveform** on each screen is a graph showing how the air pressure varies with time as the sound waves enter the microphone. The horizontal line is the time axis.

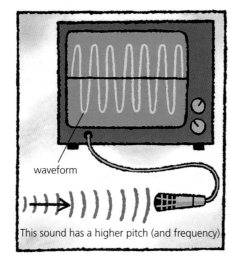

waveform

This sound has a higher pitch (and frequency)

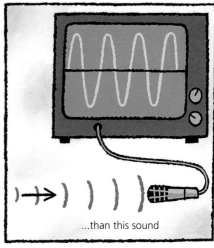

...than this sound

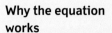

The wave equation

This equation applies to sound waves:

$$\text{speed} = \text{frequency} \times \text{wavelength}$$

In symbols: $v = f\lambda$ (λ = Greek letter *lambda*)

For example, if the speed of sound in air is 330 m/s:
sound waves of frequency 110 Hz have a wavelength of 3 m;
sound waves of frequency 330 Hz have a wavelength of 1 m;
so the *higher* the frequency, the *shorter* the wavelength.

Why the equation works

If 110 waves are sent out in one second, and each wave is 3 m long, then the waves must travel 330 metres in one second. In other words, if the frequency is 110 Hz and the wavelength is 3 m, the speed is 330 m/s.

Amplitude and loudness

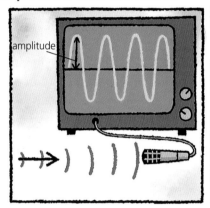 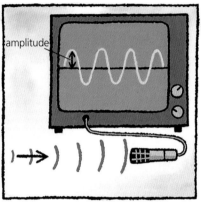

The sounds displayed on the oscilloscope screens above have the same frequency, but one is *louder* than the other. The oscillations in the air are bigger and the **amplitude** of the waveform is greater.

Sound waves carry energy. *Doubling* the amplitude means that *four* times as much energy is delivered per second.

Quality*

Middle C on a guitar does not sound quite the same as middle C on a piano, and its waveform looks different. The two sounds have a different **quality** or **timbre**. Each sound has a strong **fundamental frequency**, giving middle C. But other weaker frequencies are mixed in as well, as shown on the right. These are called **overtones**, and they differ from one instrument to another. With a synthesizer, you can select which frequencies you mix together, and produce the sound of a guitar, piano, or any other instrument.

fundamental frequency...

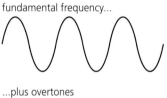

...plus overtones

...gives the final waveform

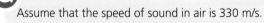

Assume that the speed of sound in air is 330 m/s.

1 Here are the frequencies of four sounds:
   A: 400 Hz   B: 150 Hz   C: 500 Hz   D: 200 Hz
   a Which sound has the highest pitch?
   b Which sound has the longest wavelength?
   c* Which two sounds are one octave apart?

2* Why does a piano not sound quite like a guitar, even if both play the same note?

3 A sound is picked up by a microphone and displayed as a waveform on an oscilloscope. How would the waveform change if
   a the sound had a higher pitch?
   b the sound was louder?

4 The lower limit of human hearing is 20 Hz; the upper limit is 20 000 Hz.
   a What is the upper limit in kHz?
   b What is the wavelength at the lower limit?
   c What is the wavelength at the upper limit?

Related topics: speed **2.1**; waves and the wave equation **6.1**; sound waves **6.3**; speed of sound **6.4**

6.6 Ultrasound

Objectives: – to know the approximate upper and lower limits of human hearing – to explain what ultrasound is, and describe some of its uses.

Sound wave essentials
Sound waves are a series of compressions ('squashes') and rarefactions ('stretches') that travel through the air or other material.

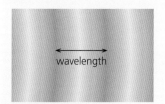

The number of waves per second is called the **frequency**. It is measured in hertz (Hz).

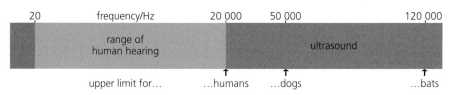

The human ear can detect sounds up to a frequency of about 20 000 Hz. Sounds above the range of human hearing are called **ultrasonic sounds**, or **ultrasound**. Here are some of the uses of ultrasound:

Cleaning and breaking*

Using ultrasound, delicate machinery can be cleaned without dismantling it. The machinery is immersed in a tank of liquid, then the vibrations of high-power ultrasound are used to dislodge the bits of dirt and grease.

In hospitals, concentrated beams of ultrasound can be used to break up kidney stones and gall stones without patients needing surgery.

Ⓔ Echo-sounding

Ships use **echo-sounders** to measure the depth of water beneath them. An echo-sounder sends pulses of ultrasound downwards towards the sea-bed, then measures the time taken for each echo (reflected sound) to return. The longer the time, the deeper the water. For example:

If a pulse of ultrasound takes 0.1 second to travel to the sea-bed and return, and the speed of sound in water is 1400 m/s:

$$\text{distance travelled} = \text{speed} \times \text{time} = 1400 \text{ m/s} \times 0.1 \text{ s} = 140 \text{ m}$$

But the ultrasound has to travel down *and back*:

So: depth of water = ½ × 140 m = 70 m

Most echo-sounders **scan** the area beneath them – they sweep their ultrasound beam backwards and forwards and from side to side. A computer displays the depth information as a picture on a screen.

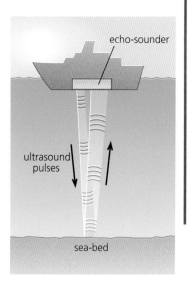

▶ This bat uses ultrasound to locate insects and other objects in front of it. It sends out a series of ultrasound pulses and uses its specially shaped ears to pick up the reflections. The process is called **echo-location**. It works like echo-sounding.

WAVES AND SOUNDS

E Metal testing

The echo-sounding principle can be used to detect flaws in metals. A pulse of ultrasound is sent through the metal as on the right. If there is a flaw (tiny gap) in the metal, *two* reflected pulses are picked up by the detector. The pulse reflected from the flaw returns first, followed by the pulse reflected from the far end of the metal. The pulses can be displayed using an oscilloscope. The trace on the screen is a graph showing how the amplitude ('strength') of the ultrasound varies with time.

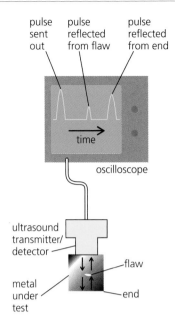

Scanning the womb

The pregnant mother in the photograph below is having her womb scanned by ultrasound. Again, the echo-sounding principle is being used. A transmitter sends pulses of ultrasound into the mother's body. The transmitter also acts as a detector and picks up pulses reflected from the baby and different layers inside the body. The signals are processed by a computer, which puts an image on the screen.

Using ultrasound is much safer than using X-rays because X-rays can cause cell damage inside a growing baby. Also, ultrasound can distinguish between different layers of soft tissue, which an ordinary X-ray machine cannot.

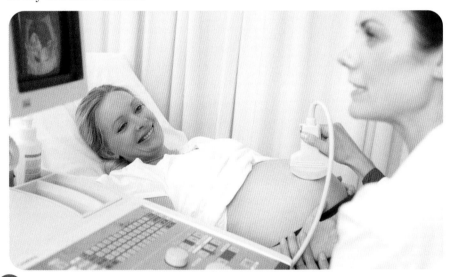

◀ An ultrasound scan of the womb. The nurse is moving an ultrasound transmitter/detector over the mother's body. A computer uses the reflected pulses to produce an image.

Q

1. What is *ultrasound*?
2. Give *two* examples of the medical use of ultrasound.
3. **a** What is an *echo-sounder* used for?
   b How does an echo-sounder work?
4. To answer this question, you will need the information on the right. A boat is fitted with an echo-sounder which uses ultrasound with a frequency of 40 kHz.
   a What is the frequency of the ultrasound in Hz?
   b If ultrasound pulses take 0.03 seconds to travel from the boat to the sea-bed and return, how deep is the water under the boat?
   c What is the wavelength of the ultrasound in water?

$$\text{speed} = \frac{\text{distance travelled}}{\text{time taken}}$$

speed of sound in water = 1400 m/s

speed = frequency × wavelength
(m/s)   (Hz)   (m)

1 kilohertz (kHz) = 1000 Hz

Related topics: sound waves **6.3**; speed of sound and echoes **6.4**; frequency **6.5**

Check-up on waves and sounds

Further questions

1 Lee and Sam are playing with a ball in the park. Unfortunately the ball finishes up in the middle of a pond, out of reach.

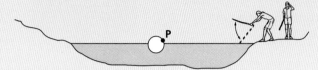

Lee thinks that hitting the water with a stick will make waves that will push the ball to the other side.

a Which **two** of these words best describe the waves that are created on the water surface?

circular   longitudinal   plane   pressure transverse [2]

b Lee hits the water surface regularly so that waves travel out to the ball and beyond it.
  i What happens to the ball? [1]
  Sam throws a stick which hits the ball at **P**.
  ii Sam is successful at moving the ball across the pond. Lee is not. Explain why. [2]

c **i** Lee hits the water surface regularly with the stick 20 times in 10 seconds. Calculate the frequency of the waves. [2]
  ii The waves travel across the pond at 0.5 m/s. Calculate the wavelength. [4]

2 a The wave in the shallow tank of water shown in the figure moves at 0.08 m/s towards the left.

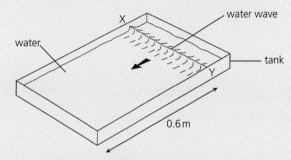

How long does it take for the wave to return to the position XY, but moving to the right? [3]

b Someone is cutting down a tree with an axe. They hear the echo of the impact of the axe hitting the tree after 1.6 s.
  i What sort of obstacle could have caused the echo? [1]
  ii The speed of sound is 330 m/s. How far is the tree from the obstacle?
c What is the difference between the sound wave in **b** and the water wave in **a**. [2]

3 The figure shows an oscilloscope trace for a sound wave produced by a loudspeaker.

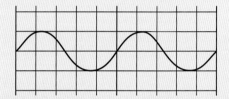

a Copy the figure and draw the trace for a louder sound of the same pitch. [2]

b It takes 1/50th of a second (0.02 s) for the whole trace to be produced.
  i Show that the frequency of the sound produced by the loudspeaker is 100 Hz.
  ii Determine the wavelength in air of the sound produced by the loudspeaker. (The speed of sound in air is 330 m/s.) [3]

4 a A sound wave travelling through air can be represented as shown in the diagram.

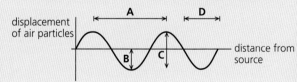

Which distance, **A**, **B**, **C**, or **D**, represents:
  i one wavelength?
  ii the amplitude of the wave? [2]

b The cone of a loudspeaker is vibrating. The diagram shows how the air particles are spread out in front of the cone at a certain time.

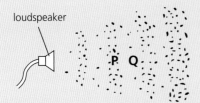

P is a compression, Q is a rarefaction.
  i Describe how the pressure in the air changes from **P** to **Q**. [2]
  ii Describe the motion of the air particles as the sound wave passes. [2]
  iii Copy the diagram of air particles above and mark and label a distance equal to one wavelength of the sound wave. [1]

5 The following are all examples of wave motion:

light waves	X-rays
sound waves	seismic P-waves
waves in ripple tank	seismic S-waves

a Write out the above in the form of a table like this:

Longitudinal waves	Transverse waves

[6]

b How are seismic waves produced? [2]
c When someone fires a starting pistol, they hears an echo 0.4 s later because some of the sound waves are reflected by the wall of a building. If the speed of sound is 330 m/s, how far away is the wall? [3]

6

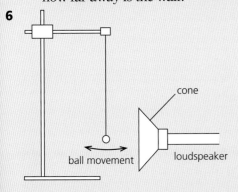

A light polystyrene ball is shown hanging very close to a loudspeaker. The loudspeaker gives out a sound of low frequency and the ball is seen to vibrate.
  a Explain how the sound from the loudspeaker causes the ball to move as described. [2]
  b Explain what will happen to the motion of the cone of the loudspeaker when:
    i the sound is made louder [1]
    ii the pitch of the sound is increased. [1]
  c Calculate the frequency of a sound which has a wavelength of 0.5 m and travels at a speed of 340 m/s in air. Write down the formula that you use and show your working. [3]

7 The figure shows a metal rod, 2.4 m long, being struck a sharp blow at one end using a light hammer. The time interval between the impact of the hammer and the arrival of the sound wave at the other end of the rod is measured electronically.

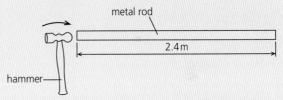

Four measurements of the time interval are 0.44 ms, 0.50 ms, 0.52 ms and 0.46 ms.
  a Determine the average value of the four measurements.
  b Hence calculate a value for the speed of sound in the rod. [4]

8 a A microphone is connected to an oscilloscope. When different sounds, A, B, and C, are made, these are the waveforms seen on the screen:

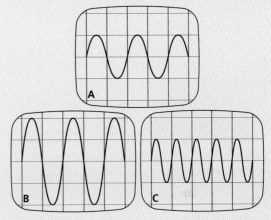

a Comparing sounds A and B, how would they sound different? [2]
b Comparing sounds A and C, how would they sound different? [2]
c Which sound has the highest amplitude? [1]
d Which sound has the highest frequency? [1]
e The speed of sound is 330 m/s. If sound A has a frequency of 220 Hz, what is its wavelength? [2]
(E) f What is the frequency of sound C? [2]

(E) 9 Ultrasound waves are high frequency longitudinal waves. X-rays are high frequency transverse waves.
a Explain the difference between transverse and longitudinal waves. [2]
b The diagram shows an ultrasound probe used to obtain an image of an unborn baby.

— mother's abdominal wall
— ultrasound probe

Give **two** reasons why ultrasound and not X-rays are used for this investigation. [2]
c Describe **one** industrial use of ultrasonic waves. [2]

Use the list below when you revise for your IGCSE examination. The spread number, in brackets, tells you where to find more information.

Revision checklist

Core Level
☐ Wave motion and wavefronts. (6.1)
☐ Waves transfer energy. (6.1)
☐ The difference between transverse and longitudinal waves, with examples of each. (6.1)
☐ The meaning of wavelength. (6.1)
☐ The meaning of amplitude. (6.1)
☐ The meaning of frequency. (6.1)
☐ The hertz, unit of frequency. (6.1)
☐ The equation linking speed, frequency, and wavelength. (6.1 and 6.5)
☐ Demonstrating these wave effects in a ripple tank:
 – reflection
 – refraction
 – diffraction through a gap and at an edge. (6.2)
☐ How refraction is caused by a change of wave speed. (6.2)
☐ How, in a ripple tank, a change in wave speed is caused by a change of depth. (6.2)
☐ Sound waves are produced by vibrations. (6.3)
☐ Sound waves are longitudinal waves. (6.3)
☐ Why sound waves need a medium (material) to travel through. (6.3)
☐ Displaying waveforms on an oscilloscope. (6.3 and 6.5)
☐ Measuring the speed of sound (in air). (6.4)
☐ How the reflection of sound causes echoes. (6.4)
☐ The frequency range of sound waves. (6.5)
☐ The link between frequency and pitch. (6.5)
☐ The link between amplitude and loudness. (6.5)
☐ What ultrasound is. (6.6)

Extended Level
As for Core Level, plus the following:
☐ How wavelength and gap size affect diffraction through a gap. (6.2)
☐ How wavelength affects diffraction at an edge. (6.2)
☐ What compressions and rarefactions are. (6.3)
☐ How the speed of sound is different in solids, liquids, and gases. (6.4)
☐ Some medical and industrial uses of ultrasound. (6.6)
☐ Echo-sounding: measuring depth or distance by timing ultrasound pulses. (6.6)

7

Rays and waves

- LIGHT RAYS AND WAVES
- REFLECTION
- REFRACTION
- PRISMS
- SPECTRUM OF LIGHT
- TOTAL INTERNAL REFLECTION
- OPTICAL FIBRES
- LENSES
- ELECTROMAGNETIC WAVES
- SENDING SIGNALS

A rainbow forms as the Sun shines on raindrops. The raindrops, acting like tiny prisms, are splitting the white sunlight into its different spectral colours and reflecting them back. Because the Sun is behind the camera and the rainbow appears to extend to the sea, rain must be falling between the camera and the clouds in the background.

7.1 Light rays and waves

Objectives: – to describe the main properties (features) of light waves – to know what the speed of light is – to know the link between wavelength and colour.

▶ If you can see a beam of light, this is because tiny particles of dust, smoke, or mist in the air are reflecting some of the light into your eyes.

For you to see something, light must enter your eyes. The Sun, lamps, lasers, and glowing TV screens all emit (send out) their own light. They are **luminous**. However, most objects are **non-luminous**. You see them only because daylight, or other light, bounces off them. They **reflect** light, and some of it goes into your eyes.

You can see this page because it reflects light. The white parts reflect most light and look bright. However, the black letters **absorb** nearly all the light striking them. They reflect very little and look dark.

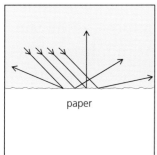

Diffuse reflection

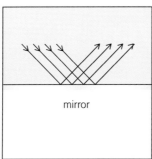

Regular reflection

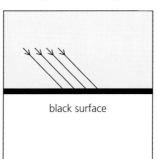

Absorption

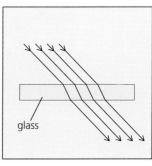

Transmission

Most surfaces are uneven, or contain particles that scatter light. As a result, they reflect light in all directions. The reflection is **diffuse**. However, mirrors are smooth and shiny. When they reflect light, the reflection is **regular**.

Transparent materials like glass and water let light pass right through them. They **transmit** light.

Features of light

Light is a form of radiation This means that light radiates (spreads out) from its source. In diagrams, lines called **rays** are used to show which way the light is going.

▲ This solar-powered car uses the energy in sunlight to produce electricity for its motor.

Light travels in straight lines You can see this if you look at the path of a sunbeam or a laser beam.

140

RAYS AND WAVES

Light transfers energy Energy is needed to produce light. Materials gain energy when they absorb light. For example, solar cells use the energy supplied by sunlight to generate electricity.

Light travels as waves Light radiates from its source rather as ripples spread across the surface of a pond. However, in the case of light, the 'ripples' are tiny, vibrating, electric and magnetic forces. Light waves have wavelengths of less than a thousandth of a millimetre (see below). Like other waves, they can be diffracted (see spread 6.2), but the effect is too small to notice unless the gaps are very narrow, for example, as in a fine mesh.

Some effects of light are best explained by thinking of light as a stream of tiny 'energy particles'. Scientists call these particles **photons**.

Light can travel through empty space Electric and magnetic ripples do not need a material to travel through. That is why light can reach us from the Sun and stars.

Light is the fastest thing there is In a vacuum (in space, for example), the speed of light is 300 000 kilometres *per second*. Nothing can travel faster than this. The speed of light seems to be a universal speed limit.

Wavelength and colour

When light enters the eye, the brain senses different wavelengths as different colours. The wavelengths range from 0.000 4 mm (violet light) to 0.000 7 mm (red light), and white light is made up of all the wavelengths in this range. Most sources emit a mixture of wavelengths. However, **lasers** emit light of a single wavelength and colour. Light like this is called **monochromatic** light.

Wave essentials

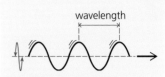

With **transverse** waves, like light, the oscillations (vibrations) are at right angles to the direction of travel.

Frequencies

Light waves have extremely high frequencies. Knowing the speed of light and the, the frequency can be calculated using the equation $v = f \times \lambda$ (See 6.1)

For example: frequency of...
red light   = 4.3×10^{14} Hz
violet light = 7.5×10^{14} Hz

In SI units, the speed of light has an exact value of 299 792 458 m/s
3×10^8 m/s is a useful approximate value.

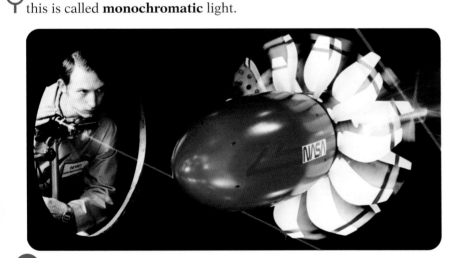

◀ Light from a laser is monochromatic (single wavelength and colour). Here, laser light is being used to measure the deflection of the rotating blades on an experimental jet engine.

Q

1. Give *two* examples each of objects which
   a. emit their own light
   b. are only visible because they reflect light from another source.
2. What evidence is there that light travels in straight lines?
3. What happens to light when it strikes
   a. white paper   b. black paper?
4. If the Moon is 384 000 km from Earth, the Sun is 150 000 000 km from Earth, and the speed of light is 300 000 km/s, calculate the time taken for light to travel from
   a. the Moon to the Earth   b. the Sun to the Earth.
5. Comparing red light with violet, which has
   a. the longer wavelength   b. the higher frequency?
6. What is meant by *monochromatic* light?

Related topics: speed 2.1; energy 4.1; colours in white light 7.4; electromagnetic waves 7.10; photons 10.10

7.2 Reflection in plane mirrors (1)

Objectives: – to know the laws of reflection – to describe how and where a plane mirror forms an image – to know that the image is virtual.

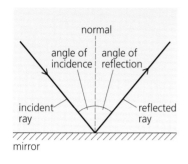

The laws of reflection

When a ray of light strikes a mirror, it is reflected as shown on the left. The incoming ray is the **incident ray**, the outgoing ray is the **reflected ray**, and the line at right angles to the mirror's surface is called a **normal**. The mirror in this case is a **plane mirror**. This just means that it is a flat mirror, rather than a curved one.

There are two **laws of reflection**. They apply to all types of mirror:

1. The angle of incidence is equal to the angle of reflection.
2. The incident ray, the reflected ray, and the normal all lie in the same plane.

Put another way, light is reflected at the same angle as it arrives, and the two rays and the normal can all be drawn on one flat piece of paper.

Definitions
Angle of incidence: this is the angle between the incident ray and the normal.
Angle of reflection: this is the angle between the reflected ray and the normal.

Image in a plane mirror

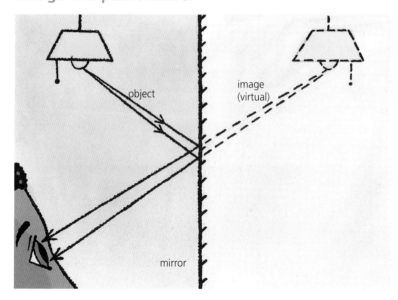

In the diagram above, light rays are coming from an **object** (a lamp) in front of a plane mirror. Thousands of rays could have been drawn but, for simplicity, only two have been shown. After reflection, some of the rays enter the girl's eye. To the girl, they seem to come from a position behind the mirror, so that is where she sees an **image** of the lamp. Dotted lines have been drawn to show the point where two of the reflected rays appear to come from. The dotted lines are *not* rays.

The image seen in the mirror looks exactly the same as the object, apart from one important difference. The image is **laterally inverted** (back to front).

Real and virtual images In a cinema, the image on the screen is called a **real image** because rays from the projector focus (meet) to form it. The image in a plane mirror is not like this. Although the rays *appear* to come from behind the mirror, no rays actually pass through the image and it cannot be formed on a screen. An image like this is called a **virtual image**.

▲ The word on this vehicle is laterally inverted so that it reads correctly when seen in a driving mirror.

Finding the position of an image in a mirror

The position of an image in a plane mirror can be found by experiment:

Put a mirror upright on a piece of paper. Put a pin (the object) in front of it. Mark the positions of the pin and the mirror.

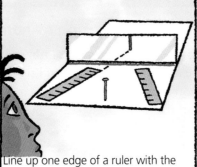
Line up one edge of a ruler with the image of the pin. Draw a line along the edge to mark its position. Then repeat with the ruler in a different position.

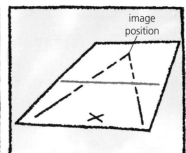
Take away the mirror, pin, and ruler. Extend the two lines to find out where they meet. This is the position of the image.

The result of the experiment can be checked like this. If a second pin is put *behind* the mirror, in the position found for the image, the pin should be in line with the image, as shown on the right. And it should stay in line when you move your head from side to side. Scientifically speaking, there should be **no parallax** (no relative movement) between the second pin and the image when you change your viewing position. If there is relative movement (parallax), then the two are not in the same position.

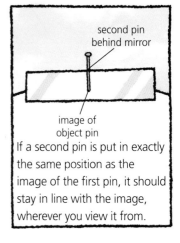
If a second pin is put in exactly the same position as the image of the first pin, it should stay in line with the image, wherever you view it from.

Rules for image size and position

When a plane mirror forms an image:

- The image is the same size as the object.
- The image is as far behind the mirror as the object is in front.
- A line joining equivalent points on the object and image passes through the mirror at right angles.

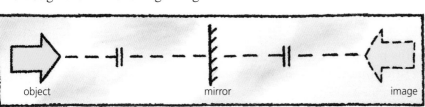

1. **a** Copy the diagram on the right. Draw in the image in its correct position.
   b The image cannot be formed on a screen. What name is given to this type of image?
   c From the object arrow's tip, A, draw two rays which reflect from the mirror and go into the person's eye.
   d Can the person see an image of the arrow's tail, B? If not, why not?

2. A man stands 10 m in front of a large, plane mirror. How far must he walk before he is 5 m away from his image?

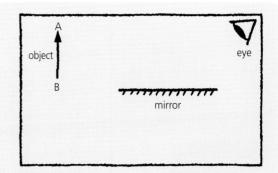

Related topics: reflection of waves **6.2**; real and virtual images formed by lenses **7.7–7.8**

7.3 Reflection in plane mirrors (2)

Objectives: – to know how to find the image position in a plane mirror using a scale drawing – to solve problems about reflected rays by calculation.

(E) Finding an image position by construction

In the diagrams below, O is a point object in front of a plane (flat) mirror. Here are two methods of finding the position of the image by geometric construction using a protractor. In Method 1, you deduce the position from the paths of two rays, but Method 2 is simpler!

Method 1

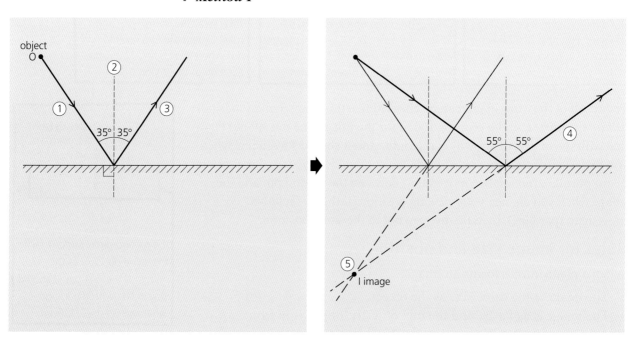

(E) 1 From the object, O, draw a ray which strikes the mirror at an angle of incidence of 35° (or value of your own choosing close to this).
2 Construct a normal (a line at right angles to the mirror's surface) at the point where the ray strikes the mirror.
3 Draw the reflected ray from this point, so that the angle of reflection is equal to the angle of incidence.
4 Repeat steps 1 to 3 for a second ray with an angle of incidence of 55° (or value of your own choosing close to this).
5 Extend the two reflected rays backwards until they intersect (meet). The point of intersection, I, is the image position.

Method 2

This method is illustrated on the left. It uses the fact that the position of the image behind the mirror matches that of the object in front.

1 From the object, O, draw a line which passes through the mirror's surface at right angles. Extend this line well beyond the mirror.
2 Measure the distance from the object to the mirror.
3 At an equal distance behind the mirror, mark a point on the extended line. This point, I, is the image position.

RAYS AND WAVES

E Reflection problem

Example A horizontal ray of light strikes a plane mirror whose surface is angled at 55° to the ground, as shown below left.
a What is the angle between the reflected ray and the ground?
b If the mirror is re-angled to reflect the ray vertically upwards, what is the new angle between the surface of the mirror and the ground?

a
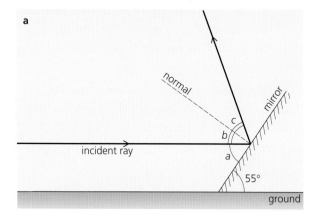

b
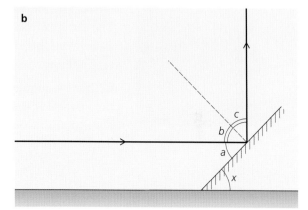

E **a** In the diagram above left, angles a, b, and c have also been labelled to help with the calculation. The incident ray is parallel to the ground, so the angle between the reflected ray and the ground is equal to $b + c$.

As the incident ray is parallel to the ground: $a = 55°$
But: $a + b = 90°$ So: $b = 35°$

As the angle of reflection = angle of incidence: $c = b$
So: $c = 35°$ Therefore: $b + c = 70°$

So, the angle between the reflected ray and the ground is 70°.

b The situation is shown above right, where angles a, b, and c all now have new values. As before: $a + b = 90°$ and $c = b$. x is the unknown angle between the surface of the mirror and the ground. It is equal to a.

As the ray is reflected vertically: $b + c = 90°$ So b and c are both 45°
But: $a + b = 90°$ So: $a = 45°$ Therefore: $x = 45°$

So, the angle between the surface of the mirror and the ground must be changed to 45°.

Reflection essentials

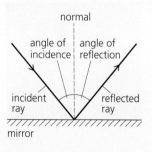

When light is reflected from a mirror, the angle of incidence is equal to the angle of reflection.

Q

You will need a ruler, protractor, and sharp pencil.

1 In the diagram on the right, two rays leave a point object O and strike a plane mirror.
  a Make an exact copy of the diagram.
  b Measure the angle of incidence of each ray.
  c Draw in the two reflected rays at the correct angles.
  d Find where the image is formed and label it.
  e By drawing or calculation, work out what angle the mirror would need to be turned through so that ray B is reflected back the way it came.

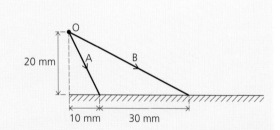

Related topics: reflection of waves **6.2**

7.4 Refraction of light

Objectives: – to describe how light is refracted by a glass block and by a prism – to know how refractive index is defined – to explain what dispersion is.

The 'broken pen' illusion on the left occurs because light is bent by the glass block. The bending effect is called **refraction**.

The diagram below shows how a ray of light passes through a glass block. The line at right angles to the side of the block is called a **normal**. The ray is refracted towards the normal when it enters the block, and away from the normal when it leaves it. The ray emerges parallel to its original direction (provided the block has parallel sides).

Refraction would also occur if the glass were replaced with another transparent material, such as water or acrylic plastic, although the angle of refraction would be slightly different. The material that light is travelling through is called a **medium**.

Definitions
Angle of incidence: this is the angle between the incident ray and the normal.
Angle of refraction: this is the angle between the refracted ray and the normal.

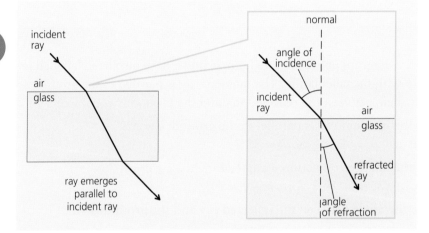

Real and apparent depth*

Because of refraction, water (or glass) looks less deep than it really is. Its apparent depth is less than its real depth. This diagram shows why:

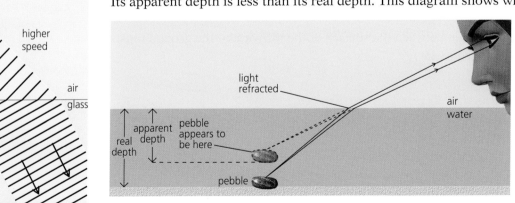

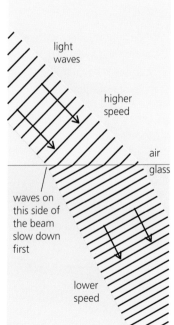

Why light is refracted

Scientists explain refraction as follows. Light is made up of tiny waves. These travel more slowly in glass (or water) than in air. When a light beam passes from air into glass, as shown on the left, one side of the beam is slowed before the other. This makes the beam 'bend'.

RAYS AND WAVES

E Refractive index

In a vacuum (empty space), the speed of light is 300 000 km/s. In air, it is effectively the same. However, in glass, light slows to 200 000 km/s.

The **refractive index** of a medium is defined like this:

$$\text{refractive index} = \frac{\text{speed of light in vacuum}}{\text{speed of light in medium}}$$

So, in the case of glass:

$$\text{refractive index} = \frac{300\ 000\ \text{km/s}}{200\ 000\ \text{km/s}} = 1.5$$

Some refractive index values are given on the right. The medium with the *highest* refractive index has the *greatest* bending effect on light because it slows the light the *most*.

medium	refractive index
diamond	2.42
glass (crown)	1.52
acrylic plastic (Perspex)	1.49
water	1.33

▲ The above figures are based on more accurate values of the speed of light than those used on the left.

The refractive index of glass varies depending on the type of glass. Refractive index also varies slightly depending on the colour of the light.

Refraction by a prism

A **prism** is a triangular block of glass or plastic. The sides of a prism are not parallel. So, when light is refracted by a prism, it comes out in a different direction. It is **deviated**.

If a narrow beam of white light is passed through a prism, it splits into a range of colours called a **spectrum**, as shown below. The effect is called **dispersion**. It occurs because white is not a single colour but a mixture of all the colours of the rainbow. The prism refracts each colour by a different amount.

Seven colours?

By tradition, there are seven 'rainbow' colours. The seventh, indigo, is between blue and violet. This idea came from the Ancient Greeks who thought that seven was a special number in the Universe – which is why we now have seven days in a week.

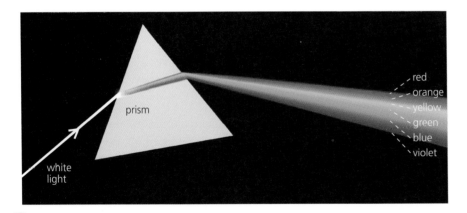

◀ Most people think that they can see about six colours in the spectrum of white light. However, the spectrum is really a continuous change of colour from beginning to end.

Red light is deviated (bent off-course) least by a prism. Violet light is deviated most. However, here the difference has been exaggerated.

Q

For questions **1b** and **3**, you will need to refer to the table at the top of the page. Assume that the speed of light in a vacuum is 300 000 km/s.

1. **a** Copy the diagram on the right. Draw in and label the *normal*, the *refracted ray*, the *angle of incidence*, and the *angle of refraction*.
   b How would your diagram be different if the ray was passing into water rather than glass?
2. **a** When white light passes through a prism, it spreads into a spectrum of colours. What is the spreading effect called?
   b Which colour is deviated most by a prism?
   c Which colour is deviated least?
3. Calculate the speed of light in water.

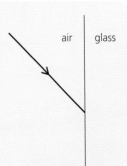

Related topics: refraction of waves **6.2**; colour and wavelength **7.1**; light waves **7.1**; refraction calculations **7.6**

147

7.5 Total internal reflection

Objectives: – to know what total internal reflection is – to describe some of the uses of total internal reflection, including in optical fibres.

Refraction essentials
The bending of light when it passes from one medium (material) to another is called **refraction**. It is caused by a change in the speed of the light.

The inside surface of water, glass, or other transparent material can act like a perfect mirror, depending on the angle at which the light strikes it.

The diagrams below show what happens to three rays leaving an underwater lamp at different angles. Angle c is called the **critical angle**. For angles of incidence greater than this, there is no refracted ray. All the light is reflected. The effect is called **total internal reflection**.

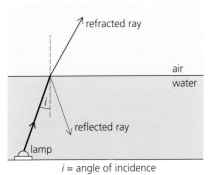

i = angle of incidence

The ray splits into a refracted ray and a weaker reflected ray.

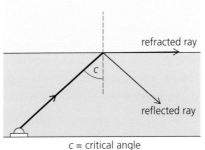

c = critical angle

The rays splits, but the refracted ray only just leaves the surface.

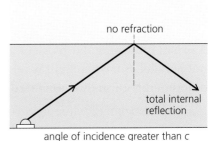

angle of incidence greater than c

There is no refracted ray. The surface of the water acts like a perfect mirror.

The value of the critical angle depends on the material. For example:

	critical angle		
water 49°	acrylic plastic 42°	glass (crown) 41°	diamond 24°

Reflecting prisms

In the diagrams below, inside faces of prisms are being used as mirrors. Total internal reflection occurs because the angle of incidence on the face (45°) is greater than the critical angle for glass or acrylic plastic.

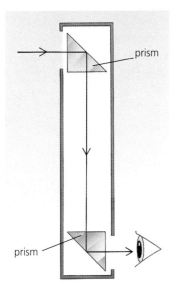

▲ **Periscope** This is an instrument for looking over obstacles. Prisms reflect the light, although they can be replaced with mirrors.

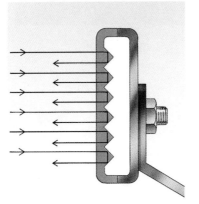

▲ **Rear reflectors** (on cars and cycles) The direction of the incoming light is reversed by two total internal reflections.

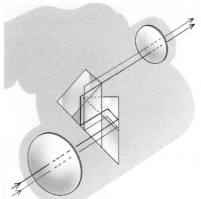

▲ **Binoculars** The lens system in each 'barrel' produces an upside-down image. Reflecting prisms are used to turn it the right way up.

E Optical fibres

Optical fibres are very thin, flexible rods made of special glass or transparent plastic. Light put in at one end is total internally reflected until it comes out of the other end, as shown below. Although some light is absorbed by the fibre, it comes out almost as bright as it goes in – even if the fibre is several kilometres long. (For more on optical fibres, see spread 7.12.)

▲ **Single optical fibre** In the type shown above, the inner glass core is coated with glass of a lower refractive index.

▲ **Bundle of optical fibres** Provided the fibres are in the same positions at both ends, a picture can be seen through them.

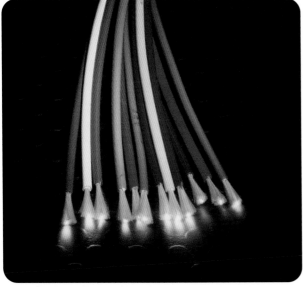

▲ Optical fibres can carry telephone calls and internet data. The signals are coded and sent along the fibre as pulses of laser light. Fewer booster stations are needed than with electrical cables.

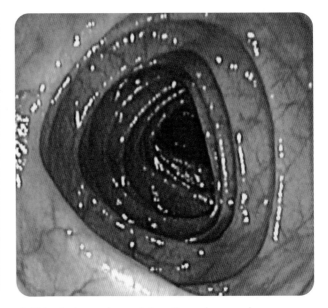

▲ This photograph was taken through an **endoscope**, an instrument used by surgeons for looking inside the body. An endoscope contains a long, thin bundle of optical fibres.

Q

1. Glass has a critical angle of 41°. Explain what this means.
2. **a** Copy and complete the diagrams on the right to show where each ray will go after it strikes the prism.
   b If the prisms on the right were transparent triangular tanks filled with water, would total internal reflection still occur? If not, why not?
3. **a** Give *two* examples of the practical use of optical fibres.
   b Give *two* other examples of the practical use of total internal reflection.

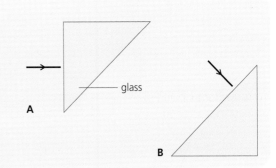

Related topics: refraction **7.4**; calculating the critical angle **7.6**; optical fibres in communications **7.12**

7.6 Refraction calculations

Objectives: – to know, and use, the equations linking refractive index, speeds of light, and ray angles during refraction - to know how to calculate the critical angle.

Refraction essentials

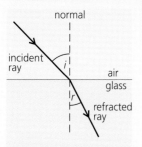

A light ray bends as it enters a glass block. The bending effect is called **refraction**. It occurs because light waves slow down when they pass from air into glass or other **medium** (see spread 7.4). Passing from glass back in to air, they would speed up again. So, if the ray in the diagram were reversed, it would pass back into the air along the same path as it came in.

Snell's law

When light is refracted, an increase in the angle of incidence i produces an increase in the angle of refraction r. In 1620, the Dutch scientist Willebrord Snell discovered the link between the two angles: their *sines* are always in proportion.

When light passes from one medium into another:

$$\frac{\sin i}{\sin r} \text{ is constant}$$

This is known as **Snell's law**. It is illustrated by these examples:

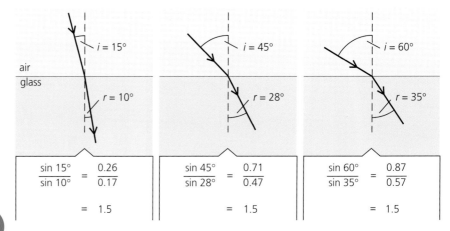

Measuring refractive index

To find the refractive index of, say, glass, you could direct a ray (from a ray box) at a glass block, mark the positions of the incident and refracted rays, measure their angles, then use the equation on the right. A semi-circular block is useful for experiments like this. If the ray passes through point O below, no bending occurs at the circular face, so it is easier to vary and measure the angles.

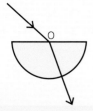

Refractive index

The refractive index of a medium is defined like this:

$$\text{refractive index} = \frac{\text{speed of light in vacuum}}{\text{speed of light in medium}}$$

In a vacuum, the speed of light is 300 000 km/s – and effectively the same in air. In glass, it drops to 200 000 km/s. So, the refractive index of glass is 300 000 km/s ÷ 200 000 km/s, which is 1.5. This is the same as the value of $\sin i \div \sin r$ in the diagrams above.

Here is an alternative definition of refractive index:

$$\text{refractive index} = \frac{\sin i}{\sin r}$$

Example Light (in air) strikes water at an angle of incidence of 45°. If the refractive index of water is 1.33, what is the angle of refraction?

Applying the above equation: $1.33 = \dfrac{\sin 45°}{\sin r}$

Rearranged, this gives $\sin r = \sin 45°/1.33$. When calculated, this gives $\sin r = 0.532$. So the angle of refraction r is 32°.

RAYS AND WAVES

Calculating the critical angle

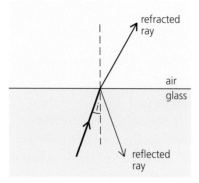

i = angle of incidence

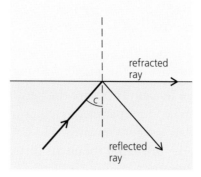

c = critical angle

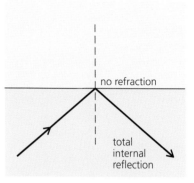

angle of incidence greater than c

 In the diagrams above, rays are travelling from glass towards air at different angles. When the angle of incidence is greater than the **critical angle**, there is no refracted ray. All the light is reflected. There is **total internal reflection**.

Knowing the refractive index of a material, the critical angle can be calculated. For example:

On the right, the middle diagram above has been redrawn with the ray direction reversed. This time, the angle of *incidence* is 90°, and angle c is now the angle of *refraction*. If the refractive index of glass is 1.5:

$$\text{refractive index} = \frac{\sin 90°}{\sin c} = \frac{1}{\sin c} \quad \text{(as } \sin 90° = 1\text{)}$$

rearranging: $\sin c = \dfrac{1}{1.5} = 0.67$

so c, the critical angle of glass, = 42°.

Note: this figure differs slightly from that in spread 7.3 because a simplified value for the refractive index of glass has been used in the calculation.

From the above calculation, it follows that the critical angle c of any medium can be calculated using this equation:

For a medium of refractive index n: $\quad \sin c = \dfrac{1}{n}$

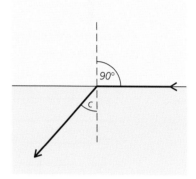

▲ Compare this with the middle diagram at the top of the page.

Q

To answer these questions, you will need a calculator (or set of tables) containing sine values.

1. The refractive index of water is 1.33. Calculate the angle of refraction if light (in air) strikes water at an angle of incidence of **a** 24° **b** 53°.
2. A transparent material has a refractive index of 2.0.
   a Calculate the critical angle.
   b If the refractive index were less than 2.0, would the critical angle be *greater* or *less* than before?
3. Diamond has a refractive index of 2.42. The speed of light in a vacuum (or in air) is 300 000 km/s. Calculate:
   a the speed of light in diamond   b the critical angle for diamond.

▲ When a diamond is cut, the facets (faces) are angled so that they produce total internal reflection. Reflected light gives the diamond its 'sparkle'.

Related topics: refraction and refractive index **7.4**; total internal reflection **7.5**

7.7 Lenses (1)

Objectives: – to know that a convex lens is a converging lens and can form a real image – to know how to draw ray diagrams showing object and image positions

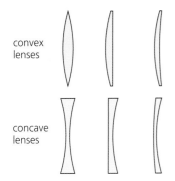

convex lenses

concave lenses

Lenses bend light and form images. There are two main types of lens. The diagram on the left shows some examples of each.

Convex lenses These are thickest in the middle and thin round the edge. When rays parallel to the principal axis pass through a convex lens, they are bent inwards. The point F where they converge (meet) is called the **principal focus**. Its distance from the centre of the lens is the **focal length**. A convex lens is known as a **converging lens**.

Rays can pass through the lens in either direction, so there is another principal focus F' on the opposite side of the lens and the same distance from it.

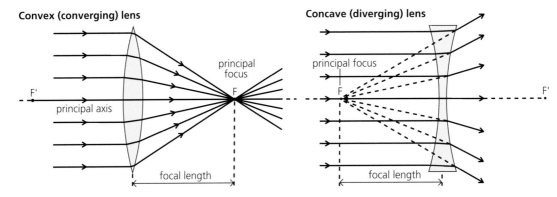

How lenses bend light

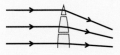

Lenses are made of glass, plastic, or other transparent material. Each section of a lens acts like a tiny prism, refracting (bending) light as it goes in and again as it comes out. Expensive lenses have special coatings to reduce the colour-spreading of the prisms.

Concave lenses These are thin in the middle and thickest round the edge. When rays parallel to the principal axis pass through a concave lens, they are bent outwards. The principal focus is the point from which the rays appear to diverge (spread out). A concave lens is a **diverging lens**.

Real images formed by convex lenses

In the diagram below, rays from a very distant object are being brought to a focus by a convex lens. Rays come from all points on the object. However, for simplicity, only a few rays from one point have been shown. Together, the rays form an image which can be picked up on a screen. An image like this is called a **real image**. It is formed in the **focal plane**.

In a camera, a convex lens is used as below to form a real image on an electronic sensor (or in older cameras, a piece of film). The image in the eye is formed in the same way.

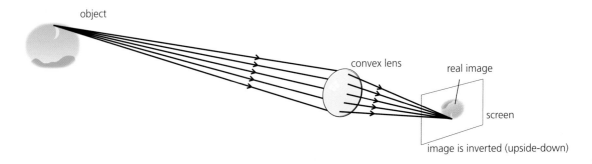

RAYS AND WAVES

The rays from a point on a very distant object are effectively parallel, so the image passes through the principal focus. However, for an object at any other distance, the image is in a different position.

You can predict where a convex lens will form an image by drawing a **ray diagram**. There are two examples below. Each has these features:
- For simplicity, rays are drawn from just one point on the object.
- The rays used are the **standard rays** described on the right. These are chosen because it is easy to work out where they go. Only two of them are needed to find where the image is.
- For simplicity, rays are shown bending at the line through the middle of the lens. In reality, bending takes place at each surface.

Standard rays
In ray diagrams, any two of the following rays are needed to fix the image position and size:

1 A ray through the centre passes straight through the lens.

2 A ray parallel to the principal axis passes through F after leaving the lens.

3 A ray through F' leaves the lens parallel to the principal axis.

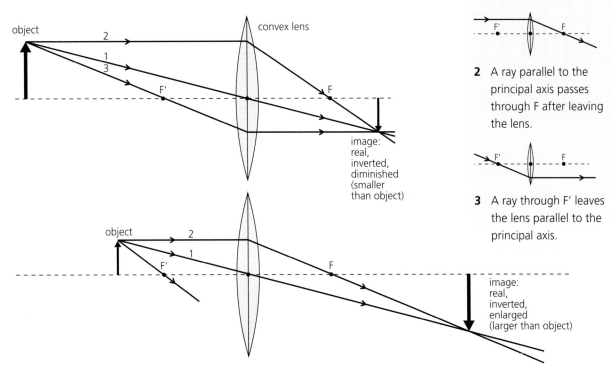

The ray diagrams above show that as the object is moved towards the lens, the image becomes bigger and further away.

A film projector uses a convex lens to form a magnified, real image on a screen a long way away from it, as in the lower diagram.

1 a Which of the lenses on the right is a convex lens?
  b Which one is a converging lens?
  c What is meant by the *principal focus* of the convex lens?
  d What is meant by the *focal length* of the convex lens?
2 a If a convex lens picks up rays from a very distant object, where is the image formed?
  b If the object is moved towards the lens, what happens to the position and size of the image?
3 Draw a ray diagram like one of those above, but with the object exactly 2 × focal length away from the lens. Draw in and describe the image.

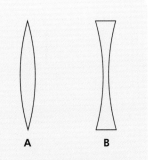

Related topics: mirrors **7.4**; refraction by a prism **7.4**; camera and eye **7.9**

7.8 Lenses (2)

Objectives: – to know how a convex lens can be used as a magnifying glass, and draw its ray diagram – to know how to draw accurate ray diagrams to scale.

Convex lens as a magnifying glass

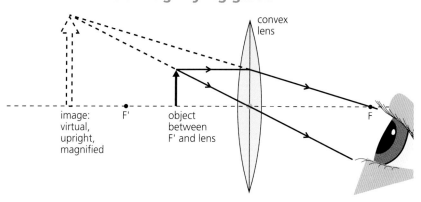

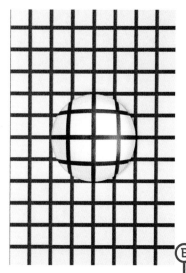

▲ Thick, bulging convex lenses have the shortest focal lengths and make the most powerful magnifying glasses.

Thin convex lenses have longer focal lengths and are much less powerful.

If an object is closer to a convex lens than the principal focus, the rays never converge. Instead, they appear to come from a position behind the lens. The image is upright and magnified. It is called a **virtual image** because no rays actually meet to form it and it cannot be picked up on a screen. Used like this, a convex lens is often called a **magnifying glass**.

Drawing accurate ray diagrams

Problems like the one below can be solved by doing a ray diagram as an accurate scale drawing on graph paper:

> *Example* An object 2 cm high stands on the principal axis at a distance of 9 cm from a convex lens. If the focal length of the lens is 6 cm, what is the image's position, height, and type?

For accuracy, you need to choose a scale that makes the diagram as large as possible. In the drawing below, 1 cm on the paper represents 2 cm of actual distance. When the final measurements are scaled up, they show that the image is 18 cm from the lens, 4 cm high, and real.

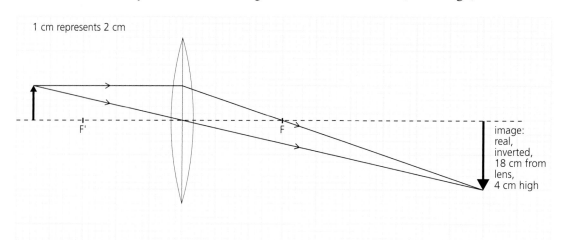

Estimating the focal length of a convex lens*

You can find an approximate value for the focal length of a convex lens by forming an image of a distant window (or other distant bright object) on a screen. Rays from the window are almost parallel, so the image is close to the principal focus of the lens. Therefore the distance from the image to the lens is approximately the same as the focal length.

Convex lenses in a telescope*

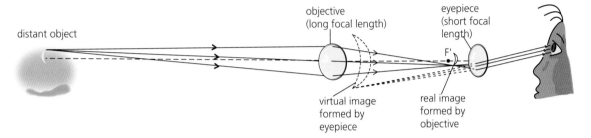

The telescope above (shown without its tube) uses two convex lenses. The **objective** forms a real image of a distant object – in this case the Moon – just inside the principal focus of the **eyepiece**. The image acts as a close object to this lens, which forms a magnified virtual image of it. The eyepiece is being used as a magnifying glass, but it is magnifying an image of the object rather than the object itself. The final image is upside down. Most binoculars – two telescopes side-by-side – have prisms in them to turn the image the right way up (see spread 7.5).

Images formed by concave lenses*

In the diagram below, two standard rays have been used to show how a concave lens forms an image. Wherever the object is positioned, the image is always small, upright, and virtual.

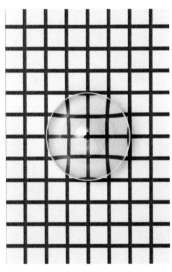

▲ A concave lens forms a small, upright, virtual image.

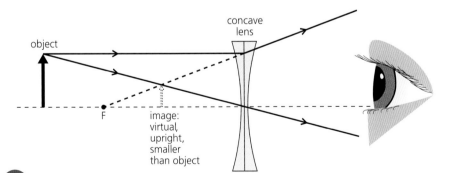

Q

1 a An object 2 cm high is placed 12 cm away from a convex lens of focal length 6 cm. By doing an accurate drawing on graph paper, find the position, height, and type of image.

b The object is moved so that it is only 10 cm away from the lens. Use another drawing to find the new position, height, and type of image.

2 Where should the object be placed if the image formed by a convex lens is to be

a virtual, and larger than the object?

b real, and the same size as the object?

c real, and larger than the object?

3* Describe how you could quickly find an approximate value for the focal length of a convex lens.

Related topics: virtual image **7.2**; binoculars **7.5**; focal length and ray diagrams **7.7**

7.9 More lenses in action

Objectives: – to describe how a real image is formed inside a camera – to explain how lenses are used to correct short-sightedness and long-sightedness.

Convex lens essentials

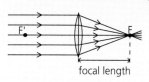

focal length

Real image An image formed by rays that converge. It can be picked up on a screen.

Focus Any point where rays leaving a lens converge. If the rays entering the lens are parallel to its axis, then they converge at the **principal focus** (F on one side of the lens, F' on the other). Rays from a point on a very distant object are effectively parallel.

The camera

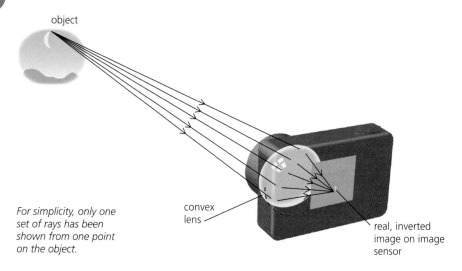

For simplicity, only one set of rays has been shown from one point on the object.

This uses a convex lens to form a small, inverted, real image on a sensor (or in older cameras, a piece of photographic film) at the back. The image sensor is a light-sensitive microchip containing millions of microscopic solar cells. When the shutter opens, these capture the image as a pattern of electric charge which can be stored as data on a memory card. This can be processed to produce the final image on a screen or in print.

The human eye*

Like a camera, this uses a convex lens system to form a small, inverted, real image at the back. The light is mainly converged by the cornea and the watery liquid behind it. The lens, which is flexible, is used to make focusing adjustments: its shape is changed by a ring of muscles. The image is formed on the retina, which contains over 100 million light-sensitive cells. Signals from these cells are sent to the brain along the optic nerve.

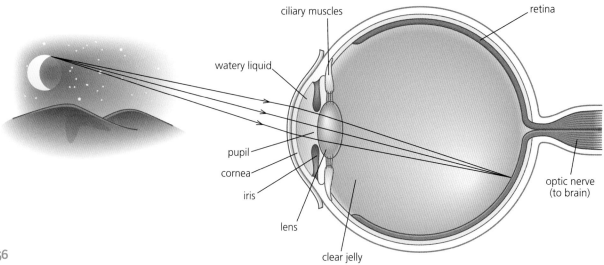

RAYS AND WAVES

Correcting defects in vision

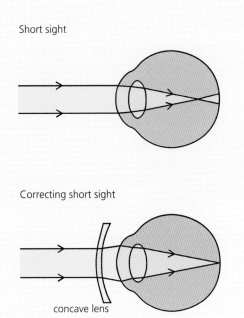

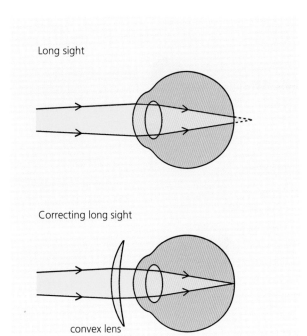

E With many people, changes in the shape of the eye are not enough to produce sharp focusing on the retina. To overcome the problem, spectacles or contact lenses have to be worn.

Short sight In a short-sighted eye, the lens cannot be made thin enough for looking at distant objects. So the rays are bent inwards too much. They converge before they reach the retina. To correct the fault, a *concave* (diverging) lens is placed in front of the eye.

Long sight In a long-sighted eye, the lens cannot be made thick enough for looking at close objects. So the rays are not bent inwards enough. When they reach the retina, they have still not met. To correct the fault, a *convex* (converging) lens is placed in front of the eye.

From middle age onwards, the eye lens becomes less flexible and loses its ability to accommodate for objects at different distances. To overcome this difficulty, some people wear **bifocals** – spectacles whose lenses have a top part for looking at distant objects and a bottom part for close ones. If the spectacles have **progressive** lenses, the changeover is gradual.

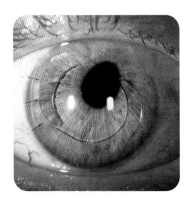

▲ This person's eye has been fitted with a plastic lens because the natural lens has developed too many cloudy patches called cataracts.

Q

You will need information from the previous spreads, 7.7 and 7.8, on how and where a convex lens forms an image.

1 In most cameras, the lens can be moved in and out to make focusing adjustments. If the camera on the opposite page is to take a picture of a object about a metre in front of it, will the lens need to moved *closer* to the sensor or *further* away?

2 A short-sighted person cannot see distant objects clearly. Why not?

3 A long-sighted person cannot see close objects clearly. Why not?

4 What type of spectacle lens or contact lens is needed to correct for
  a short sight
  b long sight?

Related topics convex lenses and ray diagrams **7.7-7.8**

7.10 Electromagnetic waves (1)

Objectives: – to know the main properties of electromagnetic waves – to know the names used for the different regions of the electromagnetic spectrum.

Light waves belong to a whole family of **electromagnetic waves**. These have several features in common. For example:

- They can travel through a vacuum (for example, space).
- They travel through a vacuum at a speed of about 300 000 kilometres per second. This is usually called the **speed of light**, although it is the speed of all electromagnetic waves. (The exact value is 299 792 458 m/s.)
- They are transverse waves – their oscillations are at right angles to the direction of travel. It is electric and magnetic fields that are oscillating, not material.
- They transfer energy. A source loses energy when it radiates electromagnetic waves. A material gains energy when it absorbs them.

The electromagnetic spectrum

The full range of electromagnetic waves is called the **electromagnetic spectrum**. It is shown in the chart on the opposite page. The range of wavelengths is huge. At one end are the longest radio waves with wavelengths of several kilometres. At the other end are the shortest gamma rays with wavelengths of less than one-billionth of a millimetre.

Where electromagnetic waves come from*

All matter is made of atoms. Atoms are themselves made up of a central **nucleus** with tiny particles called **electrons** orbiting around it. The nucleus and the electrons are electrically charged. Sometimes, electrons can escape from their atoms. For example, when an electric current passes through a wire, the current is a flow of free electrons.

Electromagnetic waves are emitted (sent out) whenever charged particles oscillate or lose energy in some way. For example, the vibrating atoms in a hot, glowing bulb filament emit infrared and light, and an oscillating electric current emits radio waves. The higher the frequency of oscillation, or the greater the energy change, the shorter the wavelength of the electromagnetic waves produced.

Wave essentials

Waves radiate (spread out) from their source. They are a form of **radiation**.

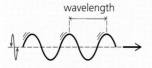

With **transverse waves** as above, the oscillations (vibrations) are at right angles to the direction of travel. The number of waves sent out per second is called the **frequency**. It is measured in hertz (Hz).

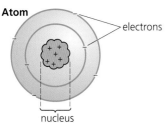

▲ In an atom, the electrons have negative (−) charge and the nucleus has positive (+) charge. Electromagnetic waves are emitted whenever charged particles oscillate or lose energy.

Wave equation

For any set of moving waves:

speed = frequency × wavelength
(m/s)   (Hz)   (m)

If the speed of the waves is unchanged, an increase in frequency means a decrease in wavelength, and vice versa.

1000 Hz = 1 kilohertz (kHz)
1 000 000 Hz = 1 megahertz (MHz)

You may need information from the next spread, 7.11.

1. Give *three* properties (features) common to all electromagnetic waves.
2. Put the following in order of wavelength, starting with the longest:
   ultraviolet   X-rays   red light   violet light   microwaves   infrared
3. Name a type of electromagnetic radiation that
   a is visible to the eye
   b is emitted by hot objects
   c is diffracted by hills
   d can cause fluorescence
   e is used for radar
   f can pass through dense metals.
4. A VHF radio station emits radio waves at a frequency of 100 MHz.
   a What is the frequency in Hz?
   b What is the wavelength? (speed of radio waves = 3×10^8 m/s)
   c What is the wavelength of radio waves from a long-wave transmitter, broadcasting at a frequency of 200 kHz?

The electromagnetic spectrum

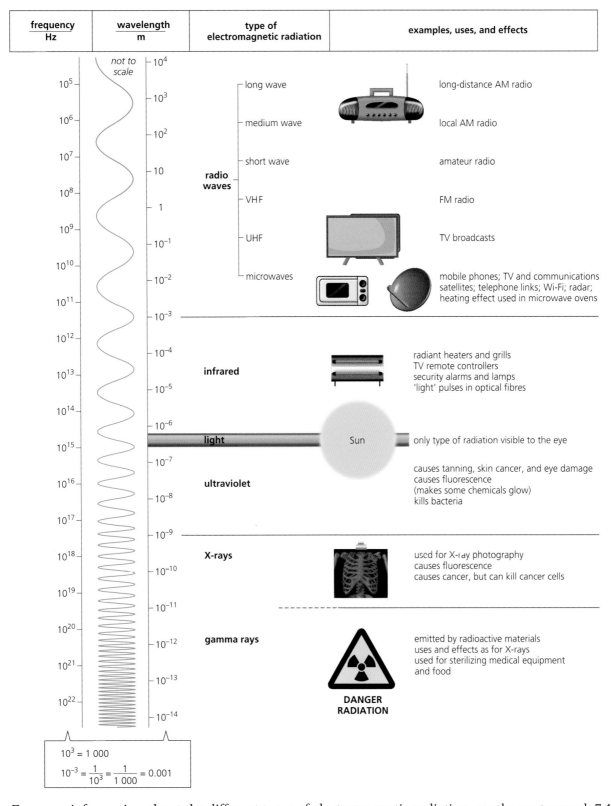

For more information about the different types of electromagnetic radiation, see the next spread, 7.11.

Related topics: thermal radiation **5.7**; transverse waves, frequency and wavelength **6.1**; radar **6.4**; light waves and speed of light **7.1**; light spectrum **7.4**; gamma rays **7.11** and **10.2**; atoms and electric charge **8.1**

7.11 Electromagnetic waves (2)

Objectives: – to know, in order of wavelength, the different types of electromagnetic waves, and some of their effects and uses.

Radio waves

Stars are natural emitters of radio waves. These can be detected by radio telescopes. However, radio waves can be produced artificially by making a current oscillate in a transmitting aerial (antenna). In a simple radio system, a microphone controls the current to the aerial so that the radio waves 'pulsate'. In the radio receiver, the incoming pulsations control a loudspeaker so that it produces a copy of the original sound.

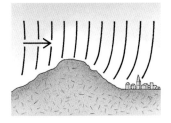

▲ Radio waves of long and medium wavelengths diffract (bend) round hills.

Long and medium waves will diffract (bend) around hills, so a radio can still receive signals even if a hill blocks the direct route from the transmitting aerial. Long waves will also diffract round the curved surface of the Earth.

VHF and UHF waves have shorter wavelengths. VHF (very high frequency) is used for stereo radio and UHF (ultra high frequency) for TV broadcasts. These waves do not diffract round hills. So, for good reception, there needs to be a straight path between the transmitting and receiving aerials.

Microwaves have the shortest wavelengths (and highest frequencies) of all radio waves. They are used by mobile phones, Wi-Fi, and for beaming TV, data, and telephone signals to and from satellites and across country.

Like all electromagnetic waves, microwaves produce a heating effect when absorbed. Water absorbs microwaves of one particular frequency. This principle is used in microwave ovens, where the waves penetrate deep into food and heat up the water in it. However, if the body is exposed to microwaves, they can cause internal heating of body tissues.

▲ This dish receives microwaves from a satellite.

Infrared radiation and light

When a radiant heater or grill is switched on, you can detect the infrared radiation coming from it by the heating effect it produces in your skin. In fact, *all* objects emit some infrared because of the motion of their atoms or molecules. Most radiate a wide range of wavelengths.

As an object heats up, it radiates more and more infrared, and shorter wavelengths. At about 700 °C, the shortest wavelengths radiated can be detected by the eye, so the object glows 'red hot'. Above about 1000 °C, the whole of the visible spectrum is covered, so the object is 'white hot'.

Short-wavelength infrared is often called 'infrared light', even though it is invisible. However, strictly speaking, light is just the part of the electromagnetic spectrum that is visible to the eye.

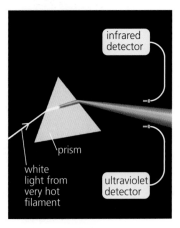

▲ Infrared and ultraviolet can be detected just beyond the two ends of the visible part of the spectrum.

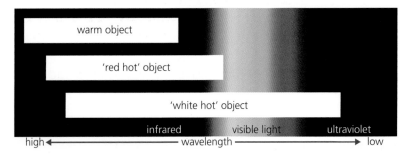

Security alarms and lamps can be switched on by motion sensors that pick up the changing pattern of infrared caused by an approaching person. At night, photographs can be taken using infrared. In telephone and data networks, signals are sent along optical fibres as pulses of infrared 'light'. And remote controllers for TVs work by transmitting infrared pulses.

Ultraviolet radiation

Very hot objects, such as the Sun, emit some of their radiation beyond the violet end of the visible spectrum. This is ultraviolet radiation, or **UV** for short. It is sometimes called 'ultraviolet light' even though it is invisible.

The Sun's UV is harmful to living cells. If too much penetrates the skin, it can cause skin cancer. If you have black or dark skin, the UV is absorbed before it can penetrate too far. But with pale skin, the ultraviolet can go deeper. UV can also damage the retina in the eye and cause blindness.

As ultraviolet is harmful to living cells, it is used in some types of sterilizing equipment to kill bacteria (germs). Water can be sterilized like this.

Fluorescence Some materials **fluoresce** when they absorb ultraviolet: they convert its energy into visible light and glow. Security marker pens contain special ink, normally invisible, which fluoresces and glows when UV light is shone on it. The same idea can be used to spot fake bank notes. Under UV, parts of a fake note glow a different colour compared with genuine note. Using a marker pen makes the difference more obvious.

▲ Sunbeds use ultraviolet to cause tanning in some types of skin.

X-rays

X-rays are given off when fast-moving electrons lose energy very quickly. For example, in an X-ray tube, the radiation is emitted when a beam of electrons hits a metal target. Short-wavelength X-rays are extremely penetrating. A dense metal like lead can reduce their strength, but not stop them. Long-wavelength X-rays are less penetrating. For example, they can pass through flesh but not bone, so bones will show up on an X-ray photograph. In engineering, X-rays can be used to take photographs that reveal flaws inside metals – for example faulty welds in pipe joints. Airport security systems also use them to detect any weapons hidden in luggage.

All X-rays are dangerous because they damage living cells deep in the body and can cause cancer or mutations (genetic change). However, concentrated beams of X-rays can be used to *treat* cancer by destroying abnormal cells.

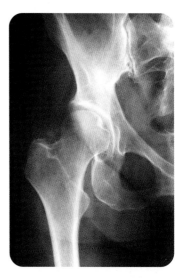

▲ An X-ray photograph.

Gamma rays

Gamma rays come from radioactive materials. They are produced when the nuclei of unstable atoms break up or lose energy. They tend to have shorter wavelengths than X-rays because the energy changes that produce them are greater. However, there is no difference between X-rays and gamma rays of the same wavelength.

Like X-rays, gamma rays can be used in the treatment of cancer, and for taking X-ray-type photographs. As they kill harmful bacteria, they are also used for sterilizing food and medical equipment.

Ionizing radiations
Ultraviolet, X-rays, and gamma rays cause **ionization** – they strip electrons from atoms in their path. The atoms are left with an electric charge, and are then known as **ions**.

Ionization is harmful because it can kill or damage living cells, or make them grow abnormally as cancers.

For questions, see the previous spread, 7.10.

Related topics: infrared and thermal radiation **5.7**; diffraction **6.2**; light spectrum **7.4**; optical fibres **7.5** and **7.12**; radioactivity, gamma rays, and ionization **10.2**

7.12 Sending signals

Objectives: – to know the difference between analogue and digital signals
– to describe the advantages of digital signals and some of their uses.

Telephone, radio, and TV are all forms of telecommunication – ways of transmitting information over long distances. The information may be sounds, pictures, or computer data. The diagram below left shows a simple telephone system. An **encoder** (the microphone) turns the incoming information (speech) into a form which can be transmitted (electrical signals). The signals pass along the **transmission path** (wires) to a **decoder** (the earphone). This turns the signals back into useful information (speech).

Other telecommunication systems use different types of signal and transmission path. The signals may be changes in voltage, changes in the intensity of a beam of light, or changes in the strength or frequency of radio waves. They may be transmitted using wires, optical fibres, or radio waves.

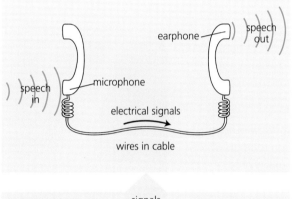

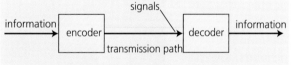

▲ Like all telecommunications systems, a simple telephone system sends signals from a coder to a decoder.

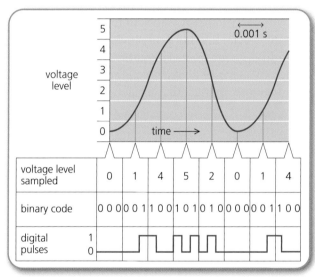

▲ How an analogue signal is converted into digital pulses. Real systems use hundreds of levels and a much faster sampling rate.

Ⓔ Analogue and digital transmission

The sound waves entering a microphone make the voltage across it vary – as shown in the graph above right. A continuous variation like this is called an **analogue signal**. The table shows how it can be converted into **digital signals** – signals represented by numbers. The original signal is **sampled** electronically many times per second. In effect, the height of the graph is measured repeatedly, and the measurements changed into **binary codes** (numbers using only 0's and 1's). These are transmitted as a series of pulses and turned back into an analogue signal at the receiving end.

Advantages of digital transmission Signals lose power as they travel along. This is called **attenuation**. They are also spoilt by **noise** (electrical interference). To restore their power and quality, digital pulses can be 'cleaned up' and amplified at different stages by **regenerators**. Analogue signals can also be amplified, but the noise is amplified as well, so the signals are of lower quality when they reach their destination.

RAYS AND WAVES

From contactless to calls

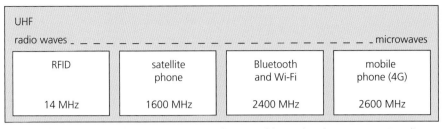

UHF radio waves			microwaves
RFID	satellite phone	Bluetooth and Wi-Fi	mobile phone (4G)
14 MHz	1600 MHz	2400 MHz	2600 MHz

▲ Contactless card payments use RFID technology.

Typical frequencies There is no commonly agreed boundary between UHF radio waves and microwaves. Some telecommunications engineers use the term 'microwaves' for both.

Wireless systems don't have a wire or cable to connect the sender and receiver. The following use UHF radio waves or microwaves to carry the signals.

RFID (Radio Frequency Identification) is used in shops and libraries for identifying which goods are being sold or taken. A tag or sticker on the product contains a tiny chip. A 'reader' sends out radio waves, causing the chip to emit data signals. Contactless debit and credit cards work in this way.

Satellite phones work in areas too remote for ordinary mobile phones. Some communicate via satellites in geostationary orbits (see spread 11.4), others via a network of satellites in low orbits around the Earth.

Bluetooth uses radio waves to link fixed and mobile devices over short distances – typically up to about 10 metres, less if walls are present. **Wi-Fi** works in a similar way to Bluetooth.

Mobile phones (cell phones) are linked by a network of masts. They use microwaves with wavelengths of a few centimetres, so only need a short aerial (antenna) inside them. However, the signals are weakened by walls.

Optical fibres

Cable TV, high-speed broadband, and telephone networks use **optical fibres** for transmission. These long, thin strands of glass carry digital signals in the form of pulses of light (or infrared). At the transmitting end, electrical signals are encoded into light by an **LED** (light-emitting diode) or a **laser diode**. At the receiving end, the signals are decoded by a **photodiode** which turns them back into electrical signals. Optical fibre cables are thinner and lighter than electric cables. They carry more signals with less attenuation over long distances. They are not affected by interference and cannot be 'tapped'.

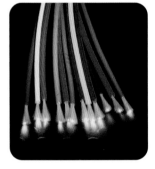

▲ Optical fibres

❶ What is the difference between a digital signal and an analogue signal?

❷ The diagram on the right shows part of a telephone system.
  a What does the laser diode do?
  b What does the photodiode do?
  c What does the regenerator do?
  d Give *two* advantages of sending digital signals rather than analogue ones.
  e Give *two* advantages of using an optical fibre link rather than a cable with wires in it.

❸ Give an example of the use of RFID.

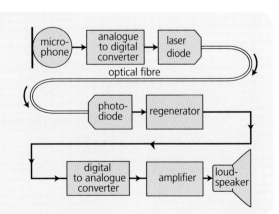

Related topics: sound waves **6.3**; optical fibres **7.5**; magnetic storage **9.4**; LEDs **8.11**

Check-up on rays and waves

Further questions

1 The diagram shows a light signal travelling through an optical fibre made of glass.

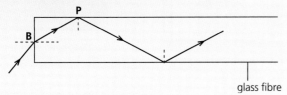

glass fibre

a State *two* changes that happen to the light when it passes from air into the glass fibre at **B**. [2]

b Explain why the light follows the path shown after hitting the wall of the fibre at **P**. [2]

2

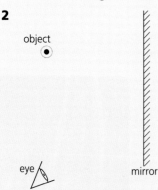

In the diagram above an object (a small bulb) has been placed in front of a plane mirror.

a Copy the diagram. Mark in the position of the image. [1]

b On your diagram, draw a single ray from the object that reflects from the mirror and goes into the eye. Include a dotted line to show where, to the eye, the ray appears to come from. [3]

c An object is 10 cm away from a plane mirror. How far is the object from its image. [1]

d If the object is moved 1 cm closer to the mirror, how far is it away from its image then? [1]

3 a Copy the diagram and draw the path of the ray of yellow light as it passes through and comes out of the glass prism. [2]

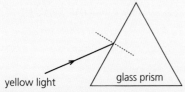

b What do we call this effect? [1]

c State why light changes direction when it enters a glass prism. [1]

4 The figure shows an object OB of height 2 cm in front of a converging lens. The principal foci of the lens are labelled F and F'. An image of OB will be formed to the right of the lens.

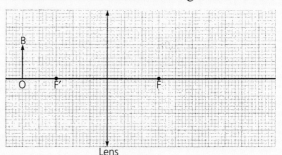

a Copy the figure and draw two rays from the top of the object B which pass through the lens and go to the image. [2]

b Draw the image formed. Label this image I and measure its size. [1]

5

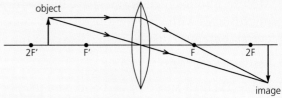

The diagram shows a converging lens forming a real image of an illuminated object. State **two** things that happen to the image when the object is moved towards F'. [2]

6

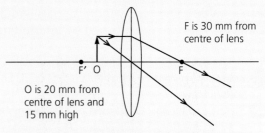

F is 30 mm from centre of lens

O is 20 mm from centre of lens and 15 mm high

The diagram shows an object **O** placed in front of a convex (converging) lens and the passage of two rays from the top of the object through the lens.

a Copy and complete the diagram (using the dimensions given) to show where the image is formed. [2]

b State **two** properties of the image. [2]

c Thin lenses can be *diverging* or *converging*. Which of these words describes
   i a convex lens   **ii** a concave lens? [2]
d Which type of lens is used to correct
   i short-sightedness   **ii** long-sightedness? [2]

7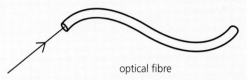
optical fibre

a In the diagram, above a monochromatic ray of light enters an optical fibre. What does *monochromatic* mean? [1]
b Why doesn't the light escape from the sides of the optical fibre? [1]
c Optical fibres are used in telecommunications. Give one other use of optical fibres (apart from decorative lamps). [1]
d The signals sent along optical fibres are *digital*. What does this mean? [2]
e Give *two* advantages of using digital signals for communications rather than analogue ones. [2]
f Give *one* example of the use of radio frequency identification (**RFID**) technology. [1]

8 A ray of light, in air, strikes one side of a rectangular glass block. The refractive index of the glass is 1.5.
a Draw a diagram to show the direction the ray will take in the glass if the angle of incidence is 0°. [2]
b Draw a diagram to show the approximate direction the ray will take in the glass if the angle of incidence is 45°, and calculate the angle of refraction. [4]
c If the speed of light in air is 3×10^8 m/s, calculate the speed of light in the glass. [2]

9 Light and gamma rays are both examples of electromagnetic radiation.
a Name three other types of electromagnetic radiation. [3]
b State two differences between light and gamma rays. [2]
c The speed of light is 3×10^8 m/s. Calculate the frequency of yellow light of wavelength 6×10^{-7} m. [2]

10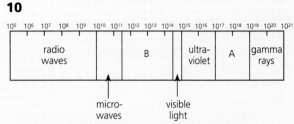

The diagram shows the main regions of the electromagnetic spectrum. The numbers show the frequencies of the waves measured in hertz (Hz).
a Name the regions
   i **A** [1]
   ii **B**. [1]
b i Write down, **in words**, the equation connecting wave speed, wavelength, and wave frequency. [1]
   ii Calculate the frequency of the radiation with a wavelength of 0.001 m (10^{-3} m), given that all electromagnetic waves travel at a speed of 300 000 000 m/s (3×10^8 m/s) in space. [2]
   iii State to which part of the electromagnetic spectrum the radiation in part **ii** belongs. [1]
c Explain how and why microwaves can cause damage to or even kill living cells. [2]

11 The figure shows a square block of glass JKLM with a ray of light incident on side JK at an angle of incidence of 60°. The refractive index of the glass is 1.50.

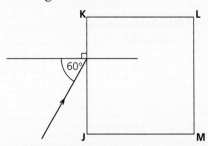

a Calculate the angle of refraction of the ray. [2]
b Calculate the critical angle for a ray of light in this glass. [2]
c Explain why the ray shown cannot emerge from side KL but will emerge from side LM. [3]

12 a | less than | the same as | greater than |

Copy the sentences below and use **one** of the three phrases above to complete each sentence. Each phrase may be used once, more than once, or not at all.

   i The wavelength of radio waves is _____ the wavelength of ultraviolet radiation. [1]

   ii In a vacuum the speed of ultraviolet radiation is _____ the speed of light. [1]

   iii The frequency of ultraviolet radiation is _____ the frequency of infrared radiation. [1]

b Name the part of the electromagnetic spectrum that is used to:

   i send information to and from satellites [1]

   ii kill harmful bacteria in food. [1]

Use the list below when you revise for your IGCSE examination. The spread number, in brackets, tells you where to find more information.

Revision checklist

Core Level
☐ The characteristics (features) of light. (7.1)
☐ The meanings of angle of incidence, angle of reflection, and normal. (7.2)
☐ A law of reflection: the angles of incidence and reflection are equal. (7.2)
☐ The image in a plane mirror, its position and how it is formed. (7.2)
☐ The image in a plane mirror is virtual. (7.2)
☐ Demonstrating the refraction of light; the meaning of angle of refraction. (7.4)
☐ How a light ray passes through a parallel-sided block of glass or plastic. (7.4)
☐ Dispersion: how a prism forms a spectrum. (7.4)
☐ The colours of the visible spectrum in order of wavelength. (7.4)
☐ Total internal reflection: what it means and how it can be used. (7.5)
☐ The meaning of critical angle. (7.5)
☐ How a convex lens focuses a beam of light. (7.7)
☐ The principal focus and focal length of a lens. (7.7)
☐ Drawing ray diagrams to show how and where a convex lens forms a real image. (7.7 and 7.8)
☐ Electromagnetic waves: the main features of the electromagnetic spectrum. (7.10)
☐ How electromagnetic waves all travel at the same speed in a vacuum. (7.10)
☐ The characteristics and properties of
  – radio waves, microwaves
  – infrared rays
  – ultraviolet rays
  – X-rays, gamma rays.   (7.10 and 7.11)
☐ The uses of electromagnetic waves:
  – domestic, industrial, medical (7.10 and 7.11)
  – in communication systems. (7.10 , 7.11, 7.12)
☐ The harmful effects of electromagnetic waves. (7.11)
☐ Using electromagnetic waves in
  – communications (radio, TV, satellite, telephone)
  – remote controllers
  – medicine
  – security systems. (7.10 and 7.11)
☐ Microwaves and X-rays: safety issues. (7.11)

Extended Level
As for Core Level, plus the following:
☐ The meaning of monochromatic. (7.1)
☐ Drawing accurate diagrams to find where a plane mirror forms an image. (7.2 and 7.3)
☐ Defining refractive index in terms of light speed. (7.4)
☐ Optical fibres and their uses. (7.5 and 7.12)
☐ The equation linking refractive index, angle of incidence, and angle of refraction (Snell's law) (7.6)
☐ Calculating the critical angle using the refractive index. (7.6)
☐ Drawing ray diagrams to show how a convex lens can form a real image. (7.8)
☐ Drawing ray diagrams to show how a convex lens can form a virtual image. (7.8)
☐ Using a convex lens as a magnifying glass. (7.8)
☐ How lenses are used to correct short-sightedness and long-sightedness. (7.9)
☐ The speed of electromagnetic waves (the speed of light). (7.1 and 7.10)
☐ The difference between analogue and digital signals. (7.12)
☐ The advantages of using digital signals (7.12)
☐ Using radio waves, microwaves, and optical fibres in communication systems (TV, satellites, mobile phones, broadband, bluetooth). (7.12)

8

Electricity

- ELECTRIC CHARGE
- CONDUCTORS AND INSULATORS
- ELECTRIC FIELDS
- CURRENT, VOLTAGE, AND RESISTANCE
- ELECTRICAL COMPONENTS
- SERIES AND PARALLEL CIRCUITS
- ELECTRICAL POWER
- CALCULATING ELECTRICAL ENERGY
- MAINS ELECTRICITY

The city of Bogotá, Colombia, at night. Like other cities, it is so bright that it can even be seen from space. Modern industrial societies rely heavily on the use of electricity – not only for lighting, as shown here, but also for running factory machinery, information and communications systems, and heating. Typically, electricity accounts for about one sixth of an industrialized country's energy use.

8.1 Electric charge (1)

Objectives: – to describe charging by rubbing – to describe the forces between charges – to know why metals are better conductors than most non-metals.

Electric charge, or 'electricity', can come from batteries and generators. But some materials become charged when they are rubbed. Their charge is sometimes called **electrostatic charge** or 'static electricity'. It causes sparks and crackles when you take off a pullover, and if you slide out of a car seat and touch the door, it may even give you a shock.

Negative and positive charges

Polythene and Perspex (acrylic resin) can be charged by rubbing them with a dry, woollen cloth.

When two charged polythene rods are brought close together, as shown below, they *repel* (try to push each other apart). The same thing happens with two charged Perspex rods. However, a charged polythene rod and a charged Perspex rod *attract* each other. Experiments like this suggest that there are two different and opposite types of electric charge. These are called **positive** (+) charge and **negative** (−) charge:

▲ This person has been charged up. Her hairs all carry the same type of charge, so they repel each other.

> Like charges repel; unlike charges attract.
> The closer the charges, the greater the force between them.

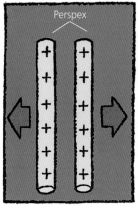 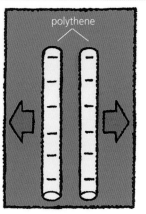 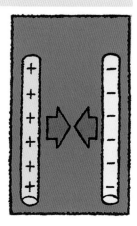

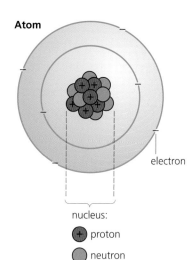

Atom

nucleus:
⊕ proton
● neutron

electron

Where charges come from

Everything is made of tiny particles called atoms. These have electric charges inside them. A simple model of the atom is shown on the left. There is a central **nucleus** made up of **protons** and **neutrons**. Orbiting the nucleus are much lighter electrons:

Electrons have a negative (−) charge.
Protons have an equal positive (+) charge.
Neutrons have no charge.

Normally, atoms have equal numbers of electrons and protons, so the *net* (overall) charge on a material is zero. However, when two materials are rubbed together, electrons may be transferred from one to the other. One material ends up with more electrons than normal and the other with less. So one has a net negative charge, while the other is left with a net positive charge. Rubbing materials together does not *make* electric charge. It just *separates* charges that are already there.

ELECTRICITY

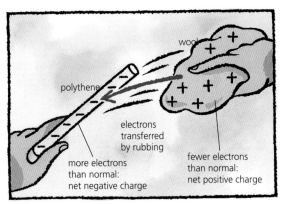

▲ When polythene is rubbed with a woollen cloth, the polythene pulls electrons from the wool.

▲ When Perspex is rubbed with a woollen cloth, the wool pulls electrons from the Perspex.

Conductors and insulators

When some materials gain charge, they lose it almost immediately. This is because electrons flow through them or the surrounding material until the balance of negative and positive charge is restored.

Conductors are materials that let electrons pass through them. Metals are the best electrical conductors. Some of their electrons are so loosely held to their atoms that they can pass freely between them. These **free electrons** also make metals good thermal conductors.

Most non-metals conduct charge poorly or not at all, although carbon is an exception.

Insulators are materials that hardly conduct at all. Their electrons are tightly held to atoms and are not free to move – although they can be transferred by rubbing. Insulators are easy to charge by rubbing because any electrons that get transferred tend to stay where they are.

Semiconductors* These are 'in-between' materials. They are poor conductors when cold, but much better conductors when warm.

Conductors

Good	Poor
metals	water
especially:	human body
silver	earth
copper	
aluminium	
carbon	

Semiconductors

silicon   germanium

Insulators

plastics	glass
e.g:	rubber
PVC	dry air
polythene	
Perspex	

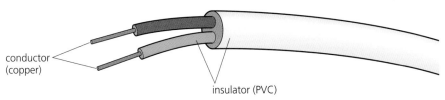

▲ The 'electricity' in a cable is a flow of electrons. Most cables have copper conducting wires with PVC plastic around them as insulation.

Q

1 Say whether the following *attract* or *repel*:
   a two negative charges
   c a negative charge and a positive charge
   b two positive charges.
2 In an atom, what kind of charge is carried by
   a protons   b electrons   c neutrons?
3 What makes copper a better electrical conductor than polythene?
4 Why is it easy to charge polythene by rubbing, but not copper?
5 Name one non-metal that is a good conductor.
6 When someone pulls a plastic comb through their hair, the comb becomes negatively charged.
   a Which ends up with more electrons than normal, the comb or the hair?
   b Why does the hair become positively charged?

Related topics: thermal conduction **5.5**; atoms **10.1**

8.2 Electric charge (2)

Objectives: – to describe some of the effects of objects being charged – to describe how electrostatic charge can be detected.

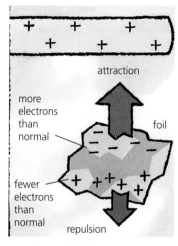

▲ A charged object attracts an uncharged one.

▶ An aircraft and its tanker must be earthed during refuelling, otherwise charge might build up as the fuel 'rubs' along the pipe. One spark could be enough to ignite the fuel vapour.

Detecting charge

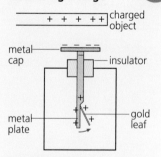

Electrostatic charge can be detected using a **leaf electroscope** as above. If a charged object is placed near the cap, charges are induced in the electroscope. Those in the gold leaf and metal plate repel, so the leaf rises.

Attraction of uncharged objects

A charged object will attract an uncharged object close to it. For example, cling film sticks to your hand because it becomes charged when pulled off the roll.

The diagram on the left shows what happens if a positively charged rod is brought near a small piece of aluminium foil. Electrons in the foil are pulled towards the rod, which leaves the bottom of the foil with a net positive charge. As a result, the top of the foil is attracted to the rod, while the bottom is repelled. However, the attraction is stronger because the attracting charges are closer than the repelling ones.

Earthing

If enough charge builds up on something, electrons may be pulled through the air and cause sparks – which can be dangerous. To prevent charge building up, objects can be **earthed**: they can be connected to the ground by a conducting material so that the unwanted charge flows away.

Induced charges*

Charges that 'appear' on an uncharged object because of a charged object nearby are called **induced charges**. In the diagram below, a metal sphere is being charged by induction. The sphere ends up with an *opposite* charge to that on the rod, which never actually touches the sphere.

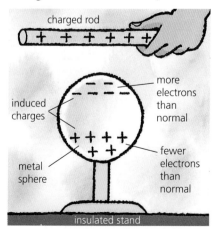
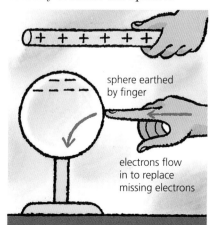

170

Unit of charge

The SI unit of charge is the **coulomb** (**C**). It is equal to the charge on about 6 billion billion electrons. For more on this, see spread 8.4. One coulomb is a relatively large quantity of charge, and it is often more convenient to measure charge in **microcoulombs**:

1 microcoulomb (μC) = 10^{-6} C   (one millionth of a coulomb)

The charge on a rubbed polythene rod is, typically, only about 0.005 μC.

Using electrostatic charge*

In the following examples, the charge comes from an electricity supply rather than from rubbing.

Electrostatic precipitators are fitted to the chimneys of some power stations and factories. They reduce pollution by removing tiny bits of ash from the waste gases. Inside the chamber of a precipitator (see right), the ash is charged by wires, and then attracted to the metal plates by an opposite charge. When shaken from the plates, the ash collects in the tray at the bottom.

Photocopiers work using the principle shown in the diagrams below.

Laser printers use the same idea except that, at stage 2, a computer-controlled laser scans the plate strip by strip to create the required image.

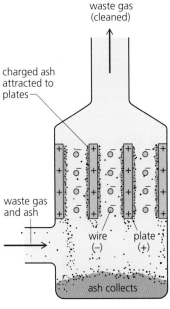

▲ An electrostatic precipitator uses charge to remove bits of ash from the waste gases produced by a factory or power station.

1 Inside the photocopier, a light-sensitive plate (or drum) is given a negative charge.

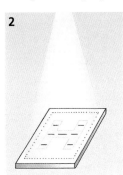

2 An image of the original document is projected onto the plate. The bright areas lose their charge but the dark

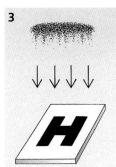

3 areas keep it. Powdered ink (called toner) is attracted to the charged (dark) areas.

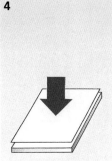

4 A blank sheet of paper is pressed against the plate and picks up powdered ink.

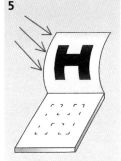

5 The paper is heated so that the powdered ink melts and sticks to it. The result is a copy of the original document.

1. Name an instrument that can be used to detect electric charge.
2. What is the SI unit of charge?
3. In the diagram on the right a negatively charged rod has been brought close to a piece of aluminium foil.
   a. Which end of the foil has an excess of electrons?
   b. Why is the foil pulled towards the rod?
4.* Give an example of where a build-up of electrostatic charge can be dangerous. How can the problem be solved?
5.* Give two examples of the practical application of electrostatic charge.

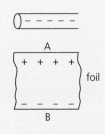

Related topics: SI units **1.2**; charge and current **8.4**; earth wires **8.13**

8.3 Electric fields

Objectives: – to know what an electric field is – to describe the field patterns around charged points and metal spheres, and between parallel plates.

Atom and charge essentials

Electric charge can be positive (+) or negative (−). Like charges repel. Unlike charges attract.

Charges come from atoms. In an atom, the charged particles are electrons (−) and protons (+). Normally, an atom has equal amounts of − and + charge, so it is uncharged. However, if an atom gains or loses electrons, it is left with a net (overall) negative or positive charge.

Most materials are made up of groups of atoms, called molecules.

A charged object will cause a redistribution of the + and − charges in uncharged objects nearby. Concentrations of + or − charge which occur because of this are called induced charges.

An electric current is a flow of charge. When a metal conducts, there is a flow of electrons.

The girl on the right has given herself an electric charge by touching the dome of a Van de Graaff generator. The dome can reach over 100 000 volts, although this is reduced when she touches it. However, the current that flows into her body (0.000 02 amperes or less) is far too small to be dangerous.

The force of repulsion between the charges on the girl's head and hairs is strong enough to make her hairs stand up. If electric charges feel a force, then, scientifically speaking, they are in an **electric field**. So there is an electric field around the dome and the girl.

Electric field patterns

In diagrams, lines with arrows on them are used to represent electric fields. There are some examples of field patterns below. In each case, the arrows show the direction in which the force on a *positive* (+) charge would act. As like charges repel, the field lines always point *away* from positive (+) charge and *towards* negative (−) charge. The force due to the field is a vector.

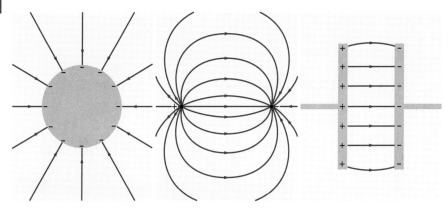

▲ Electric field close to a negatively charged sphere. The field around a Van de Graaff dome is similar to this.

▲ Electric field between two opposite, point charges.

▲ Electric field between two parallel plates with opposite charges on them.

Curves, points, and ions*

When a conductor is charged up, the charges repel each other, so they collect on the outside. The charges are most concentrated near the sharpest curve. This is where the electric field is strongest and the field lines closest together.

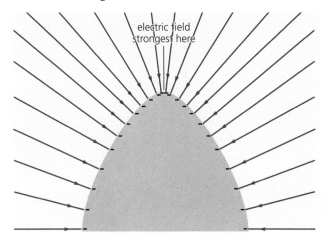

▲ The electric field is strongest where the charges are most concentrated and the field lines are closest together.

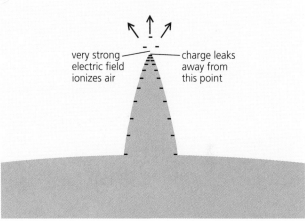

▲ At a sharp point, the electric field may be strong enough to ionize the air so that it will conduct charge away.

If a sharp spike is put on the dome of a Van de Graaff generator, any charge on the dome immediately leaks away from the point. At the point, the metal is very sharply curved. Here, the charge is so concentrated that the electric field is strong enough to ionize the air (see above). Ionized air conducts, so the dome loses its charge through the air.

Ions are electrically charged atoms (or groups of atoms). Atoms become ions if they lose (or gain) electrons. A stream of ions is a flow of charge, so it is another example of a current.

Most of the molecules in air are uncharged, but not all, as shown on the right. Flames, air movements, and natural radiation from space or rocks can all remove electrons from molecules in air so that ions are formed. Although these soon recombine with any free electrons around, more are being formed all the time. With no ions in it, air is a good electrical insulator. But with ions present, it has charges that are free to move, so the air becomes a conductor.

In a thunderstorm, the concentrations of different ions may be so great that a very high current may flow through the air, causing a flash of lightning.

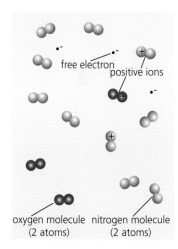

▲ Air is mainly a mixture of nitrogen and oxygen molecules. The charged ones are called ions.

1 The diagram on the right shows electric field lines round a charged metal sphere (in air).
   a Copy the diagram. Draw in the direction of the electric field on each field line.
   b If a positive charge were placed at X, in which direction would it move?
   c If a negative charge were placed at X, in which direction would it move?
   d* If a sharp spike were placed on top of the sphere, what would happen to the charge on the sphere?

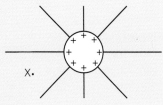

Related topics: vectors **2.1**; atoms and molecules **5.1**; charges and conductors **8.1**; induced charges **8.2**; ionizing radiation **10.2**

8.4 Current in a simple circuit

Objectives: – to know how to measure an electric current – to know the rule for the current in a simple circuit - to know the link between current and charge.

Charge essentials

Electric charge can be positive (+) or negative (−). Like charges repel, unlike charges attract.

Charges come from atoms. In atoms, the charged particles are protons (+) and **electrons** (−).

Electrons can move through some materials, called **conductors**. Copper is the most commonly used conductor.

The unit of charge is the **coulomb (C)**.

An electric **cell** (commonly called a **battery**) can make electrons move, but only if there is a conductor connecting its two terminals. Then, chemical reactions inside the cell push electrons from the negative (−) terminal round to the positive (+) terminal.

The cell below is being used to light a lamp. As electrons flow through the lamp, they make a filament (thin wire) heat up so that it glows. The conducting path through the lamp, wires, switch, and battery is called a **circuit**. There must be a *complete* circuit for the electrons to flow. Turning the switch OFF breaks the circuit and stops the flow.

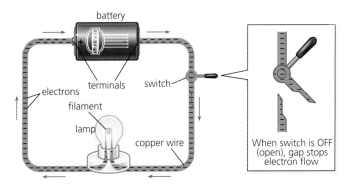

The above circuit can be drawn using **circuit symbols**:

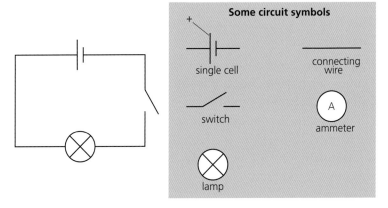

▲ Ammeter

To measure a current, you need to choose a meter with a suitable range on its scale. This ammeter cannot measure currents above 1 A. Also to measure, say, 0.1 A accurately, it would be better to use a meter with a lower range.

When connecting up a meter, the red (+) terminal should be on the same side of the circuit as the + terminal of the battery.

Measuring current

A flow of charge is called an electric **current**. The higher the current, the greater the flow of charge.

The SI unit of current is the **ampere (A)**. About 6 billion billion electrons flowing round a circuit every second would give a current of 1 A. However, the ampere is not defined in this way (see next page).

Currents of about an ampere or so can be measured by connecting an **ammeter** into the circuit. For smaller currents, a **milliammeter** is used. The unit in this case is the **milliampere (mA)**. 1000 mA = 1 A

ELECTRICITY

Some typical current values

Current in a small torch lamp ..... 0.2 A (200 mA)
Current in a car headlight lamp ..... 4 A
Current in an electric kettle element ..... 10 A

Putting ammeters (or milliammeters) into a circuit has almost no effect on the current. As far as the circuit is concerned, the meters act just like pieces of connecting wire.

The circuit on the right has two ammeters in it. Any electrons leaving the battery must flow through both, so both give the same reading:

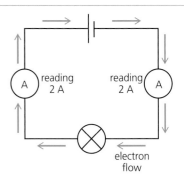

> The current is the same at all points in a simple circuit.

E Charge and current

There is a link between charge and current:

If charge flows at this rate...	then the current is...
1 coulomb per second	1 ampere
2 coulombs per second	2 amperes ...and so on.

The link can also be expressed as an equation:

$$\text{current} = \frac{\text{charge}}{\text{time}} \quad \text{In symbols:} \quad I = \frac{Q}{t}$$

For example, if a charge of 6 coulombs (C) is delivered in 3 seconds, the current is 2 A

Definitions

In 2019, some SI units were redefined to improve precision. The coulomb (C) and the ampere (A) are now linked to the charge on the proton, which is defined as exactly

$1.602\,176\,634 \times 10^{-19}$ C

The coulomb and the ampere are linked like this:

1 C is the charge passing when a current of 1 A flows for 1 s.

Current direction

Some circuit diagrams have arrowheads marked on them. These show the **conventional current direction**: the direction from + to − round the circuit. Electrons actually flow the other way. Being negatively charged, they are repelled by negative charge, so are pushed out of the negative terminal of the battery.

The conventional current direction is equivalent to the direction of transfer of positive charge. It was defined before the electron was discovered and scientists realized that positive charge did not flow through wires. However, it isn't 'wrong'. Mathematically, a transfer of positive charge is the same as a transfer of negative charge in the opposite direction.

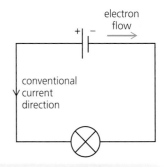

Q

1. Convert these currents into amperes: **a** 500 mA **b** 2500 mA
2. Convert these currents into milliamperes: **a** 2.0 A **b** 0.1 A
3. **a** Draw the circuit on the right using circuit symbols.
   b On your diagram, mark in and label the conventional current direction and the direction of electron flow.
   c The current reading on one of the ammeters is shown. What is the reading on the other one?
   d Which lamp(s) will go out if the switch contacts are moved apart? Give a reason for your answer.
4. What charge is delivered if
   a a current of 10 A flows for 5 seconds
   b a current of 250 mA flows for 40 seconds?

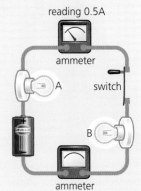

Related topics: SI units **1.2**; electrons, charge, coulombs and conductors **8.1–8.2**

8.5 Potential difference

Objectives: – to explain what is meant by p.d. and e.m.f. – to know the rule for the p.d.s around a simple circuit.

Circuit essentials

A cell can make electrons flow round a circuit. The flow of electrons is called a current. Electrons carry a negative (−) charge. As like charges repel, electrons are pushed out of the negative (−) terminal of the cell.

Charge is measured in coulombs (C).

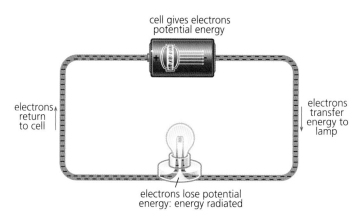

Energy and work essentials

Energy is measured in joules (J).

Energy can be transferred from one store to another, but the total quantity always stays the same.

Work is also measured in joules.

Work is done whenever energy is transferred:

work done = energy transferred

P.d. (voltage) across a cell

A cell normally has a voltage marked on it. The higher the voltage, the more work is done in pushing out each coulomb. In other words, the more energy per coulomb is transferred by the electrons to the lamp. The scientific name for voltage is **potential difference (p.d.)**. It can be measured by connecting a **voltmeter** across the terminals. The SI unit of p.d. is the **volt (V)**.

p.d. (V), charge (Q), and work done (W) are linked by this equation:

$$V = \frac{W}{Q}$$

For example, if the p.d. across the cell is 1.5 V:

1.5 J of work is done in pushing out 1 C of charge.

3 J of work is done in pushing out 2 C of charge... and so on.

The p.d across a cell is highest when it isn't connected in a circuit. This maximum p.d. is called the **electromotive force (e.m.f.)** of the cell. When the cell is supplying a current, the p.d. across it drops because the cell heats up and energy is wasted. For example, a car battery labelled 12 V might only produce 9 V when being used to turn a starter motor.

Cells in series

To produce a higher p.d., several cells can be connected in **series** (in line) as shown below. The word 'battery' really means a collection of joined cells, although it is commonly used for a single cell as well.

▲ Voltmeter and symbol. (For information about range and connection, see note under ammeter in previous spread.)

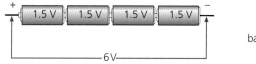

P.d.s around a circuit

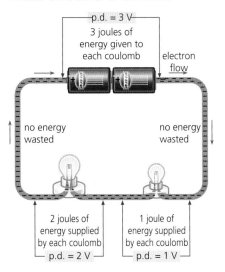
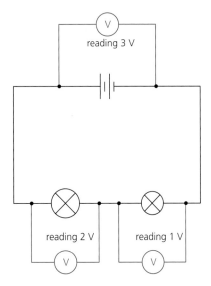

E In the circuit above, the electrons flow through two lamps. Some energy is transferred to the first lamp and the rest to the second. (If the connecting wires are thick enough not to heat up, no energy is wasted in the wires.)

Like the battery, each lamp has a p.d. across it:

If a lamp (or other component) has a p.d. of 1 volt across it, then 1 joule of work is done in pushing each coulomb of charge through it. As a result, 1 joule of energy is transferred to it.

The second diagram shows the same circuit with voltmeters connected across different sections (the voltmeters do not affect how the circuit works). The readings illustrate a principle which applies in any circuit:

> Moving round a circuit, from one battery terminal to the other, the sum of the p.d.s across the components is equal to the p.d. across the battery.

Definitions
The electromotive force (e.m.f.) of a cell (or other source) is the work done per unit of charge by the cell in driving charge round a complete circuit (including the cell itself).

The potential difference (p.d.) across a component is the work done per unit of charge in driving charge through the component.

Q

1 In what unit is each of these measured?
   a p.d.   b e.m.f.   c charge   d current   e energy
2 In the circuit on the right, the two lamps are of different sizes and brightnesses.
   a What type of meter is meter X?
   b What type of meter is meter Y?
   c What is the reading on meter Y?
   d How much energy is transferred to each coulomb of charge as it is pushed from the battery?
   e How much work is done on each coulomb in pushing it through lamp A?
   f How much charge passes through lamp A every second?

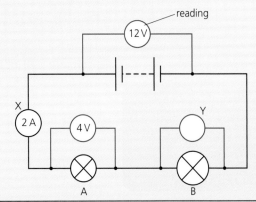

Current is measured in amperes (A) using an ammeter. If 1 ampere flows for 1 second, the charge passing is 1 coulomb.

Related topics: SI units **1.2**; energy **4.1–4.2**; electrons and charge **8.1–8.2**; charge and current **8.4**; cell arrangements **8.9**

8.6 Resistance (1)

Objectives: – to know the definition of resistance – to know that resistance has a heating effect – to describe some devices and components that have resistance.

Circuit essentials

A battery pushes electrons round a circuit. The flow of electrons is called a current. Current is measured in amperes (A).

Potential difference (p.d.), or voltage, is measured in volts (V). The greater the p.d. across a battery, the more energy each electron is given. The greater the p.d. across a lamp or other component, the more energy each electron transfers to it as it passes through.

To make a current flow in a conductor, there must be a potential difference (voltage) across it. Copper connecting wire is a good conductor and a current passes through it easily. However, a similar piece of nichrome wire is not so good and less current flows for the same p.d. The nichrome wire has more **resistance** than the copper.

Resistance is calculated using the equation below. The SI unit of resistance is the **ohm** (Ω). (The symbol Ω is the Greek letter *omega*.)

$$\text{resistance } (\Omega) = \frac{\text{p.d. across conductor (V)}}{\text{current through conductor (A)}}$$

For example, if a p.d. of 6 V is needed to make a current of 3 A flow in a wire: resistance = 6 V/3 A = 2 Ω.

With a *lower* resistance, a *lower* p.d. would be needed to give the same current. Even copper connecting wire has some resistance. However, it is normally so low that only a very small p.d. is needed to make a current flow in it, and this can be neglected in calculations.

Some factors affecting resistance

The resistance of a conductor depends on several factors:

- **Length** Doubling the length of a wire doubles its resistance.
- **Cross-sectional area** Halving the 'end on' area of a wire doubles its resistance. So a thin wire has more resistance than a thick one.
- **Material** A nichrome wire has more resistance than a copper wire of the same size.
- **Temperature** For metal conductors, resistance increases with temperature. For semiconductors, it decreases with temperature.

Resistance and heating effect

There is a heating effect whenever a current flows in a resistance. This principle is used in heating elements, and also in light bulbs with filaments. The heating effect occurs because electrons collide with atoms as they pass through a conductor. The electrons lose energy. The atoms gain energy and vibrate faster. Faster vibrations mean a higher temperature.

▲ The filament of this lamp is made of very thin tungsten wire. Tungsten has a high melting point.

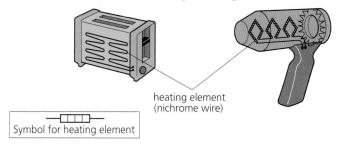

Heating elements are normally made of nichrome.

ELECTRICITY

Resistance components

Resistors are specially made to provide resistance. In simple circuits, they reduce the current. In more complicated circuits, such as those in radios, TVs, and computers, they keep currents and p.d.s at the levels needed for other components (parts) to work properly.

Resistors can have values ranging from a few ohms to several million ohms. For measuring higher resistances, these units are useful:

1 kilohm (k Ω) = 1000 Ω   1 megohm (MΩ) = 1 000 000 Ω

Like all resistances, resistors heat up when a current flows in them. However, if the current is small, the heating effect is slight.

Variable resistors (rheostats) are used for varying current. The one on the right is controlling the brightness of a lamp. In hi-fi equipment, rotary (circular) variable resistors are used as volume controls.

Thermistors have a high resistance when cold but a much lower resistance when hot. They contain semiconductor materials. Some electrical thermometers use a thermistor to detect temperature change.

Light-dependent resistors (LDRs) have a high resistance in the dark but a low resistance in the light. They can be used in electronic circuits which switch lights on and off automatically.

Diodes have an extremely high resistance in one direction but a low resistance in the other. In effect, they allow current to flow in one direction only. They are used in electronic circuits.

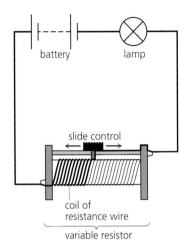

▲ Moving the slide control of the variable resistor to the right increases the length of resistance wire in the circuit. This reduces the current and dims the lamp.

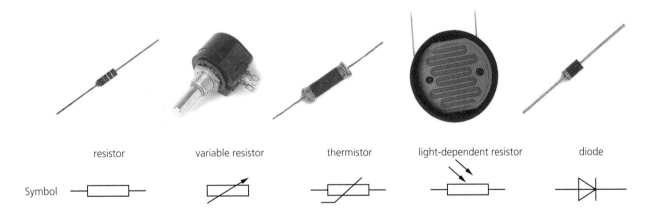

resistor   variable resistor   thermistor   light-dependent resistor   diode

Symbol

Q

1. When a kettle is plugged into the 230 V mains, the current in its element is 10 A.
   a What is the resistance of its element?
   b Why does the element need to have resistance?
2. In the diagram at the top of this page, a variable resistor is controlling the brightness of a lamp. What happens if the slide control is moved to the left? Give a reason for your answer.
3. Which of the components in the photographs above has each of these properties?
   a A high resistance in the dark but a low resistance in the light.
   b A resistance that falls sharply when the temperature rises.
   c A very low resistance in one direction, but an extremely high resistance in the other.

Related topics: SI units **1.2**; temperature, vibrating atoms, and thermometers **5.2**; conductors and semiconductors **8.1**; current, circuits and symbols **8.4**; potential difference **8.5**; diodes, LDRs, and thermistors **8.11**; resistor colour code **page 321**

8.7 Resistance (2)

Objectives: – to know, and use, the equation linking p.d., current, and resistance – to interpret current-p.d. graphs for wires and a semiconductor diode.

Resistance equation

resistance $= \dfrac{\text{potential difference}}{\text{current}}$

Units:
resistance: ohm (Ω)
potential difference (p.d.): volt (V)
current: ampere (A)

V, I, R equations

The resistance equation can be written using symbols:

$$R = \dfrac{V}{I}$$

where R = resistance, V = p.d. (voltage), and I = current

(Note the difference between the symbol V for p.d. and the symbol V for volt.)

The above equation can be rearranged in two ways:

$$V = IR \quad \text{and} \quad I = \dfrac{V}{R}$$

These are useful if the p.d. across a known resistance, or the current in it, is to be calculated.

Example A 12 Ω resistor has a p.d. of 6 V across it. What is the current in the resistor?

In this case: $V = 6$ V, $R = 12$ Ω, and I is to be found. So:

$$I = \dfrac{V}{R} = \dfrac{6}{12} = 0.5 \quad \text{(omitting units for simplicity)}$$

So the current is 0.5 A.

▲ This triangle gives the V, I, and R equations. To find the equation for I, cover up the I, ...and so on.

How current varies with p.d. for a metal conductor

The circuit below left can be used to investigate how the current in a conductor depends on the p.d. across it. The conductor in this case is a coiled-up length of nichrome wire, kept at a steady temperature by immersing it in a large amount of water. The p.d. across the nichrome can be varied by adjusting the variable resistor. Typical results are shown in the table below. The experiment is also one method of measuring resistance.

The results can also be shown in the form of a graph, as below.

p.d.	current	p.d. / current
1.0 V	0.2 A	5.0 Ω
2.0 V	0.4 A	5.0 Ω
3.0 V	0.6 A	5.0 Ω
4.0 V	0.8 A	5.0 Ω
5.0 V	1.0 A	5.0 Ω

resistance

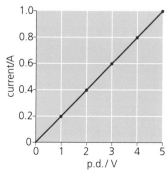

180

ELECTRICITY

(E) Ohm's law

In the experiment on the opposite page, the results have these features:
- A graph of current against p.d. is a straight line through the origin.
- If the p.d. doubles, the current doubles, ...and so on.
- p.d. ÷ current always has the same value (5 Ω in this case).

Mathematically, these can be summed up as follows:

> The current is proportional to the p.d.

This is known as **Ohm's law**, after George Ohm, the 19th century scientist who first investigated the electrical properties of wires.

Metal conductors obey Ohm's law, provided their temperature does not change. Put another way, a metal conductor has a constant resistance, provided its temperature is constant. This is not always the case with other types of conductor.

Current–p.d. graphs

Here are two more examples of current–p.d. graphs. In both, the resistance varies depending on the p.d. In the case of the diode, the negative part of the graph is for readings obtained when the p.d. is reversed (i.e. when the diode is connected into the test circuit the opposite way round).

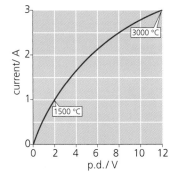

▲ **Tungsten filament** As the current increases, the temperature rises and the resistance goes up. So the current is not proportional to the p.d.

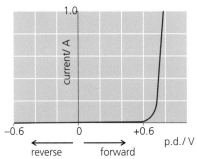

▲ **Semiconductor diode** The current is not proportional to the p.d. And if the p.d. is reversed, the current is almost zero. In effect, the diode 'blocks' current in the reverse direction.

1 A resistor has a steady resistance of 8 Ω.
  a If the current in the resistor is 2 A, what is the p.d. across it?
  b What p.d. is needed to produce a current of 4 A?
  c If the p.d. falls to 6 V, what is the current?
2 The graph lines A and B on the right are for two different conductors. Which conductor has the higher resistance?
3 Using the left-hand graph above, calculate the resistance of the tungsten filament when its temperature is a 1500 °C b 3000 °C.
4 In the right-hand graph above, does the diode have its highest resistance in the forward direction or the reverse? Explain your answer.

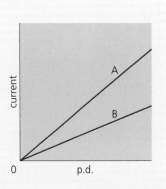

Related topics: conductors and semiconductors 8.1; current 8.4; potential difference 8.5; diodes 8.6 and 8.11

8.8 More about resistance factors

Objectives: – to know the factors that affect the resistance of a wire, and how each of them increases or decreases the resistance.

Resistance essentials

To make a current flow in a conductor, there must be a p.d. (voltage) across it. The resistance of the conductor is calculated like this:

$$\text{resistance} = \frac{\text{p.d.}}{\text{current}}$$

Units:
resistance: ohms (Ω)
p.d.: volts (V)
current: amperes (A)

Even copper connecting wires have some resistance, although this is usually very small. Resistors and heating elements are designed to have resistance.

The resistance of a wire depends on its length and cross-sectional area. It also depends on the material and its temperature (although for metals, the change of resistance with temperature is small).

The effects of length and area

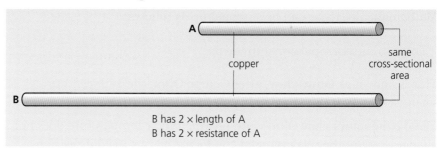

B has 2 × length of A
B has 2 × resistance of A

The copper wires above have the same cross-sectional area and temperature. But B is *twice* as long as A. As a result, it has *twice* the resistance of A. If B were *three* times as long as A, it would have three *times* the resistance, and so on. Results like this can be summed up as follows:

(E) Provided other factors do not change:

resistance ∝ length   (the symbol ∝ means 'directly proportional to')

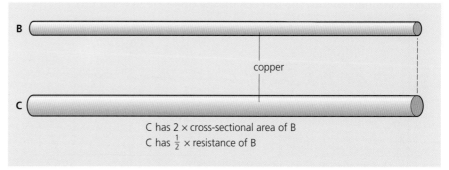

C has 2 × cross-sectional area of B
C has $\frac{1}{2}$ × resistance of B

The copper wires above have the same length and temperature. But C has *twice* the cross-sectional area of B. As a result, it has *half* the resistance of B. If C had *three* times the cross-sectional area of B, it would have *one third* of the resistance, and so on. Results like this can be summed up as follows:

(E) Provided other factors do not change:

$$\text{resistance} \propto \frac{1}{\text{area}}$$ ('area' means 'cross-sectional area')

The above proportionalities are true for other types of wire, although the resistances will differ. For example, nichrome has much more resistance than copper of the same length, cross-sectional area, and temperature.

The results can be combined as follows:

For any given conducting material at constant temperature:

$$\text{resistance} \propto \frac{\text{length}}{\text{area}}$$

ELECTRICITY

E Proportionality problems

When there are mathematical problems to solve, equations are much more useful than proportionalities. Fortunately, the proportionality linking resistance (R), length (l), and area (A) can be converted into an equation like this:

$$R = \rho \times \frac{l}{A}$$ (ρ = Greek letter 'rho')

where ρ is a constant for the material at a particular temperature. ρ is called the **resistivity** of the material (Table 1). Rearranging the above equation gives:

$$\rho = \frac{R \times A}{l}$$

This is useful when comparing different wires, A and B, made from the same material. As ρ is the same for each wire (at a particular temperature):

$$\frac{\text{resistance}_A \times \text{area}_A}{\text{length}_A} = \frac{\text{resistance}_B \times \text{area}_B}{\text{length}_B}$$

Example Wire A has a resistance of 12 Ω. If wire B is twice the length of A and twice the diameter, what is its resistance? (Assume that both wires are at the same temperature.)

As wire B has twice the diameter of A, it has four times the cross-sectional area (see the box above right).

The resistance of wire B is to be found: call it R_B. As no measurements are given, use letters to represent these as well, as in the diagram on the right.

If length$_A$ = x, then length$_B$ = $2x$

If area$_A$ = A, then area$_B$ = $4A$

Also, resistance$_A$ = 12 Ω and resistance$_B$ = R_B

Substituting the above values in the previous equation gives:

$$\frac{12 \times A}{x} = \frac{R_B \times 4A}{2x}$$ (omitting units for simplicity)

Rearranging and cancelling gives: $R_B = 6$

So, the resistance of wire B is 6 Ω.

Diameter and area

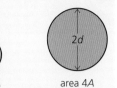

If one wire has *twice* the diameter of another, as above, then it has *four times* the cross-sectional area. That follows from the equation for the area of a circle: $A = \pi r^2$. Doubling the diameter doubles the radius. So, replacing r in the equation with $2r$ gives:

new area
$= \pi(2r)^2 = 4\pi r^2 = 4A$.

Similarly *three* times the diameter gives *nine times* the area, and so on. So:

area $\propto$ diameter^2

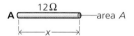

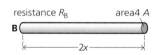

Typical resistivity values/ Ω m

Constantan	49×10^{-8}
Manganin	44×10^{-8}
Nichrome	100×10^{-8}
Tungsten	55×10^{-8}

Table 1

Q

1 Wire X has a resistance of 18 Ω. Wire Y is made of the same material and is at the same temperature. If Y is the same length as X, but 3 times the diameter, what is its resistance?

2 Wires A and B are made of the same material and are at the same temperature. The chart on the right gives some information about them.

  a If you were to use part of wire A to make an 18 Ω resistor, what length would you need?

  b What is the resistance of wire B?

  c What length of wire B would you need to make a 20 Ω resistor?

	wire A	wire B
length	1000 mm	2500 mm
area	2.0 mm²	0.5 mm²
resistance	25 Ω	

Related topics: resistance and resistors 8.6–8.7

8.9 Series and parallel circuits (1)

Objectives: – to know how series circuits and parallel circuits compare
– to know the rules that apply in each arrangement.

Circuit essentials
Potential difference (p.d.), or voltage, is measured in volts (V). The greater the p.d. aross a lamp or other component, the greater the current flowing in it. Current is measured in amperes (A).

Lamps, resistors, and other components have resistance to a flow of current. Resistance is measured in ohms (Ω).

The lamps above have to get their power from the same supply. There are two basic methods of connecting lamps, resistors, or other components together. The circuits below demonstrate the differences between them.

Lamps in series and parallel

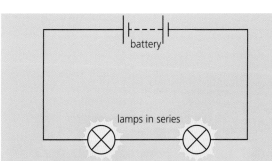

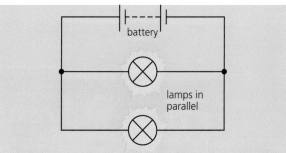

These lamps are connected in **series**.
- The lamps share the p.d. (voltage) from the battery, so each glows dimly.
- If one lamp is removed, the other goes out because the circuit is broken.

These lamps are connected in **parallel**.
- Each gets the full p.d. from the battery because each is connected directly to it. So each glows brightly.
- If one lamp is removed, the other keeps working because it is still part of an unbroken circuit.

Circuits and switches

If two or more lamps have to be powered by one battery, as in a car lighting system, they are normally connected in parallel. Each lamp gets the full battery p.d. Also, each can be switched on and off independently:

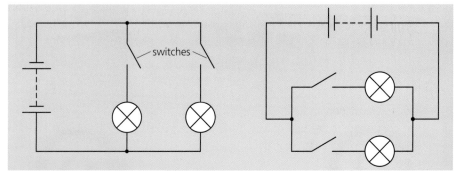

▶ These diagrams show two different ways of drawing the same circuit for independently switched lamps.

ELECTRICITY

Basic circuit rules

There are some basic rules for all series and parallel circuits. They are illustrated by the examples below. The particular current values depend on the resistances and p.d.s. However, the equation on the right *always* applies to *every* resistor.

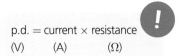

p.d. = current × resistance
(V)   (A)   (Ω)

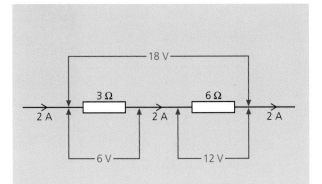

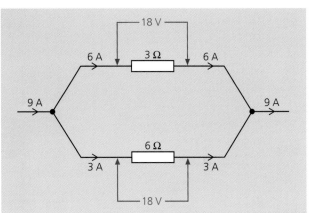

When resistors or other components are in **series**:
- the current in each of the components is the same
- (E) the total p.d. (voltage) across all the components is the sum of the p.d.s across each of them.

When resistors or other components are in **parallel**:
- the p.d. (voltage) across each of the component is the same
- (E) the total current in the main circuit is the sum of the currents in the branches.

(E) Cell arrangements

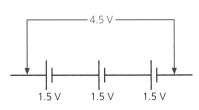

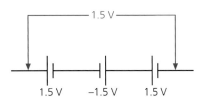

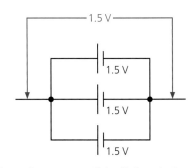

(E) These cells are connected in series. The total p.d. (voltage) across them is the sum of the individual p.d.s.

Here, a mistake has occurred. One of the cells is the wrong way round, so it cancels out one of the others.

The p.d. across parallel cells is only the same as from one cell. But together, the cells can deliver a higher current.

Q

1. When one of the lamps on a string of lights breaks, the others go out as well. What does this tell you about the way the lamps are connected?
2. Give *two* advantages of connecting lamps to a battery in parallel.
3. Redraw either of the circuits on the left so that it has a single switch which turns both lamps on and off together.
4. This question is about the circuit on the right:
   a. The readings on two of the ammeters are labelled. What are the readings on ammeters X and Y?
   b. If the p.d. across the battery is 6 V, what is the p.d. across each of the lamps? (Note: you can neglect the p.d. across an ammeter.)

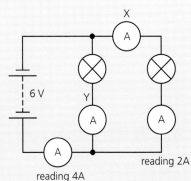

Related topics: current and circuits 8.4; potential difference (voltage) 8.5; resistance 8.6–8.7

8.10 Series and parallel circuits (2)

Objectives: – to know how to calculate the combined resistance of resistors in series, and in parallel – to solve problems involving these.

Combined resistance of resistors in series

If two (or more) resistors are connected in series, they give a *higher* resistance than any of the resistors by itself. The effect is the same as joining several lengths of resistance wire to form a longer length.

If resistors R_1 and R_2 are in series, their combined resistance R is given by this equation:

$$R = R_1 + R_2$$

There is an example on the left. For three or more resistors, the above equation can be extended by adding R_3 ... and so on.

These resistors...

3 Ω   6 Ω

...are equivalent to this resistor

9 Ω

resistance = 3 Ω + 6 Ω = 9 Ω

Combined resistance of resistors in parallel

If two (or more) resistors are connected in parallel, they give a *lower* resistance than any of the resistors by itself. The effect is the same as using a thick piece of resistance wire instead of a thin one. There is a wider conducting path than before.

(E) If two resistors R_1 and R_2 are in parallel, their combined resistance R is given by this equation (there is a proof at the bottom of the page):

$$\frac{1}{R} = \frac{1}{R_1} + \frac{1}{R_2}$$

These resistors...

3 Ω

6 Ω

...are equivalent to this resistor

2 Ω

For three or more resistors, the equation can be extended by adding $1/R_3$, ... and so on.

If the above equation for two resistors is rearranged, it becomes: $R = \dfrac{R_1 \times R_2}{R_1 + R_2}$

Omitting units for simplicity:

$\dfrac{1}{\text{resistance}} = \dfrac{1}{3} + \dfrac{1}{6}$

$= \dfrac{2}{6} + \dfrac{1}{6}$

$= \dfrac{3}{6}$

$= \dfrac{1}{2}$

So: resistance = 2 Ω

In words: combined resistance = $\dfrac{\text{resistances multiplied}}{\text{resistances added}}$

For example, if 3 Ω and 6 Ω resistors are in parallel:

$$\text{combined resistance} = \frac{6 \times 3}{6 + 3} = 2 \ \Omega$$

Note: this method of calculation works only for *two* resistors in parallel.

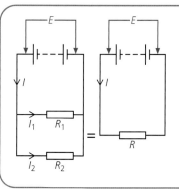

Proving the parallel resistor equation

In the circuit on the left, R_1 has the full battery p.d. of E across it. So does R_2.

As current = $\dfrac{\text{p.d.}}{\text{resistance}}$: $I_1 = \dfrac{E}{R_1}$ and $I_2 = \dfrac{E}{R_2}$

But $I = I_1 + I_2$ so $I = \dfrac{E}{R_1} + \dfrac{E}{R_2}$

If resistor R is equivalent to R_1 and R_2 in parallel, it must take the same current I from the battery:

$$I = \frac{E}{R}$$ Therefore: $\dfrac{E}{R} = \dfrac{E}{R_1} + \dfrac{E}{R_2}$ so $\dfrac{1}{R_2} = \dfrac{1}{R_1} + \dfrac{1}{R_2}$

ELECTRICITY

E Solving circuit problems

To solve problems about circuits, you need to know the basic circuit rules on the previous spread. You also need to know the link between p.d. (voltage), current, and resistance. This is given on the right.

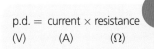

p.d. = current × resistance
(V)   (A)   (Ω)

In symbols: $V = IR$

Example 1 Calculate the p.d.s. across the 3 Ω resistor and the 6 Ω resistor in the circuit on the right.

The first stage is to calculate the total resistance in the circuit, and then use this information to find the current:

total resistance = 3 Ω + 6 Ω = 9 Ω

so: current $I = \dfrac{\text{p.d.}}{\text{resistance}} = \dfrac{18V}{9Ω} = 2$ A

Knowing that the 3 Ω resistor has a current of 2 A in it, you can calculate the p.d. across it:

p.d. = current × resistance = 2 A × 3 Ω = 6 V

The p.d. across the 6 Ω resistor can be worked out in the same way. However, it can also be deduced from the fact that the p.d.s across the two resistors must add up to 18 V, the p.d. across the battery. By either method, the p.d. across the 6 Ω resistor is 12 V.

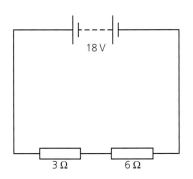

Example 2 Calculate the currents I, I_1, and I_2 in the circuit on the right.

The 3 Ω resistor has the full battery p.d. of 18 V across it. So:

$I_1 = \dfrac{\text{p.d.}}{\text{resistance}} = \dfrac{18V}{3\,Ω} = 6$ A

Using the same method: $I_2 = 3$ A

The current I is the total of the currents in the two branches. So:

$I = I_1 + I_2 = 6$ A $+ 3$ A $= 9$ A

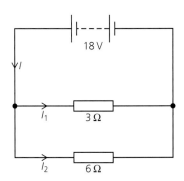

Q

1 In circuit A on the right:
  a What does the ammeter read?
  b What is the p.d. across each of the resistors?
2 In circuit B on the right:
  a What does the ammeter read when the switch is open (OFF)?
  b What is the current in each of the 4 Ω resistors when the switch is closed (ON)?
  c What does the ammeter read when the switch is closed?
  d What is the combined resistance of the two resistors when the switch is closed?
3 a Which resistor arrangement, C or D, on the right has the lower resistance?
  b Check your answer by calculation.

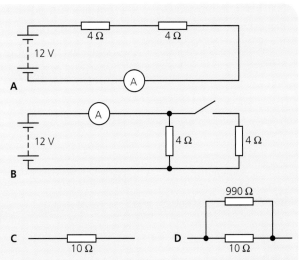

Related topics: current and circuits **8.4**; potential difference (voltage) **8.5**; resistance **8.6–8.7**

8.11 More on components

Objectives: – to know what the properties of a diode and LED are – to explain what a potential divider is and how to calculate its output voltage.

▲ Diode

E Diodes

Diodes allow current to flow in them in one direction only. The circuits below show what happens when a diode is connected into a circuit one way round and then the other:

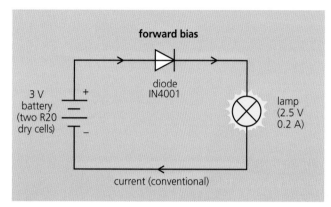
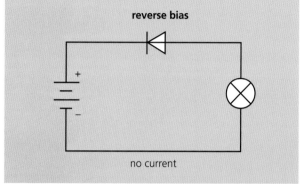

▲ When the diode is **forward biased**, it has an extremely *low* resistance, so a current flows in it and the lamp lights up. In this case, the arrowhead in the symbol points the same way as the conventional (plus-to-minus) current direction.

▲ When the diode is **reverse biased**, it has an extremely *high* resistance and the lamp does not light. In effect, the diode blocks the current.

Circuit essentials

A.c. (alternating current) flows alternately backwards and forwards. D.c. (direct current) flows one way only.

When resistors are in series, each has the same current in it. The resistor with the highest resistance has the greatest p.d. (voltage) across it.

Rectification* This is the process of changing a.c. to d.c. It is done using diodes which, doing this job, are known as **rectifiers**. A simple rectifier circuit is shown below. The diode lets the forward parts of the alternating current through but blocks the backwards part. So the current in the resistor flows one way only. It has become a rather jerky form of d.c.

An oscilloscope can be used to show how the circuit changes the a.c. input. The bottom part of the output waveform is missing. The current is flowing one way, in surges, with short periods of no current between.

With extra components added to the circuit, the pulsing current from a rectifier can be smoothed out. The result is a steady d.c. output.

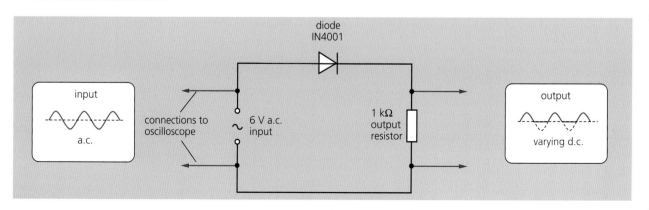

188

ELECTRICITY

E LEDs

LED stands for **Light-Emitting Diode**. As the name suggests, it is a diode that gives off light when a current is passed through it. Red, green, and blue LEDs are used as indicator lights on electronic equipment. Arrays of white LEDs are the source of light in some low-energy bulbs and torches.

▲ LED and symbol

Potential divider

A **potential divider** is an arrangement that delivers only a proportion of the voltage from a battery (or other source). Circuit A shows the principle:

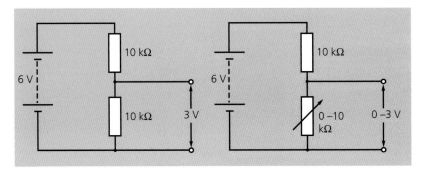

A In this potential divider, the lower resistor has half the total resistance of the two resistors, so its share of the battery's voltage is also a half.

B Using a variable resistor as above, the output voltage can be changed. Here it can range from 0 to 3 V, depending on the setting on the variable resistor.

P.d. and resistance

In a potential divider, the p.d.s across the two resistors are in proportion to their resistances. If the resistances are R_1 and R_2 and these have p.d.s of V_1 and V_2 across them:

$$\frac{R_1}{R_2} = \frac{V_1}{V_2}$$

Some electronic circuits are designed to switch on when a voltage reaches a set value. If the variable resistor in circuit B were replaced by an **LDR** (light-dependent resistor), then the circuit controlling a lamp could be switched on when it got dark. Similarly, a fire alarm could be switched on by a potential divider containing a **thermistor** (temperature-dependent resistor).

Reed switch

A **reed switch** is operated by a magnetic field. In the example on the right, the contacts close if a magnet is brought near, then open again if it is moved away. Burglar alarm circuits often contain reed switches. The magnets are attached to the moving parts of windows and doors.

With a coil round it, a reed switch becomes a **reed relay**. The current in one circuit (through the coil) switches on another circuit (through the contacts).

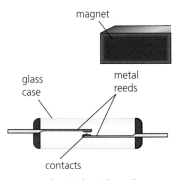

▲ A reed switch. When the magnet is moved near, the reeds become magnetized and attract each other.

Q

1. What does a diode do?
2. What is the purpose of a rectifier?
3. Look at circuits X and Y on the right. In which one
   a does the lamp light up
   b does the diode have a very high resistance?
4. If, in circuit A above, the lower resistor were replaced with one of 5 kΩ, how would this affect the output voltage of the potential divider?
5* How would you close the contacts in a reed switch?

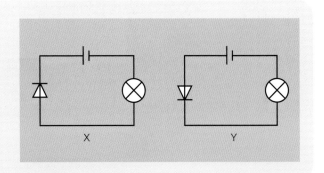

Related topics: current direction **8.4**; resistance **8.6**; relay **9.4**; a.c. and d.c. **9.9**

8.12 Electrical energy and power

Objectives: – to know how to calculate the power of an electrical device – to know the link between energy and power, and that energy can be measured in kWh.

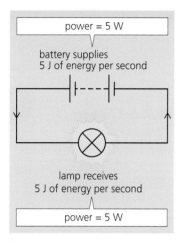

In the circuit on the left, the battery supplies energy which is transferred to the lamp. Energy is transferred from the lamp by radiation.

Power is the rate at which energy is transferred (moved from one store to another). The SI unit of power in the watt (W):

$$\text{power} = \frac{\text{energy transferred}}{\text{time taken}}$$

The battery on the left is supplying 5 joules of energy per second, so the power it transfers is 5 watts. The lamp is transferring power to its surroundings at the same rate, 5 watts.

Appliances such as toasters, irons, and TVs have a **power rating** marked on them, either in watts or in kilowatts:

1 kilowatt (kW) = 1000 watts

Some typical power ratings are shown below. Each figure tells you the power the appliance will take *if connected to a supply of the correct voltage*. For any other voltage, the actual power would be different.

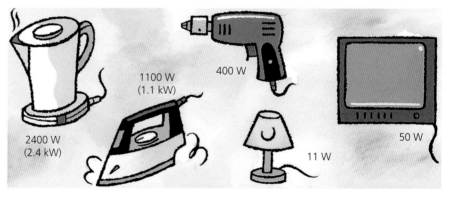

Circuit essentials

In a circuit like the one above, the charge is carried by electrons. Charge is measured in coulombs (C). The flow is called a current and is measured in amperes (A). 1 A = 1 C/s

Energy is measured in joules (J) Potential difference (p.d.) or voltage, is measured in volts (V). The greater the p.d. across a battery, the more energy per coulomb it supplies. 1 V = 1 J/C

The greater the p.d. across a lamp or other component, the more energy per coulomb is being transferred to it.

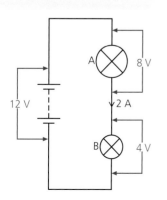

Electrical power equation

For circuits, there is a more useful version of the power equation. If a battery, lamp, or other component has a p.d. (voltage) across it and a current in it, the power is given by this equation:

$$\text{power} + \text{p.d.} \times \text{current}$$
$$(\text{W}) \quad (\text{V}) \quad (\text{A})$$

In symbols: $P = VI$

Example In the circuit on the left, what is the power of the battery and each of the lamps?

For the battery: power = p.d. × current = 12 V × 2 A = 24 W
For lamp A: power = p.d. × current = 8 V × 2 A = 16 W
For lamp B: power = p.d. × current = 4 V × 2 A = 8 W
The lamps are the only items getting power from the battery, so their total power (16 W + 8 W) is the same as that supplied by the battery (24 W).

> **Why the electrical power equation works***
> The equation power = p.d. × current is a result of how the volt, ampere, coulomb, joule, and watt are related. The following example should explain why.
> Here are two ways of describing what is happening on the right:
>
> *General description*
> Battery transfers 12 joules of energy to each coulomb of charge
> 12 joules of energy from the battery
> 2 coulombs of charge are pushed out from the battery every second
> So 12 × 2 joules of energy are transferred from the battery every second
>
> *Scientific description*
> p.d. = 12 volts
> current = 2 amperes
> power = 24 watts

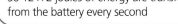

Calculating electrical energy

If the power of an appliance is known, the energy transferred in any given time can be calculated by rearranging the first equation on the opposite page like this:

energy transferred = power × time taken

For example, if a 1000 W heating element is switched on for 5 seconds (s): energy transferred = 1000 W × 5 s × 5000 J. So the heating element gives off 5000 J of thermal energy.

As power = p.d. × current, the above equation can also be written like this:

energy transferred = p.d. × current × time taken
(J)   (V)   (A)   (s)

In symbols: $E = VIt$

Example A 12 V water heater takes a current of 2 A. If it is switched on for 60 seconds, how much energy is transferred to thermal energy?

energy transferred = p.d. × current × time = 12 V × 2 A × 60 s = 1440 J
All of this is transferred to thermal energy, so:
Energy transferred to thermal energy = 1440 J

1 In 5 seconds, a hairdryer receives 10 000 joules of energy from the mains supply. What is the power transferred **a** in watts **b** in kilowatts?
2 If an electric heater takes a current of 4 A when connected to a 230 V supply, what power is being transferred?
3 If a lamp has a power of 24 W when connected to a 12 V supply, what is the current through it?
4 Calculate the energy transferred to an 11 W lamp
   a in 1 second   **b** in 1 minute.
5 A lamp takes a current of 3 A from a 12 V battery.
   a What power is being transferred to the lamp?
   b How much energy is transferred in 10 minutes?

Related topics: SI units **1.2**; energy **4.1**; power **4.4**; current **8.4**; p.d. **8.5**

The kilowatt-hour
Electricity supply companies use the **kilowatt-hour (kWh)** rather than the joule as their unit of energy measurement:

One kilowatt hour (kWh) is the energy supplied when an appliance of power 1 kW is used for 1 hour.

1 kW is 1000 W, and 1 hour is 3600 s. So, if a 1 kW appliance is used for 1 hour:
energy = power × time
= 1000 W × 3600 s
= 3 600 000 J

Therefore: 1 kWh = 3 600 000 J

Counting the cost
Electricity supply companies charge a set amount per kWh for the energy they supply. For example, if the cost is 20p per hour (where 'p' stands or the local currency unit):
Leaving a 2 kW heater on for 3 hours would require 6 kWh of energy, so the cost would be 6 × 20p, which is 120p.

8.13 Living with electricity

Objectives: – to know that mains electricity is a.c. To describe when mains electricity could be dangerous and the methods used to make it safe.

Circuit essentials

A p.d. (potential difference) is needed to make a current flow round a circuit. p.d. is measured in volts (V) and is more commonly called voltage. Current is measured in amperes (A).

When you plug a kettle into a mains socket, you are connecting it into a circuit, as shown below. The power comes from a generator in a power station. The supply voltage depends on the country. For household circuits, some countries use a voltage in the range 220–240 V, others in the range 110–130 V.

Mains current is **alternating current** (a.c.). It flows backwards and forwards, backwards and forwards... 50 times per second, in some countries. The **mains frequency** is 50 hertz (Hz). In other countries, the mains frequency is 60 Hz. A.c. is easier to generate than one-way direct current (d.c.) like that from a battery.

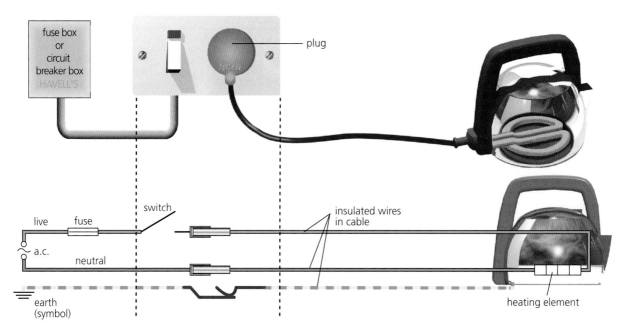

▲ This table lamp has an insulating body and does not need an earth wire.

Live (or line) wire This goes alternately negative and positive, making the current flow backwards and forwards in the circuit.

Neutral (or cold) wire This completes the circuit. In many systems, it is kept at zero voltage by the electricity supply company.

Switch This is fitted in the live wire. It would work equally well in the neutral, but wire in the cable would still be live with the switch OFF. This would be dangerous if, for example, the cable was accidentally cut.

Fuse This is a thin piece of wire which overheats and melts if the current is too high. Like the switch, it is placed in the live wire, often as a cartridge. If a fault develops, and the current gets too high, the fuse 'blows' and breaks the circuit before the cable can overheat and catch fire. Many circuits use a **circuit breaker (trip switch)** instead of a fuse (see spread 9.4 and next page).

Earth (grounded) wire This is a safety wire. It connects the metal body of the kettle to earth and stops it becoming live. For example, if the live wire comes loose and touches the metal body, a current immediately flows to earth and blows the fuse. This means that the kettle is then safe to touch.

ELECTRICITY

Double insulation Some appliances – radios for example – do not have an earth wire. This is because their outer case is made of plastic rather than metal. The plastic acts as an extra layer of insulation around the wires.

For extra safety, circuits may be fitted with a type of breaker called a **residual current device** (**RCD**). This compares the currents in the live and neutral wires. If they are not the same, then current must be flowing to earth – perhaps through someone touching an exposed wire. The RCD senses the difference and switches off the current before any harm can be done.

Plugs

Plugs are a safe and simple way of connecting appliances to the mains. Over a dozen different types of plug are in use around the world. You can see an example on the right.

A few countries use a three-pin plug with a fuse inside. The fuse value is typically either 3 A or 13 A. This tells you the current needed to blow the fuse. It must be *greater* than the normal current in the appliance, but *as close to it as possible*, so that the fuse will blow as soon as the current gets too high. For example:

- If a kettle takes a current of 10 A, then a 13 A fuse is needed.
- If a TV takes a current of 0.2 A, then a 3 A fuse is needed. The TV would still work with a 13 A fuse. But if a fault developed, its circuits might overheat and catch fire without the fuse blowing.

▲ This two-pin plug has earth connections in grooves at the edge.

Electrical hazards

Mains electricity can be dangerous. Here are some of the hazards:
- Old, frayed wiring. Broken strands mean that a wire will have a higher resistance at one point. When a current flows in it, the heating effect may be enough to melt the insulation and cause a fire.
- Long extension leads. These may overheat if used when coiled up. The current warms the wire, but the heat has less area to escape from a tight bundle.
- Water in sockets or plugs. Water will conduct a current, so if electrical equipment gets wet, there is a risk that someone might be electrocuted.
- Accidentally cutting cables. With lawnmowers and hedgetrimmers, a plug-in RCD can be used to avoid the risk of electrocution.

If an accident happens, and someone is electrocuted, you must switch off at the socket and pull out the plug before giving any help.

▲ A plug-in RCD gives protection against the risk of electrocution.

1. What is a fuse, and how does it work?
2. In a mains circuit, why should the switch always be in the live wire rather than the neutral?
3. In mains appliances, what is the purpose of the earth wire?
4. Some countries use plugs with a fuse in. For each appliance on the right, decide whether its plug should be fitted with a 3 A or a 13 A fuse.
5. Why should a 13 A fuse *not* be used for a TV taking a current of 0.2 A?
6. If an accident occurs and someone is electrocuted, what **two** things must you do before giving help?

appliance	current
hairdryer	6 A
food mixer	2 A
iron	10 A
lamp	0.1 A

Related topics: current and circuits **8.4**; voltage (p.d.) **8.5**; resistance and heating effect **8.6**; power **8.12**; how a circuit breaker works **9.4**; generators **9.9**; electricity supply system **9.12**

Check-up on electricity

Further questions

1. **a** When a balloon is rubbed in your hair, the balloon becomes negatively charged.
   i Explain how the balloon becomes negatively charged. [2]
   ii State what you know about the size and sign of the charge left on your hair. [2]
   b The negatively charged balloon is brought up to the surface of a ceiling. The balloon sticks to the ceiling. Explain how and why this happens. [3]

2. **Read the following passage carefully** before answering the questions.
   Spraying crops with insecticides has become more efficient. A portable high voltage generator gives the drops of liquid insecticide a small positive charge. This makes the liquid break up into smaller drops and causes the spray to become finer and spread out more.
   The plants, which are all reasonable conductors, are in contact with the earth. As the droplets of spray get near the plants, the plants themselves become slightly charged and attract the droplets.
   a **i** Explain why the positive charge on the droplets makes the spray spread out. [1]
   ii State what charge appears on the plants as the droplets come near to them. [1]
   b Draw a diagram to show the electric field pattern around one of the droplets (assume it is spherical and that there are no other droplets nearby). [2]
   c How would the field be different if the droplet had a negative rather than positive charge? [1]

3.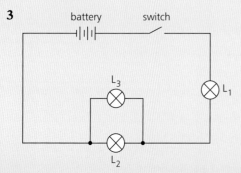

The circuit shows a battery connected to a switch and three identical lamps, L_1, L_2 and L_3.
   a Copy the diagram and add:
   i an arrow to show the conventional current direction in the circuit when the switch is closed [1]
   ii a voltmeter V, to measure the voltage across L_1 [1]
   iii a switch, labelled S, that controls L_3 only. [1]
   b State and explain what effect adding another cell to the battery would have on the lamps in the circuit. [2]

4. The circuit diagram shows a battery connected to five lamps. The currents in lamps **A** and **B** are shown.

   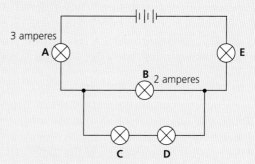

   Write down the current flowing in
   a lamp **C**, [1]
   b lamp **E**. [1]

5. **a** How much energy is transferred by a battery of e.m.f. 4.5 V when 1.0 C of charge passes through it? [1]
   b How much power is transferred by a battery of e.m.f. 4.5 V when a current of 1.0 A is passing through it? [1]

6. The diagram shows a circuit which contains two resistors.

   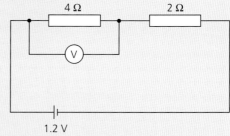

Calculate
a the total resistance of the two resistors
  in series, (Ω) [1]
b the current flowing in the cell, (A) [1]
c the current flowing in the 4 Ω
  resistor, (A) [1]
d the reading of the voltmeter, (V) [1]
e the power transferred by the 4 Ω
  resistor. (W) [1]

7 A small electric hairdryer has an outer case made of plastic. The following information is printed on the case:

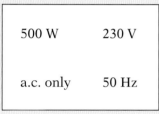

a Explain the meaning of these terms:
  i a.c. only [1]
  ii 50 Hz [1]
b The hairdryer does not have an earth wire. Instead, it is **double insulated**. Explain what this means. [2]
c What current does the hairdryer take? [2]
d The hairdryer is protected by its own fuse.
  i What is the purpose of the fuse? [1]
  ii Given a choice of a 3 A or a 13 A fuse for the hairdryer, which would you select, and why? [2]
e* If the hairdryer were used in a country where the mains voltage was only 110 V, what difference would this make, and why? [3]

8 A small generator is labelled as having an output of 2 kW, 230 V a.c. (at constant frequency). It is used to provide emergency lighting for a large building in the event of a breakdown of the mains supply. The circuit is shown below.

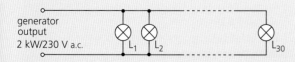

There are 30 light fittings on the circuit, each with a 230 V, 28 W halogen lamp.

a Calculate the maximum current which the generator is designed to supply. [2]
b i Calculate the power needed when all the lamps are turned on at the same time.
  ii Explain why this generator is suitable for supplying the power required but would not be suitable if all the lamps were exchanged for 100 W lamps. [4]
c Write down two reasons why all the lamps are connected in parallel rather than in series. In each answer, you should refer to both types of circuit. [4]
d Calculate the resistance of the filament of each 28 W lamp. [4]
e The figure below shows the current output of the generator when it is supplying all 30 of the 28 W lamps.

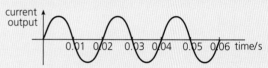

  i* Calculate the frequency of the supply from the generator.
  ii Copy the diagram and sketch another graph to show the approximate current output of the generator when 15 lamps are removed from their fittings. [4]

9 A student investigates how the current in a lamp varies with the voltage (p.d.) across it. She uses the circuit shown below.

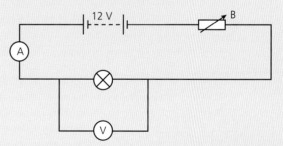

a Three of the components are labelled, A, V, and B. Write down what each one is. [3]
b Describe how the student should carry out the experiment. [3]
From her results, the student plots this graph:

MEASUREMENTS AND UNITS

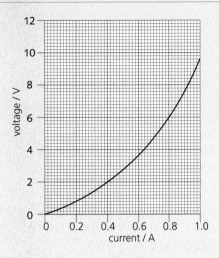

c What is the current when the voltage across the lamp is 2.0 V? [1]

d What is the resistance of the lamp when the voltage across it is 2.0 V? [2]

e What is the resistance of the lamp when the voltage across it is 6.0 V? [2]

f What happens to the resistance of the lamp as the voltage across it is increased? [1]

10 A small electric heater takes a power of 60 W from a 12 V supply.
  a What is the current in the heater? [2]
  b What is the resistance of the heater? [2]
  (E) c How much charge (in C) passes through the heater in 20 seconds? [2]
  d How much energy (in J) is transferred by the heater in 20 seconds? [2]

Use the list below when you revise for your IGCSE examination. The spread number, in brackets, tells you where to find more information.

Revision checklist
Core Level
☐ Two types of electric charge and the attractions and repulsions between them. (8.1)
☐ Charging by friction: adding or removing electrons. (8.1)
☐ Electrical conductors and insulators. (8.1)
☐ Why metals are good conductors, while most other materials are insulators. (8.1)
☐ Detecting charge. (8.2)
☐ Current as a flow of charge. (8.4)
☐ Current in a metal is a flow of electrons. (8.4)
☐ The ampere, unit of current; measuring current with an ammeter. (8.4)
☐ Using circuit diagrams and symbols (excluding the diode). (8.4, 8.5 and page 321)
☐ Using switches, resistors, thermistors, LDRs, and other components. (8.4, 8.6, and 8.9)
☐ How the current is the same at all points round a series circuit. (8.4 and 8.9)
☐ The volt, unit of p.d. and e.m.f.; measuring p.d. with a voltmeter. (8.5)
☐ Defining p.d. and e.m.f. (8.5)
☐ The ohm, unit of resistance. (8.6)
☐ The equation linking resistance, p.d., and current. (8.6 and 8.7)
☐ Factors affecting the resistance of a wire. (8.6)
☐ How current is split in a parallel circuit. (8.9)
☐ Advantages of connecting lamps in parallel. (8.9)
☐ Calculating the combined resistance of resistors in series. (8.10)
☐ The combined resistance of two resistors in parallel is less than that of either resistor by itself. (8.10)

☐ The equation linking power, p.d. (voltage), and current. (8.12)
☐ The equation linking energy, p.d., current, and time. (8.12)
☐ The kilowatt-hour. (8.12)
☐ The difference between a.c. and d.c. (8.13)
☐ The key features of a mains circuit. (8.13)
☐ Using fuses and circuit breakers. (8.13)
☐ The hazards of mains electricity; the importance of using earthing or double insulation. (8.13)

Extended Level
As for Core Level, plus the following:
☐ Why charged objects attract uncharged ones. (8.2)
☐ The coulomb, unit of charge. (8.2)
☐ What an electric field is. (8.3)
☐ Electric field is a vector; the direction of an electric field. (8.3)
☐ Electric fields patterns around a point charge, a charged conducting sphere, and between two oppositely-charged parallel plates. (8.3)
☐ The equation linking current, charge, and time. (8.4)
☐ Electron flow and conventional current. (8.4)
☐ The rule linking the p.d.s round a circuit. (8.5 and 8.9)
☐ current-p.d. characteristics (graphs) for a metal wire at constant temperature, a filament as it heats up, and a diode. (8.7)
☐ The relationship between the resistance, length, and cross-sectional area of a wire. (8.8)
☐ The rule for currents in a parallel circuit. (8.9)
☐ Calculating the combined resistance of two resistors in parallel. (8.10)
☐ Diodes and LEDs. (8.11)
☐ The action of a potential divider. (8.11)

9 Magnets and currents

- MAGNETS
- MAGNETIC FIELDS
- MAGNETIC EFFECT OF A CURRENT
- ELECTROMAGNETS
- MAGNETIC FORCE ON A CURRENT
- ELECTRIC MOTORS
- ELECTROMAGNETIC INDUCTION
- GENERATORS
- TRANSFORMERS
- POWER TRANSMISSION AND DISTRIBUTION

Computer model of the magnetic field inside the doughnut-shaped chamber of a nuclear fusion reactor. Like the Sun, fusion reactors release energy by smashing hydrogen atoms together to form helium. One day, they may provide the energy to run power stations on Earth.

In the reactor, the magnetic field is used to trap the charged particles from hydrogen at a temperature of over 100 million °C.

9.1 Magnets

Objectives: – to describe the forces between magnetic poles – to know what induced magnetism is – to know about magnetic and non-magnetic materials.

Magnetic poles

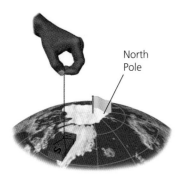

If a small bar magnet is dipped into iron filings, the filings are attracted to its ends, as shown in the photograph on the opposite page. The magnetic force seems to come from two points, called the **poles** of the magnet.

The Earth exerts forces on the poles of a magnet. If a bar magnet is suspended as on the left, it swings round until it lies roughly north–south. This effect is used to name the two poles of a magnet. These are called:
- the **north-seeking pole** (or **N pole** for short)
- the **south-seeking pole** (or **S pole** for short).

If you bring the ends of two similar bar magnets together, there is a force between the poles as shown below:

> Like poles repel; unlike poles attract.
> The closer the poles, the greater the force between them.

Properties of magnets
A magnet:
- Has a magnetic field around it (see the next spread).
- Has two opposite poles (N and S) which exert forces on other magnets. Like poles repel; unlike poles attract.
- Will attract magnetic materials by inducing magnetism in them. In some materials (e.g steel) the magnetism is permanent. In others (e.g. iron) it is temporary.
- Will exert little or no force on a non-magnetic material.

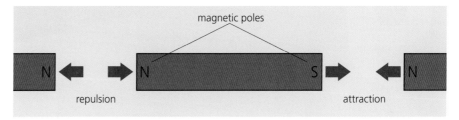

Induced magnetism

Materials such as iron and steel are attracted to magnets because they themselves become magnetized when there is a magnet nearby. The magnet **induces** magnetism in them, as shown below. In each case, the induced pole nearest the magnet is the *opposite* of the pole at the end of the magnet. The attraction between unlike poles holds each piece of metal to the magnet.

The steel and the iron behave differently when pulled right away from the magnet. The steel keeps some of its induced magnetism and becomes a **permanent magnet**. However, the iron loses virtually all of its induced magnetism. It was only a **temporary magnet**.

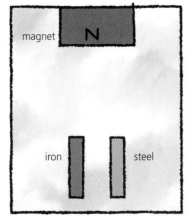
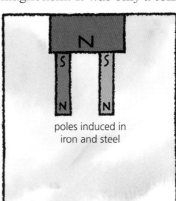
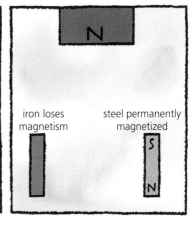

MAGNETS AND CURRENTS

Making a magnet

A piece of steel becomes permanently magnetized when placed near a magnet, but its magnetism is usually weak. It can be magnetized more strongly by stroking it with one end of a magnet, as on the right. However, the most effective method of magnetizing it is to place it in a long coil of wire and pass a large, direct (one-way) current in the coil. The current has a magnetic effect which magnetizes the steel.

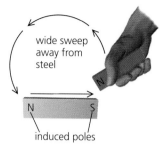

▲ Magnetizing a piece of steel by stroking it with a magnet.

Magnetic and non-magnetic materials

A **magnetic material** is one which can be magnetized and is attracted to magnets. All strongly magnetic materials contain iron, nickel, or cobalt. For example, steel is mainly iron. Strongly magnetic metals like this are called **ferromagnetics**. They are described as *hard* or *soft* depending on how well they keep their magnetism when magnetized:

Hard magnetic materials such as steel, and alloys called Alcomax and Magnadur, are difficult to magnetize but do not readily lose their magnetism. They are used for permanent magnets.

Soft magnetic materials such as iron and Mumetal are relatively easy to magnetize, but their magnetism is only temporary. They are used in the cores of electromagnets and transformers because their magnetic effect can be 'switched' on or off or reversed easily.

Non-magnetic materials include metals such as brass, copper, zinc, tin, and aluminium, as well as non-metals.

Ferrous and non-ferrous
Iron and alloys (mixtures) containing iron are called **ferrous** metals (*ferrum* is Latin for iron). Aluminium, copper, and the other non-magnetic metals are **non-ferrous**.

Where magnetism comes from*

In an atom, tiny electrical particles called electrons move around a central nucleus. Each electron has a magnetic effect as it spins and orbits the nucleus.

In many types of atom, the magnetic effects of the electrons cancel, but in some they do not, so each atom acts as a tiny magnet. In an unmagnetized material, the atomic magnets point in random directions. But as the material becomes magnetized, more and more of its atomic magnets line up with each other.

Together, billions of tiny atomic magnets act as one big magnet.

If a magnet is hammered, its atomic magnets are thrown out of line: it becomes **demagnetized**. Heating it to a high temperature has the same effect.

▲ Magnetic materials are attracted to magnets and can be made into magnets.

1 What is meant by the *N pole* of a magnet?
2 Magnetic materials are sometimes described as *hard* or *soft*.
   a What is the difference between the two types?
   b Give one example of each type.
3 Name *three* ferromagnetic metals.
4 Name *three* non-magnetic metals.
5 The diagram on the right shows three metal bars. When different ends are brought together, it is found that A and B attract, A and C attract, but A and D repel. Decide whether each of the bars is a permanent magnet or not.

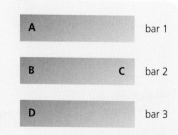

Related topics: atoms and electrons **8.1**; the Earth's magnetism **9.2**; electromagnets **9.4**; transformers **9.10–9.11**

9.2 Magnetic fields

Objectives: – to know how to plot the field pattern around a bar magnet – to know how the direction of a magnetic field is defined.

In the photograph below, iron filings have been sprinkled on paper over a bar magnet. The filings have become tiny magnets, pulled into position by forces from the poles of the magnet. Scientifically speaking, there is a **magnetic field** around the magnet, and this exerts forces on magnetic materials in it.

Magnetic field patterns

Magnetic fields can be investigated using a small **compass**. The 'needle' is a tiny magnet which is free to turn on its spindle. When near a magnet, the needle is turned by forces between its poles and the poles of the magnet. The needle comes to rest so that the turning effect is zero.

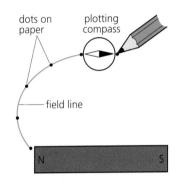

The diagram on the left shows how a small compass can be used to plot the field around a bar magnet. Starting with the compass near one end of the magnet, the needle position is marked using two dots. Then the compass is moved so that the needle lines up with the previous dot... and so on. When the dots are joined up, the result is a magnetic **field line**. More lines can be drawn by starting with the compass in different positions.

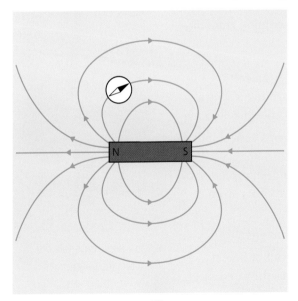
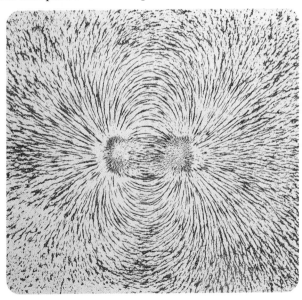

In the diagram above, a selection of field lines has been used to show the magnetic field around a bar magnet:

- The field lines run from the N pole to the S pole of the magnet. The field direction, shown by an arrowhead, is defined as the direction in which the force on a N pole would act. It is the direction in which the N end of a compass needle would point.
- The magnetic field is strongest where the field lines are closest together.

Magnet essentials
A magnet has a north-seeking (N) pole at one end and a south-seeking (S) pole at the other. When two magnets are brought together:

like poles repel, unlike poles attract.

If two magnets are placed near each other, their magnetic fields combine to produce a single field. Two examples are shown at the top of the next page. At the **neutral point**, the field from one magnet exactly cancels the field from the other, so the magnetic force on anything at this point is zero.

MAGNETS AND CURRENTS

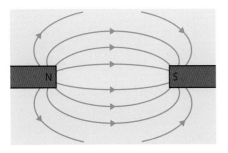
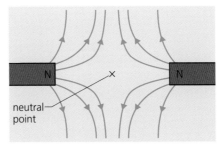

◀ Between magnets with unlike poles facing, the combined field is almost uniform (even) in strength. However, between like poles, there is a neutral point where the combined field strength is zero.

The Earth's magnetic field*

The Earth has a magnetic field. No one is sure of its cause, although it is thought to come from electric currents generated in the Earth's core. The field is rather like that around a large, but very weak, bar magnet.

With no other magnets near it, a compass needle lines up with the Earth's magnetic field. The N end of the needle points north. But an N pole is always attracted to an S pole. So it follows that the Earth's magnetic S pole must be in the north! It lies under a point in Canada called **magnetic north**.

Magnetic north is over 1200 km away from the Earth's geographic North Pole. This is because the Earth's magnetic axis is not quite in line with its north–south axis of rotation.

Magnetic screening

Some electronic equipment is easily upset by magnetic fields from nearby generators, motors, transformers, or the Earth. The equipment can be screened (shielded) by enclosing it in a layer of a soft magnetic material, such as iron or nickel. This redirects the field so that it does not pass through the equipment.

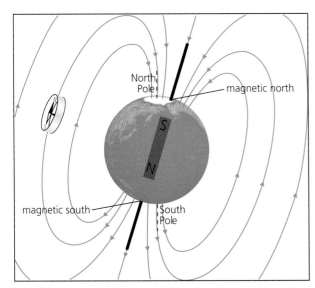

▲ The Earth behaves as if it has a large but very weak bar magnet inside it.

▲ A compass is of no use in polar regions because the Earth's magnetic field lines are vertical.

1 In the diagrams on the right, the same compass is being used in both cases.
   a Copy diagram A. Label the N and S ends of the compass needle.
   b Copy diagram B. Mark in the poles of the magnet to show which is N and which is S. Then draw an arrowhead on the field line to show its direction.
   c In diagram B, at which position, X or Y, would you expect the magnetic field to be the stronger?

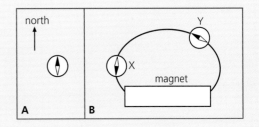

Related topics: magnetic poles and the Earth's magnetic effect **9.1**

9.3 Magnetic effect of a current

Objectives: – to know that a current produces a magnetic field – to describe the field patterns around the current in a straight wire, coil, and solenoid.

Magnet essentials
Like poles repel; unlike poles attract. Magnetic field lines show the direction of the force on a N pole.

Magnetic field around a wire

If an electric current is passed through a wire, as shown below left, a weak magnetic field is produced. The field has these features:
- the magnetic field lines are circles
- the field is strongest close to the wire
- increasing the current increases the strength of the field.

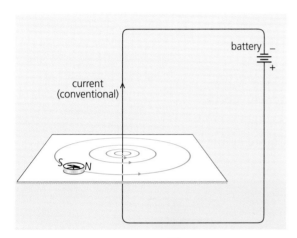

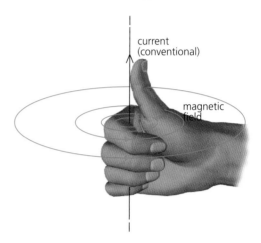

Current essentials
In a circuit the current is a flow of electrons: tiny particles which come from atoms.

The current arrows shown on circuit diagrams run from + to −. This is the **conventional current direction**. Electrons, being negatively charged, flow the other way.

A rule for field direction The direction of the magnetic field produced by a current is given by the **right-hand grip rule** shown above right. Imagine gripping the wire with your right hand so that your thumb points in the conventional current direction. Your fingers then point in the same direction as the field lines.

Magnetic fields from coils

A current produces a stronger magnetic field if the wire it flows in is wound into a coil. The diagrams below show the magnetic field patterns produced by two current-carrying coils. One is just a single turn of wire. The other is a long coil with many turns. A long coil is called a **solenoid**.

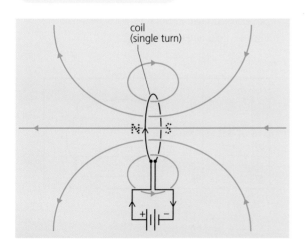

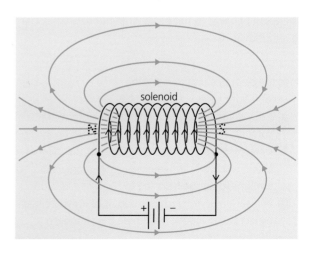

MAGNETS AND CURRENTS

The magnetic field produced by a current-carrying coil has these features:
- the field is similar to that from a bar magnet, and there are magnetic poles at the ends of the coil
- increasing the current increases the strength of the field
- increasing the number of turns on the coil increases the strength of the field.

A rule for poles* To work out which way round the poles are, you can use another **right-hand grip rule**, as shown on the right. Imagine gripping the coil with your right hand so that your fingers point in the conventional current direction. Your thumb then points towards the N pole of the coil.

*Magnets are made – and demagnetized – using coils as shown below. In video recorders and hard drives, tiny coils are used to put magnetic patterns on a disc (see next spread). The patterns store pictures, sounds, and data.

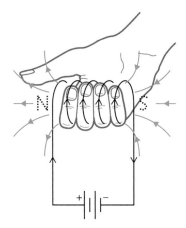

Right-hand grip rule for poles

Making a magnet

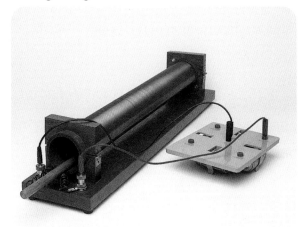

Above, a steel bar has been placed in a solenoid. When a current is passed through the solenoid, the steel becomes magnetized and makes the magnetic field much stronger than before. And when the current is switched off, the steel stays magnetized. Nearly all permanent magnets are made in this way.

Demagnetizing a magnet

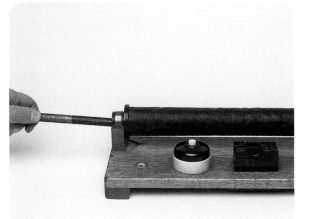

Above, a magnet is slowly being pulled out of a solenoid through which an alternating current is passing. Alternating current (a.c.) flows backwards, forwards, backwards, forwards... and so on. It produces a magnetic field which changes direction very rapidly and throws the atoms in the magnet out of line.

1 The coil in diagram A is producing a magnetic field.
  a Draw a diagram to show the shape of the magnetic field around the coil.
  b Give *two* ways in which the strength of the field could be increased.
  c How could the direction of the field be reversed?
2 Redraw diagram B to show which way the compass needles point when a current flows in the wire. (Assume that the black end of each compass needle is a N pole, the conventional current direction is away from you, into the paper, and that the only magnetic field is that due to the current.)

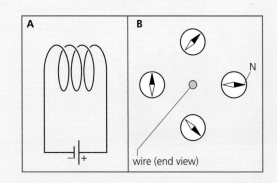

Related topics: current in a circuit **8.4**; alternating current **8.12**; magnetic poles **9.1**; magnetic fields **9.2**; magnetic storage **9.4**

9.4 Electromagnets

Objectives: – to know the factors affecting the strength of the magnetic field around an electromagnet – to describe some uses of electromagnets.

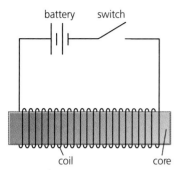

▲ A simple electromagnet

Unlike an ordinary magnet, an **electromagnet** can be switched on and off. In a simple electromagnet, a **coil**, consisting of several hundred turns of insulated copper wire, is wound round a **core**, usually of iron or Mumetal. When a current flows in the coil, it produces a magnetic field. This magnetizes the core, creating a magnetic field about a thousand times stronger than the coil by itself. With an iron or Mumetal core, the magnetism is only temporary, and is lost as soon as the current in the coil is switched off. Steel would not be suitable as a core because it would become permanently magnetized.

(E) The strength of the magnetic field is increased by:
- increasing the current
- increasing the number of turns in the coil.

Reversing the current reverses the direction of the magnetic field.

The following all make use of electromagnets.

The magnetic relay

A magnetic relay is a switch operated by an electromagnet. With a relay, a small switch with thin wires can be used to turn on the current in a much more powerful circuit – for example, one with a large electric motor in it:

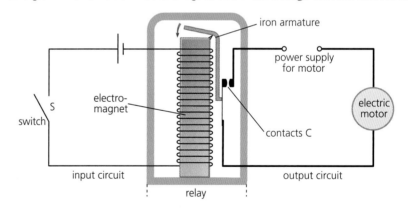

! **Magnetic essentials**
A **hard** magnetic material (for example, steel) is one which, when magnetized, does not readily lose its magnetism.
A **soft** magnetic material (for example, iron) quickly loses its magnetism when the magnetizing field is removed.

When the switch S in the input circuit is closed, a current flows in the electromagnet. This pulls the iron armature towards it, which closes the contacts C. As a result, a current flows in the motor.

The relay above is of the 'normally open' type: when the input switch is OFF, the output circuit is also OFF. A 'normally closed' relay works the opposite way: when the input switch is OFF, the output circuit is ON. In practice, most relays are made so that they can be connected either way.

▲ With a relay, a small switch can be used to turn on a powerful starter motor.

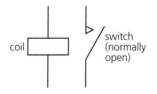

Normally open relay (symbol)

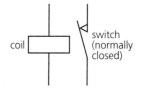

Normally closed relay (symbol)

MAGNETS AND CURRENTS

The circuit breaker

A circuit breaker is an automatic switch which cuts off the current in a circuit if this rises above a specified value. It has the same effect as a fuse but, unlike a fuse, can be reset (turned ON again) after it has tripped (turned OFF).

In the type shown on the right, the current flows in two contacts and also in an electromagnet. If the current gets too high, the pull of the electromagnet becomes strong enough to release the iron catch, so the contacts open and stop the current. Pressing the reset button closes the contacts again.

Magnetic storage*

Some recording studios use magnetic tape on reels or in cassettes for recording sounds. The tape consists of a long, thin plastic strip, coated with a layer of iron oxide or similar material. Magnetically, iron oxide is between soft and hard. Once magnetized it keeps its magnetism, but is relatively easy to demagnetize, ready for another recording. The diagram below shows a simple system for recording sound on tape. The hard drive in a computer also stores data as a pattern of varying magnetism. In both examples, an electromagnet creates the varying magnetic field needed for recording. Later, a playback head can read the pattern to give a varying current.

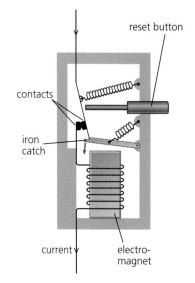

▲ Circuit breaker

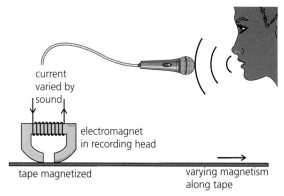

▲ **Recording on magnetic tape** The incoming sound waves are used to vary the current in a tiny electromagnet in the recording head. As the tape moves past the head, a track of varying magnetism is created along the tape.

▲ **Computer hard drive** The recording head is at the end of the arm. It contains a tiny electromagnet which is used to create tracks of varying magnetism on a spinning disc. The disc is made of aluminium or glass, and is coated with a layer of magnetic material similar to that on a tape.

1 An electromagnet has a core.
   a What is the purpose of the core?
   b Why is iron a better material for the core than steel?
   c Write down *two* ways of increasing the strength of the magnetic field from an electromagnet.
2 In the diagram on the opposite page, an electric motor is controlled by a switch connected to a relay.
   a What is the advantage of using a relay, rather than a switch in the motor circuit itself?
   b Why does the motor start when switch S is closed?
3 The diagram at the top of the page shows a circuit breaker.
   a What is the purpose of the circuit breaker?
   b How do you think the performance of the circuit breaker would be affected if the coil of the electromagnet had more turns?
4* Sounds can be recorded on tape.
   a Why is an electromagnet needed for this?
   b Why must the coating on the tape be between soft and hard magnetically?

Related topics: using circuit breakers 8.13; magnetic materials 9.1; fields from coils 9.3

9.5 Magnetic force on a current

Objectives: – to know that there is a force on a current-carrying conductor in a magnetic field – to know the rule for working out the direction of the force.

Magnet essentials

The N and S poles of one magnet exert forces on those of another:

like poles repel, unlike poles attract.

The magnetic field around a magnet can be represented by field lines. These show the direction in which the force on an N pole would act.

In the experiment shown below, a length of copper wire has been placed in a magnetic field. Copper is non-magnetic, so it is feels no force from the magnet. However, with a current passing through it, there is a force on the wire. The force arises because the current produces its own magnetic field which acts on the poles of the magnet. In this case, the force on the wire is upwards (see box below left). It would be downwards if either the magnetic field or the current were reversed. Whichever way the experiment is done, the wire moves *across* the field. It is *not* attracted to either pole.

The force is increased if:
- the current is increased
- a stronger magnet is used
- the length of wire in the field is increased.

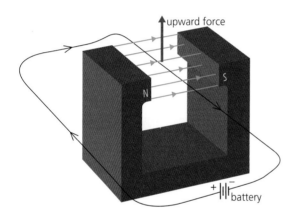

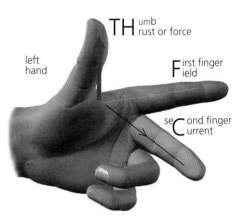

▲ Fleming's left-hand rule

Field and force

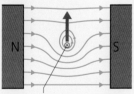

current direction out of paper

By itself, the current in a straight wire produces a circular magnetic field pattern. However, when the wire is between the poles of a magnet, the combined field is as above. In situations like this, the field lines tend to straighten. So, in this case, the wire gets pushed upwards.

Fleming's left-hand rule

In the above experiment, the direction of the force can be predicted using **Fleming's left-hand rule**, as illustrated above right. If you hold the thumb and first two fingers of your left hand at right angles, and point the fingers as shown, the thumb gives the direction of the force.

In applying the rule, it is important to remember how the field and current directions are defined:
- The field direction is from the N pole of a magnet to the S pole.
- The current direction is from the positive (+) terminal of a battery round to the negative (−). This is called the *conventional* current direction.

Fleming's left-hand rule only applies if the current and field directions are at right angles. If they are at some other angle, there is still a force, but its direction is more difficult to predict. If the current and field are in the *same* direction, there is *no* force.

If a beam of charged particles (such as electrons) passes through a magnetic field, there is a force on it, just as for a current in a wire: see spread 10.2.

MAGNETS AND CURRENTS

The moving-coil loudspeaker

Most loudspeakers are of the moving-coil type shown on the right. The cylindrical magnet produces a strong radial ('spoke-like') magnetic field at right angles to the wire in the coil. The coil is free to move backwards and forwards and is attached to a stiff paper or plastic cone.

The loudspeaker is connected to an amplifier which gives out alternating current. This flows backwards, forwards, backwards... and so on, causing a force on the coil which is also backwards, forwards, backwards.... As a result, the cone vibrates and gives out sound waves. The sound you hear depends on how the amplifier makes the current alternate.

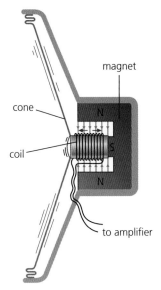

▲ Moving-coil loudspeaker

Turning effect on a coil

The coil below lies between the poles of a magnet. The current flows in opposite directions along the two sides of the coil. So, according to Fleming's left-hand rule, one side is pushed *up* and the other side is pushed *down*. In other words, there is a turning effect on the coil. With more turns on the coil, the turning effect is increased.

The meter in the photograph uses the above principle. Its pointer is attached to a coil in the field of a magnet. The higher the current in the meter, the further the coil turns against the springs holding it, and the further the pointer moves along the scale.

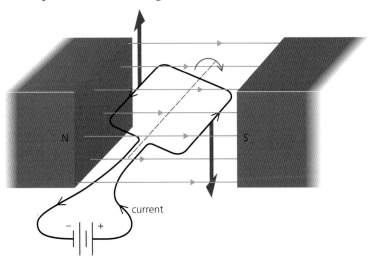

▲ Moving-coil meter

Q

1 There is a force on the wire in the diagram on the right.
   a Give *two* ways in which the force could be increased.
   b Use Fleming's left-hand rule to work out the direction of the force.
   c Give *two* ways in which the direction of the force could be reversed.
2 Explain why the cone of a loudspeaker vibrates when alternating current passes through its coil.
3 The diagram above shows a current-carrying coil in a magnetic field. What difference would it make if
   a there were more turns of wire in the coil
   b the direction of the current were reversed?

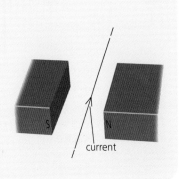

Related topics: sound waves **6.3**; current in a circuit **8.4**; magnetic fields **9.2**; field around a wire **9.3**; force on particle beam **10.2**

9.6 Electric motors

Objectives: – to know that if a current-carrying coil is in a magnetic field, the forces on it have a turning effect, and how this is used in an electric motor.

Turning effect on a coil

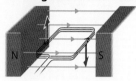

When a current flows in this coil, there is an upward force on one side and a downward force on the other. The direction of each force is given by Fleming's left-hand rule, explained on the previous spread.

If a coil is carrying a current in a magnetic field, as on the left, the forces on it produce a turning effect. Many electric motors use this principle.

A simple d.c. motor

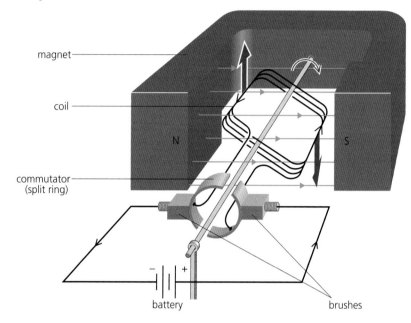

The diagram above shows a simple electric motor. It runs on direct current (d.c.), the 'one-way' current that flows from a battery.

E The coil is made of insulated copper wire. It is free to rotate between the poles of the magnet. The **commutator**, or split-ring, is fixed to the coil and rotates with it. Its action is explained below and in the diagrams on the left. The **brushes** are two contacts which rub against the commutator and keep the coil connected to the battery. They are usually made of carbon.

When the coil is horizontal, the forces are furthest apart and have their maximum turning effect (leverage) on the coil. With no change to the forces, the coil would eventually come to rest in the vertical position. However, as the coil overshoots the vertical, the commutator changes the direction of the current in it. So the forces change direction and push the coil further round until it is again vertical... and so on. In this way, the coil keeps rotating clockwise, half a turn at a time. If either the battery or the poles of the magnet were the other way round, the coil would rotate anticlockwise.

The action of the commutator

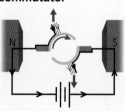

When the coil is nearly vertical, the forces cannot turn it much further...

...but when the coil overshoots the vertical, the commutator changes the direction of the current in it, so the forces change direction and keep the coil turning.

The turning effect on the coil can be increased by:
- increasing the current
- using a stronger magnet
- increasing the number of turns on the coil
- * increasing the area of the coil. (A longer coil means higher forces because there is a greater length of wire in the magnetic field; a wider coil gives the forces more leverage.)

Practical motors*

The simple motor on the opposite page produces a low turning effect and is jerky in action, especially at low speeds. Practical motors give a much better performance for these reasons:

- Several coils are used, each set at a different angle and each with its own pair of commutator segments (pieces), as shown on the right. The result is a greater turning effect and smoother running.
- The coils contain hundreds of turns of wire and are wound on a core called an **armature**, which contains iron. The armature becomes magnetized and increases the strength of the magnetic field.
- The pole pieces are curved to create a radial ('spoke-like') magnetic field. This keeps the turning effect at a maximum for most of the coil's rotation.

In some motors, the field is provided by an electromagnet rather than a permanent magnet. One advantage is that the motor can be run from an alternating current (a.c.) supply. As the current flows backwards and forwards in the coil, the field from the electromagnet changes direction to match it, so the turning effect is always the same way and the motor rotates normally. The mains motors in drills and food mixers work like this.

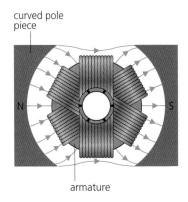

▲ Practical motors have curved pole pieces, and several coils wound on an iron armature.

Brushless motors

Many electric motors don't have brushes. For example, those used in electric and hybrid cars work in a different way. An electronic unit feeds current to a set of fixed coils in such a way that a rotating magnetic field is created. This pulls on a set of magnets so that they spin round.

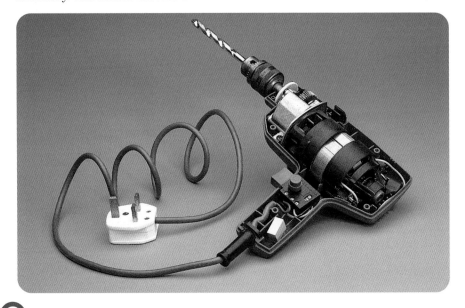

◀ In this electric drill, the motor is in the centre. Note the commutator segments at the right hand end, and the electromagnet.

1 Which part(s) of an electric motor
  a connect the power supply to the split-ring and coil
  b changes the current direction every half-turn?

2 On the right, there is an end view of the coil in a simple electric motor.
  a Redraw the diagram to show the position of the coil when the turning effect on it is **i** maximum **ii** zero.
  b Give *three* ways in which the maximum turning effect on the coil could be increased.
  c Use Fleming's left-hand rule to work out which way the coil will turn.

3 What is the advantage of using an electromagnet in an electric motor, rather than a permanent magnet?

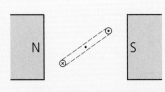

⊗ = current into paper
⊙ = current out of paper

Related topics: current **8.4**; a.c. and d.c. **8.13**; magnetic fields **9.2**; electromagnets **9.4**; Fleming's left-hand rule and turning effect **9.5**

9.7 Electromagnetic induction

Objectives: – to know that a voltage is induced in a wire if it is moved through a magnetic field – to know the factors affect the size of the voltage.

A current produces a magnetic field. However, the reverse is also possible: a magnetic field can be used to produce a current.

Induced e.m.f. and current in a moving wire

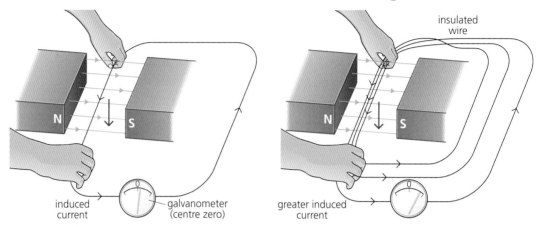

When a wire is moved across a magnetic field, as shown above left, a small e.m.f. (voltage) is generated in the wire. The effect is called **electromagnetic induction**. Scientifically speaking, an e.m.f. is **induced** in the wire. If the wire forms part of a complete circuit, the e.m.f. makes a current flow. This can be detected by a meter called a **galvanometer**, which is sensitive to very small currents. The one shown in the diagram is a centre-zero type. Its pointer moves to the left or right of the zero, depending on the current direction.

The induced e.m.f. (and current) can be increased by:
- moving the wire faster
- using a stronger magnet
- increasing the length of wire in the magnetic field – for example, by looping the wire through the field several times, as shown above right.

The above results are summed up by **Faraday's law of electromagnetic induction**. In simplified form, this can be stated as follows:

> The e.m.f. induced in a conductor is proportional to the rate at which magnetic field lines are cut by the conductor.

In applying this law, remember that field lines are used to represent the strength of a magnetic field as well as its direction. The closer together the lines, the stronger the field.

Either of the following will reverse the direction of the induced e.m.f. and current:
- moving the wire in the opposite direction
- turning the magnet round so that the field direction is reversed.

If the wire is not moving, or is moving parallel to the field lines, there is no induced e.m.f. or current.

Circuit essentials
For a current to flow in a circuit, the circuit must be complete, with no breaks in it. Also, there must a source of e.m.f. (voltage) to provide the energy. A battery is one such source. Others include a wire moving through a magnetic field, as explained on the right.

E.m.f. stands for electromotive force. It is measured in volts.

Magnet essentials
The N and S poles of one magnet exert forces on those of another:
like poles repel, unlike poles attract.

The magnetic field around a magnet can be represented by field lines. These show the direction in which the force on an N pole would act.

Induced e.m.f. and current in a coil

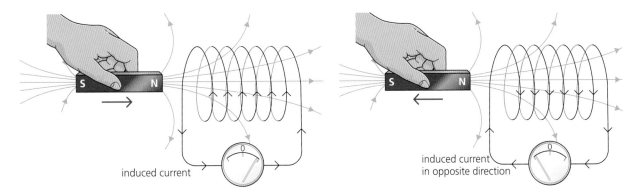

induced current

induced current in opposite direction

If a bar magnet is pushed into a coil, as shown above left, an e.m.f. is induced in the coil. In this case, it is the magnetic field that is moving rather than the wire, but the result is the same: field lines are being cut. As the coil is part of a complete circuit, the induced e.m.f. makes a current flow.

The induced e.m.f. (and current) can be increased by:
- moving the magnet faster
- using a stronger magnet
- increasing the number of turns on the coil (as this increases the length of wire cutting through the magnetic field).

E Experiments with the magnet and coil also give the following results.
- If the magnet is pulled *out of* the coil, as shown above right, the direction of the induced e.m.f. (and current) is reversed.
- If the S pole of the magnet, rather than the N pole, is pushed into the coil, this also reverses the current direction.
- If the magnet is held still, no field lines are cut, so there is no induced e.m.f. or current.

The playback heads in video recorders and hard drives contain tiny coils. A tiny, varying e.m.f. is induced in the coil as the magnetized tape passes over it and field lines are cut by the coil. In this way, the magnetized patterns on a disc are changed into electrical signals which can be used to recreate the original pictures, sounds, or data. For more about magnetic recording, see spreads 9.3 and 9.4.

▲ The pick-ups under the strings of this guitar are tiny coils with magnets inside them. The steel strings become magnetized. When they vibrate, current is induced in the coils, boosted by an amplifier, and used to produce sound.

1. The wire on the right forms part of a circuit. When the wire is moved downwards, a current is induced in it. What would be the effect of
   a moving the wire upwards through the magnetic field
   b holding the wire still in the magnetic field
   c moving the wire parallel to the magnetic field lines?
2. In the experiment at the top of the page, what would be the effect of
   a moving the magnet faster
   b having more turns on the coil?
   c turning the magnet round, so that the S pole is pushed into the coil

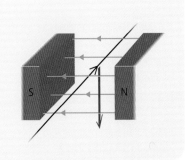

Related topics: current **8.4**; e.m.f. **8.5**; magnetic fields **9.2**; magnetic recording **9.3** and **9.4**; direction of induced current (Lenz's law) **9.8**

9.8 More about induced currents

Objectives: – to know about the factors affecting the direction of an induced current – to know the rules for working out the direction of this current.

Magnetic essentials

Like magnetic poles repel; unlike ones attract. Magnetic field lines run from the N pole of a magnet to the S pole.

In diagrams, the conventional current direction is used. This runs from the + of the supply to the −.

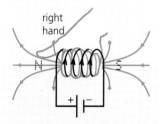

A current-carrying coil produces a magnetic field. The **right-hand grip rule** above tells you which end is the N pole. It is the end your thumb points at when your fingers point the same way as the current.

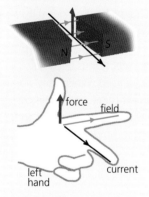

If a current-carrying wire is in a magnetic field as above, the direction of the force is given by **Fleming's left-hand rule**.

If a conductor is moving through a magnetic field, or in a changing field, an e.m.f. (voltage) is induced in it.

Induced current direction: Lenz's law

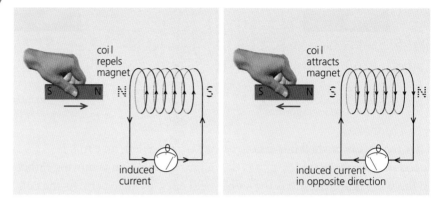

If a magnet is moved in or out of a coil, a current is induced in the coil. The direction of this current can be predicted using **Lenz's law**:

> An induced current always flows in a direction such that it opposes the change which produced it.

Above, for example, the induced current turns the coil into a weak electromagnet whose N pole *opposes* the approaching N pole of the magnet. When the magnet is pulled *out of* the coil, the induced current alters direction and the poles of the coil are reversed. This time, the coil attracts the magnet as it is pulled away. So, once again, the change is opposed.

Lenz's law is an example of the law of conservation of energy. Energy is spent when a current flows round a circuit, so energy must be spent to induce the current in the first place. In the example above, you have to spend energy to move the magnet against the opposing force.

Induced current direction: Fleming's right-hand rule

If a straight wire (in a complete circuit) is moving at right angles to a magnetic field, the direction of the induced current can be found using **Fleming's right-hand rule**, as shown below:

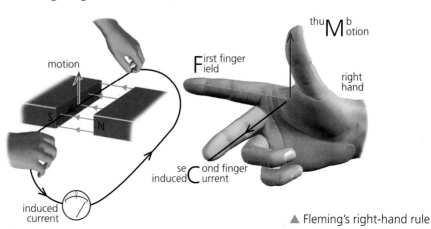

▲ Fleming's right-hand rule

MAGNETS AND CURRENTS

E) On the opposite page, there is information about Fleming's right-hand and left-hand rules. The two rules apply to different situations:
- when a *current* causes *motion*, the *left*-hand rule applies
- when *motion* causes a *current*, the *right*-hand rule applies.

Fleming's right-hand rule follows from the left-hand rule and Lenz's law. The diagram on the right illustrates this. Here, the upward motion induces a current in the wire. The induced current is in the magnetic field, so there is a force on it whose direction is given by the *left*-hand rule. The force must be downwards to *oppose* the motion, so you can use this fact and the left-hand rule to work out which way the current must flow. However, the *right*-hand rule gives the same result – without you having to reason out all the steps!

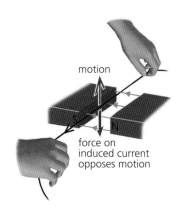

Eddy currents*

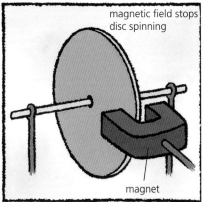

If the aluminium disc above is set spinning, it may be many seconds before frictional force finally brings it to rest. However, if it spinning between the poles of a magnet, it stops almost immediately. This is because the disc is a good conductor and currents are induced in it as it moves through the magnetic field. These are called **eddy currents**. They produce a magnetic field which, by Lenz's law, opposes the motion of the disc. Eddy currents occur wherever pieces of metal are in a changing magnetic field – for example, in the core of a transformer.

Metal detectors rely on eddy currents. Typically, a pulse of current through a flat coil produces a changing magnetic field. This induces eddy currents in any metal object underneath. The eddy currents give off their own changing field which induces a second pulse in the coil. This is detected electronically.

▲ A metal detector creates eddy currents in metal objects and then detects the magnetic fields produced.

Q

1. Look at the diagrams on the opposite page, illustrating Fleming's right-hand rule. If the directions of the magnetic field and the motion were both reversed, how would this affect the direction of the induced current?
2. On the right, a magnet is being moved towards a coil.
   a As current is induced in the coil, what type of pole is formed at the left end of the coil? Give a reason for your answer.
   b* In which direction does the (conventional) current flow in the meter, AB or BA?
3* Aluminium is non-magnetic. Yet a freely spinning aluminium disc quickly stops moving if a magnet is brought close to it. Explain why.

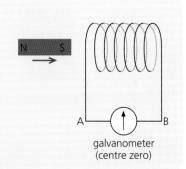

Related topics: law of conservation of energy **4.2**; right-hand grip rule **9.3**; Fleming's left-hand rule **9.5**; induced current **9.7**

9.9 Generators

Objectives: – to describe how an a.c. generator works – to give the factors affecting its output, and how the output can be shown as a graph.

Most of our electricity comes from huge **generators** in power stations. There are smaller generators in cars and on some bicycles. These generators, or dynamos, all use electromagnetic induction. When turned, they induce an e.m.f. (voltage) which can make a current flow. Most generators give out alternating current (a.c.). A.c. generators are also called **alternators**.

Ⓔ A simple a.c. generator

The diagram below shows a simple a.c. generator. It is providing the current for a small lamp. The coil is made of insulated copper wire and is rotated by turning the shaft. The **slip rings** are fixed to the coil and rotate with it. The **brushes** are two contacts which rub against the slip rings and keep the coil connected to the outside part of the circuit. They are usually made of carbon.

When the coil is rotated, it cuts magnetic field lines, so an e.m.f. is generated. This makes a current flow. As the coil rotates, each side travels upwards, downwards, upwards, downwards... and so on, through the magnetic field. So the current flows backwards, forwards... and so on. In other words, it is a.c. The graph shows how the current varies through one cycle (rotation). It is a maximum when the coil is horizontal and cutting field lines at the fastest rate.
It is zero when the coil is vertical and cutting no field lines.

The following all increase the maximum e.m.f. (and the current):
- increasing the number of turns on the coil
- increasing the area of the coil
- using a stronger magnet
- rotating the coil faster.

Faster rotation also increases the frequency of the a.c. Mains generators must keep a steady frequency – for example, 50 Hz (cycles per second) in the UK.

Electromagnetic induction

If a conductor is moved through a magnetic field so that it cuts field lines, an e.m.f. (voltage) is induced in it. In a complete circuit, the induced e.m.f. makes a current flow.

Alternating current

Alternating current (a.c.) flows alternately backwards and forwards. Mains current is a.c.

With a.c. circuits, giving voltage and current values is complicated by the fact that these vary all the time, as the graph on this page shows. To overcome the problem, a type of average called a **root mean square** (**RMS**) value is used. For example, Europe's mains voltage, 230 V, is an RMS value. It is equivalent to the steady voltage which would deliver energy at the same rate.

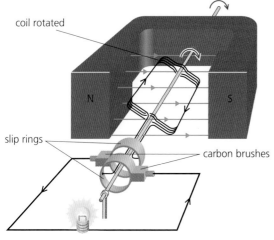

▲ Simple a.c. generator, connected to a lamp

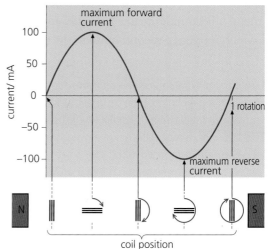

▲ Graph showing the generator's a.c. output

MAGNETS AND CURRENTS

Practical generators*

▲ Alternator from a car

◀ One of the alternators (a.c. generators) in a large power station. It is turned by a turbine, blown round by the force of high-pressure steam.
It generates an e.m.f. of over 20 000 volts, although consumers get their supply at a much lower voltage than this.

Unlike the simple generator on the opposite page, most a.c. generators have a fixed set of coils arranged around a rotating electromagnet. The various coils are made from many hundreds of turns of wire. To create the strongest possible magnetic field, they are wound on specially shaped cores containing iron. Slip rings and brushes are still used, but only to carry current to the spinning electromagnet. As the other coils are fixed, the current delivered by the generator does not have to flow through sliding contacts. (Sliding contacts can overheat if the current is very high.)

Direct current (d.c.) is 'one-way' current like that from a battery. D.c. generators are similar in construction to d.c. motors, with a fixed magnet, rotating coil, brushes, and a commutator to reverse the connections to the outside circuit every half-turn. When the coil is rotated, alternating current is generated. However, the action of the commutator means that the current in the outside circuit always flows the same way – in other words, it is d.c.

Cars need d.c. for recharging the battery and running other circuits. In petrol and diesel cars, the engine turns a generator. However, an alternator is used, rather than a d.c. generator, because it can deliver more current. A device called a **rectifier** changes its a.c. output to d.c.

Moving-coil microphone

Like generators, some microphones use the principle of electromagnetic induction.

In a moving-coil microphone, incoming sound waves strike a thin metal plate called a diaphragm and make it vibrate. The vibrating diaphragm moves a tiny coil backwards and forwards in a magnetic field. As a result, a small alternating current is induced in the coil. When amplified (made larger), the current can be used to drive a loudspeaker.

Q

1. The diagram on the right shows the end view of the coil in a simple generator. The coil is being rotated. It is connected through brushes and slip rings to an outside circuit.
   a What type of current is generated in the coil, a.c. or d.c.? Explain why it is this type of current being generated.
   b Give *three* ways in which the current could be increased.
   c The current varies as the coil rotates. What is the position of the coil when the current is a maximum? Why is the current a maximum in this position?
   d What is the position of the coil when the current is zero? Why is the current zero in this position?
2* Give *three* differences between the simple a.c. generator on the opposite page and most practical a.c. generators.

Related topics: e.m.f. **8.5**; rectifiers **8.11**; mains a.c. **8.13**; electromagnets **9.4**; d.c. motors **9.6**; electromagnetic induction **9.7**

9.10 Coils and transformers (1)

Objectives: – to know how a changing current in one coil can induce a voltage in another – to know what a transformer does, and the equation for its output voltage.

A *moving* magnetic field can induce an e.m.f. (voltage) in a conductor, as on the left. A *changing* magnetic field can have the same effect.

Mutual induction

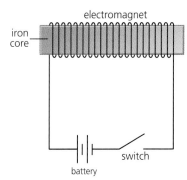
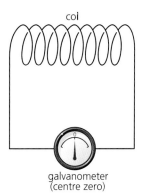

Electromagnetic induction

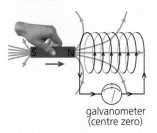

If a magnet is pushed in or out of a coil, the coil cuts through magnetic field lines, so an e.m.f. (voltage) is induced in it. This is an example of electromagnetic induction. If the coil is in a complete circuit, the induced voltage makes a current flow.

(E) As the electromagnet above is switched on, an e.m.f. is induced in the other coil, but only for a fraction of a second. The effect is equivalent to pushing a magnet towards the coil very fast. With a steady current in the electromagnet, no e.m.f. is induced because the magnetic field is not changing. As the electromagnet is switched off, an e.m.f. is induced in the opposite direction. The effect is equivalent to pulling a magnet away from the coil very fast.

The induced e.m.f. at switch-on or switch-off is increased if:
- the core of the electromagnet goes right through the second coil
- the number of turns on the second coil is increased.

When coils are magnetically linked, as above, so that a changing current in one causes an induced e.m.f. in the other, this is called **mutual induction**.

▲ Using mutual induction, 40 000 volts (or more) for spark plugs is produced from a 12 volt supply. The high voltage is induced in a coil by switching an electromagnet on and off electronically.

▲ In an induction hob, each 'plate' contains a coil that gives off a strong, alternating magnetic field. This generates a high current in the metal base of the saucepan, which heats up as a result.

MAGNETS AND CURRENTS

A simple transformer

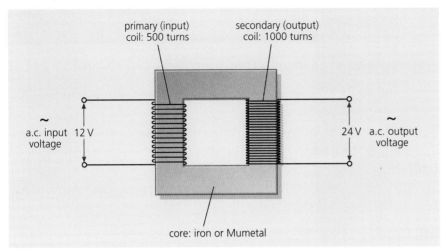

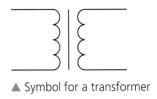

▲ Symbol for a transformer

A.c. voltages can be increased or decreased using a **transformer**. A simple transformer is shown in the diagram above. It works by mutual induction.

When alternating current flows in the **primary** (input) coil, it sets up an alternating magnetic field in the core and, therefore, in the **secondary** (output) coil. This changing field induces an alternating voltage in the output coil. Provided all the field lines pass through both coils, and the coils waste no energy because of heating effects, the following equation applies:

$$\frac{\text{output voltage}}{\text{input voltage}} = \frac{\text{turns on output coil}}{\text{turns on input coil}}$$

In symbols: $\dfrac{V_s}{V_p} = \dfrac{N_s}{N_p}$ where s means secondary and p means primary

For the transformer above, $N_s/N_p = 1000/500 = 2$. The transformer has a **turns ratio** of 2. The same ratio links the voltages: $V_s/V_p = 24/12 = 2$. Put in words, the output coil has twice the number of turns of the input coil, so the output voltage is twice the input voltage.

A transformer does not give you something for nothing. If it increases voltage, it reduces current. This is explained in the next spread.

D.c. and a.c.
Direct current (d.c.) flows one way only.

Alternating current (a.c.) flows alternately backwards and forwards.

P.d. e.m.f. and voltage
P.d. (potential difference) is the scientific name for voltage. The p.d. produced within a battery or other source is called the e.m.f. (electromotive force).

For convenience, engineers often use the word voltage rather than p.d. or e.m.f. especially when dealing with a.c.

Voltages in a.c. circuits are commonly called a.c. voltages, although, strictly speaking, an 'alternating current voltage' doesn't make much sense!

1 In the experiment on the right, what happens when
  a the switch is closed (turned ON)
  b the switch is left in the closed (ON) position
  c the switch is then opened (turned OFF)?

2 In the experiment on the right, what would be the effect of
  a extending the iron core so that it goes through both coils
  b replacing the battery and switch by an a.c. supply?

3 A transformer has a turns ratio (N_s/N_p) of 1/4. Its input coil is connected to a 12 volt a.c. supply. Assuming there are no energy or field line losses:
  a What is the output voltage?
  b What turns ratio would be required for an output voltage of 36 volts?

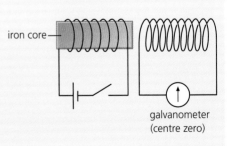

Related topics: p.d. and e.m.f. **8.5**; magnetic field lines **9.2**; electromagnets **9.4**; electromagnetic induction **9.7**; d.c. and a.c. **9.9**

9.11 Coils and transformers (2)

Objectives: – to know the difference between a step-up and step-down transformer – to know the equation for the power transferred by a transformer.

Step-up and step-down transformers

Depending on its turns ratio, a transformer can increase or decrease an a.c. voltage.

Step-up transformers have more turns on the output coil than on the input coil, so their output voltage is *more* than the input voltage. The transformer in the diagram below is a step-up transformer. Large step-up transformers are used in power stations to increase the voltage to the levels needed for overhead power lines. The next spread explains why.

Step-down transformers have fewer turns on the output coil than on the input coil, so the output voltage is *less* than the input voltage. In battery chargers, computers, and other electronic equipment, they reduce the voltage of the a.c. mains to the much lower levels needed for other circuits.

Both types of transformer work on a.c., but not on d.c. Unless there is a *changing* current in the input coil, no voltage is induced in the output coil. Connecting a transformer to a d.c. supply can damage it. A high current flows in the input coil, which can make it overheat.

▲ A transformer connected to local power lines

Power essentials

Energy is measured in joules (J).

Power is measured in watts (W).

An appliance with a power output of 1000 W delivers energy at the rate of 1000 joules per second.

In circuits, power can be calculated using this equation:

power = voltage × current
(watts)   (volts)   (amperes)
(W)   (V)   (A)

(E) Power through a transformer

If no energy is wasted in a transformer, the power (energy per second) delivered by the output coil will be the same as the power supplied to the input coil. So:

input voltage × input current = output voltage × output current

In symbols:
$$V_p I_p = V_s I_s$$

As voltage × current is the same on both sides of a transformer, it follows that a transformer which *increases* the voltage will *reduce* the current in the same proportion, and vice versa. The figures in the diagram below illustrate this.

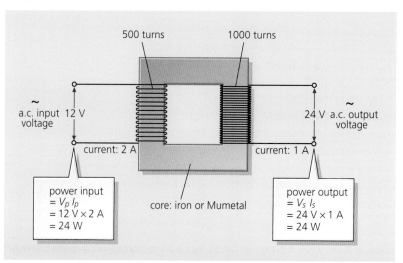

MAGNETS AND CURRENTS

Practical transformers*

The diagram on the right shows two ways of arranging the coils and core in a practical transformer. Both methods are designed to trap the magnetic field in the core so that all the field lines from one coil pass through the other.

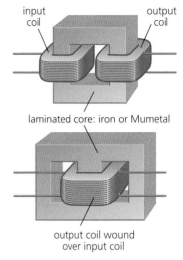

All transformers waste some energy because of heating effects in the core and coils. Here are two of the causes:
- The coils are not perfect electrical conductors and heat up because of their resistance. To keep the resistance low, thick copper wire is used where possible.
- The core is itself a conductor, so the changing magnetic field induces currents in it. These circulating **eddy currents** have a heating effect. To reduce them, the core is laminated (layered): it is made from thin, insulated sheets of iron or Mumetal, rather than a solid block.

Large, well-designed transformers can have efficiencies as high as 99%. In other words, their useful power output is 99% of their power input.

▲ Practical transformers

Solving problems

Example Assuming that the transformer on the right has an efficiency of 100%, calculate **a** the supply voltage **b** the current in the input coil.

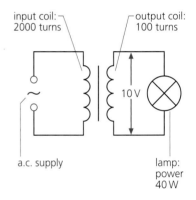

a This is solved using the transformer equation: $\dfrac{V_s}{V_p} = \dfrac{N_s}{N_p}$

where V_1 is the supply voltage to be calculated.

Substituting values: $\dfrac{10\ \text{V}}{\text{supply voltage}} = \dfrac{100}{2000}$

Rearranged, this gives: supply voltage = 200 V

b This is solved using the power equation: $V_p I_p = V_s I_s$

where $V_s I_s$ is already known to be 40 W.

Substituting values: 200 V × input current = 40 W

Rearranged, this gives: input current = 0.2 A

1. How does a step-up transformer differ from a step-down transformer?
2. Explain each of the following:
   a. a transformer will not work on d.c.
   b.* the core of a transformer needs to be laminated
   c. if a transformer increases voltage, it reduces current.
3. In the circuit on the right, a transformer connected to the 230 V a.c. mains is providing power for a low-voltage heater. Using the information in the diagram, and assuming that the efficiency is 100%, calculate
   a. the voltage across the heater
   b. the power supplied by the mains
   c. the power delivered to the heater
   d. the current in the heater.

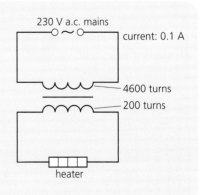

Related topics: resistance 8.6; power calculations 8.11; eddy currents 9.8; d.c. and a.c. 9.9; power transmission 9.12

219

9.12 Power across the country

Objectives: – to describe the main features of a cross-country power transmission system – to explain, using calculations, why very high voltages are used.

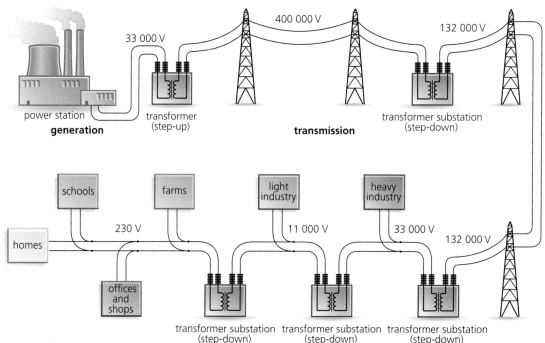

▲ A typical mains supply system. Actual voltages may differ, depending on the country.

Power essentials

An appliance with a power output of 1000 watts (W) delivers energy at the rate of 1000 joules per second.

In circuits

power = voltage × current
(watts)   (volts)   (amperes)
(W)   (V)   (A)

Transformer essentials

Transformers are used to increase or decrease a.c. voltages. If a transformer is 100% efficient, its power output and input are equal. So if it increases voltage, it reduces current in the same proportion so that 'voltage × current' stays the same.

Power for the a.c. mains is *generated* in power stations, *transmitted* (sent) through long-distance cables, and then *distributed* to consumers.

Typically, a large power station might contain four generators, each producing a current of 20 000 amperes at a voltage of 33 000 volts. The current from each generator is fed to a huge step-up transformer which transfers power to overhead cables at a greatly increased voltage (275 000 V or 400 000 V in the UK). The reason for doing this is explained on the next page. The cables feed power to a nationwide supply network called a **grid**. Using the grid, power stations in areas where the demand is low can be used to supply areas where the demand is high. Also power stations can be sited away from heavily populated areas.

Power from the grid is distributed by a series of **substations**. These contain step-down transformers which reduce the voltage in stages to the level needed by consumers. Depending on the country, this might be between 110 V and 230 V for home consumers, although industry normally uses a higher voltage.

Transmission issues

A.c. or d.c.? Alternating current (a.c.) is used for the mains. On a large scale, it can be generated more efficiently than 'one-way' direct current (d.c.). However, the main advantage of a.c. is that voltages can be stepped up or down using transformers. Transformers will not work with d.c.

MAGNETS AND CURRENTS

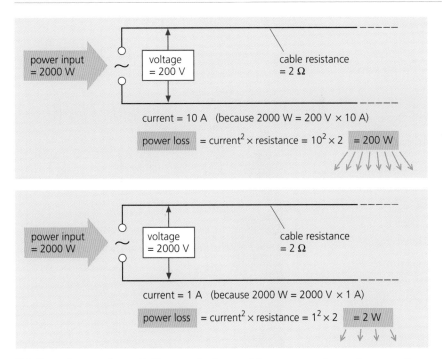

Calculating power loss
When current flows in a cable, the resistance causes a drop in voltage along the cable and a loss of power.

power loss
= voltage drop × current

But: voltage drop
= current × resistance

So: power loss
= current × resistance × current
= current^2 × resistance

In symbols: $P = I^2 R$

◀ These calculations show the power losses in a cable when the same amount of power is sent at two different voltages (for simplicity, some units have been omitted).

High or low voltage? Transmission cables are good conductors, but they still have significant resistance – especially when they are hundreds of kilometres long. This means that energy is wasted because of the heating effect of the current. The calculations above demonstrate why less power is lost from a cable if power is transmitted through it at high voltage. By using a transformer to increase the voltage, the current is reduced, so thinner, lighter, and cheaper cables can be used.

Overhead or underground?* There are two ways of running high-voltage transmission cables across country. They can be suspended overhead from tall towers called pylons, or they can be put underground.

In countries where power has to be transmitted very long distances, overhead cables are more common because they are cheaper. They are easier to insulate because, over most of their length, the air acts as an insulator. Also, costly digging operations are avoided. However, pylons and overhead cables spoil the environment. They are often not allowed in densely populated areas or in areas of outstanding natural beauty. So underground cables (called land lines) are used instead.

▲ Pylons and overhead cables are not usually permitted in areas like this.

Q

1 In a mains supply system, how are voltage changes made?

2 Explain each of the following.
  a A.c. rather than d.c. is used for transmitting mains power.
  b The voltage is stepped up before power from a generator is fed to overhead transmission cables.

3* Give an example of where underground transmission cables might be used instead of overhead ones, despite the extra cost.

4 The second paragraph on the opposite page describes the output of the four generators in a typical, large power station. Calculate the power station's total power output in MW. (1 MW = 1 000 000 W)

5 The diagram at the top of this page compares power losses from a cable at two different voltages. Calculate the power loss if the same power is sent at 20 000 V.

6 4 kW of power is fed to a transmission cable of resistance 5 Ω. Calculate the power loss in the cable if the power is transmitted at **a** 200 V **b** 200 000 V.

Related topics: power stations **4.5–4.6**; resistance **8.6–8.7**; mains electricity **8.13**; generators **9.9**; transformers **9.10–9.11**

Check-up on magnets and currents

Further questions

1 An electromagnet is made by winding wire around an iron core.

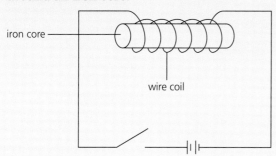

The diagram shows an electromagnet connected to a circuit.

a State **two** ways of making the strength of the electromagnet weaker. [2]

b Explain why the core is made of iron instead of steel. [1]

2 **A**, **B**, **C** and **D** are small blocks of different materials. The table below shows what happens when two of the blocks are placed near one another.

Arrangement of blocks		Effect
A	B	attraction
B	C	attraction
A	C	no effect
B	D	no effect

| a magnet | a magnetic material | a non-magnetic material |

Use one of the phrases in the above boxes to describe the magnetic property of each block. Each phrase may be used once, more than once or not at all.

a Block **A** is _____
b Block **B** is _____
c Block **C** is _____
d Block **D** is _____ [4]

3 The figure below shows a circuit, which includes an electrical relay, used to switch on a motor M.

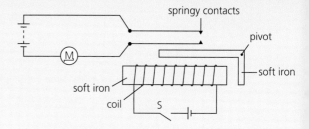

Explain, in detail, how closing switch S causes the motor M to start. [4]

4 The diagram shows a long wire placed between the poles of a magnet. When current *I* flows in the wire, a force acts on the wire causing it to move.

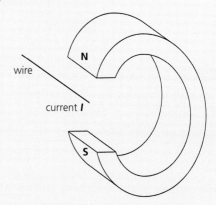

a Use Fleming's left-hand rule to find the direction of the force on the wire. Copy the diagram and show the direction of the force on your copy with an arrow labelled **F**. [1]

b State what happens to the force on the wire when
  i the size of the current in the wire is increased, [1]
  ii a weaker magnet is used, [1]
  iii the direction of the current is reversed. [1]

c Name **one** practical device which uses this effect. [1]

5 The diagram below shows a permanent magnet being moved towards a coil whose ends are connected to a sensitive ammeter. As the magnet approaches, the ammeter needle gives a **small** deflection to the **left**.

MAGNETS AND CURRENTS

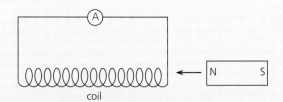

coil

a State what you would expect the ammeter to show if, in turn,

(E) **i** the magnet was pulled away from the coil
ii the magnet was reversed so that the S pole was moved towards the coil
iii the magnet was now pulled away from the coil, at a much higher speed. [4]

b Give the name of the process which is illustrated by these experiments. [1]

6 a Many power stations have to burn fuel in order to generate electricity. Explain how the energy transferred from the burning fuel is eventually carried away by an electric current. [4]

b Power stations use transformers to increase the voltage to very high values before transmitting it to all parts of the country. Explain why electricity is transmitted at very high voltages. [1]

c A power station produces electricity at 25 000 V which is increased by a transformer to 400 000 V. The transformer has 2000 turns on its primary coil. Use the formula

$$\frac{\text{voltage across primary coil}}{\text{number of turns on primary coil}} = \frac{\text{voltage across secondary coil}}{\text{number of turns on secondary coil}}$$

to calculate the number of turns on its secondary coil. [2]

7 The diagram shows a simple transformer.

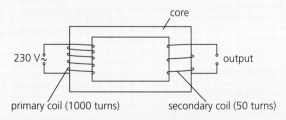

The transformer is a **step-down** transformer.
a What is a **step-down** transformer? [1]
b How can you tell from the diagram that this is a **step-down** transformer? [1]
c Calculate the output voltage of this transformer. [3]

d Explain why transformers are used when power needs to be transmitted over long distances. [3]
e What is the core of a transformer usually made of? [2]

8 The diagram shows the main parts of one type of ammeter. There are two short iron bars inside a coil of insulated wire. One bar is fixed and cannot move and the other is on the end of a pivoted pointer. The diagram shows the ammeter in use and measuring a current of 1.5 amperes (A).

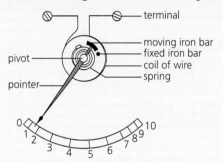

(E) **a** How much electrical charge will pass through this ammeter in one minute? Include in your answer the equation you are going to use. Show clearly how you get to your final answer and give the unit. [3]
b i Apart from a heating effect, what will be produced by the coil of wire when the electricity passes through it? [1]
ii What effect will this have on the two iron bars? What causes the effect? Draw one or more diagrams if this will help you to explain. [4]

*To answer part **a**, you will need information from Chapter 8.*

(E) **9 a** When a coil rotates in a magnetic field, an alternating voltage is produced. Explain how the voltage is produced. [2]

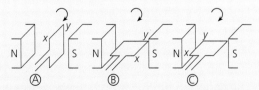

b The diagrams A B and C show three positions of a coil as it rotates clockwise in a magnetic field produced by two poles. The graph below shows how the voltage produced changes as the coil rotates.

223

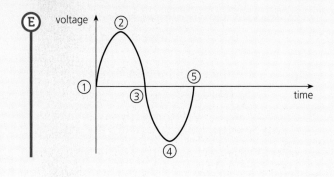

E When the coil is in the position shown by diagram A, the output voltage is zero and is marked as 1 on the voltage–time graph. State which point on the voltage–time graph corresponds to the coil position shown by
  i diagram B, [1]
  ii diagram C. [1]
 c State **one** way of increasing the size of the voltage produced by **this** coil rotating in a magnetic field. [1]

Use the list below when you revise for your IGCSE examination. The spread number, in brackets, tells you where to find more information.

Revision checklist

Core Level
☐ The two types of magnetic pole and the attractions and repulsions between them. (9.1)
☐ The properties of magnets. (9.1)
☐ Induced magnetism. (9.1)
☐ Magnetic and non-magnetic materials. (9.1)
☐ Hard and soft magnetic materials; the different magnetic properties of steel and iron. (9.1)
☐ Plotting the field around a magnet. (9.2)
☐ The field around a bar magnet, and the direction of the field lines. (9.2)
☐ The magnetic fields around a current-carrying straight wire and a solenoid (long coil). (9.3)
☐ Electromagnets and their uses. (9.4)
☐ How a magnetic relay works. (9.4)
☐ The force on a current-carrying conductor in a magnetic field; the effects of reversing the current and field directions. (9.5)
☐ How a moving-coil loudspeaker works. (9.5)
☐ The turning effect on a current-carrying coil in a magnetic field and the factors affecting it. (9.5)
☐ Electromagnetic induction: how an e.m.f. is induced in a wire or coil if it is in a changing magnetic field. (9.7)
☐ The factors affecting the size of an induced e.m.f. (9.7)
☐ The construction of a transformer. (9.10)
☐ Step-up and step-down transformers. (9.10)
☐ The equation linking a transformer's input (primary) and output (secondary) voltages. (9.10)
☐ How transformers are used in the transmission of mains power across country. (9.12)
☐ Why power is transmitted at high voltage. (9.12)

Extended Level
As for Core Level, plus the following:
☐ The variation in magnetic field strength around a current-carrying straight wire and a solenoid. (9.3)
☐ How the magnetic field from a straight wire or solenoid is affected if the current is increased or its direction changed. (9.3)
☐ How to work out the direction of the force on a current-carrying wire in a magnetic field (Fleming's left-hand rule). (9.5)
☐ How a simple d.c. motor works, and the action of the commutator. (9.6)
☐ How the direction of an induced current always opposes the change causing it (Lenz's law). (9.8)
☐ How to work out the direction of the induced current when a wire is moved through a magnetic field. (Fleming's right-hand rule). (9.8)
☐ How a simple a.c. generator works. (9.9)
☐ How the output voltage of an a.c. generator varies with time, and is related to the position of the coil. (9.9)
☐ How a transformer works. (9.10)
☐ Why a transformer uses a.c. not d.c. (9.10)
☐ The equation linking a transformer's input (primary) and output (secondary) powers. (9.11)
☐ Using the equation $P = I^2R$ to explain why power losses in transmission cables are lower when the voltage is higher. (9.12)

10
Atoms and radioactivity

- ATOMIC PARTICLES
- ISOTOPES
- IONIZATION
- ALPHA, BETA, AND GAMMA RADIATION
- RADIOACTIVE DECAY
- HALF-LIFE
- NUCLEAR FISSION
- NUCLEAR FUSION
- USING RADIOACTIVITY

The *aurora borealis* ('northern lights') in the night sky over Alaska, USA. The shimmering curtain of light is produced when atomic particles streaming from the Sun strike atoms and molecules high in the Earth's atmosphere. The Earth's magnetic field concentrates the incoming atomic particles above the north and south polar regions, so that is where aurorae are normally seen.

10.1 Inside atoms

Objectives: – to describe a simple model of the atom – to know what isotopes are – to know the terms used, and the notation, for representing different types of atom.

A simple model of the atom

> **Charge essentials**
> There are two types of electric charge: positive (+) and negative (−). Like charges repel; unlike charges attract.

Everything is made of atoms. Atoms are far too small to be seen with any ordinary microscope – there are more than a billion billion of them on the surface of this full stop. However, by shooting tiny atomic particles through atoms, scientists have been able to develop **models** (descriptions) of their structure. In advanced work, scientists use a mathematical model of the atom. However, the simple model below is often used to explain the basic ideas.

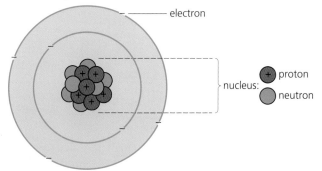

▶ A simple model of the atom. In reality, the nucleus is far too small to be shown to its correct scale. If the atom were the size of a concert hall, its nucleus would be smaller than a pill!

An atom is made up of smaller particles:

- There is a central **nucleus** made up of **protons** and **neutrons**. Around this, **electrons** orbit at high speed. The numbers of particles depends on the type of atom.
- Protons have a positive (+) charge. Electrons have an equal negative (−) charge. Normally, an atom has the same number of electrons as protons, so its total charge is zero.
- Protons and neutrons are called **nucleons**. Each is about 1800 times more massive than an electron, so virtually all of an atom's mass is in its nucleus.
- Electrons are held in orbit by the force of attraction between opposite charges. Protons and neutrons are bound tightly together in the nucleus by a different kind of force, called the **strong nuclear force**.

Elements and atomic number

element	chemical symbol	atomic number (proton number)
hydrogen	H	1
helium	He	2
lithium	Li	3
beryllium	Be	4
boron	B	5
carbon	C	6
nitrogen	N	7
oxygen	O	8
radium	Ra	88
thorium	Th	90
uranium	U	92
plutonium	Pu	94

All materials are made from about 100 basic substances called **elements**. An atom is the smallest 'piece' of an element you can have. Each element has a different number of protons in its atoms: it has a different **atomic number** (sometimes called the **proton number**). There are some examples on the left. The atomic number also tells you the number of electrons in the atom.

Isotopes and mass number

The atoms of any one element are not all exactly alike. Some may have more neutrons than others. These different versions of the element are called **isotopes**. They have identical chemical properties, although their atoms have different masses. Most elements are a mixture of two or more isotopes. You can see some examples in the chart on the opposite page.

ATOMS AND RADIOACTIVITY

The total number of protons and neutrons in the nucleus is called the **mass number** (or **nucleon number**). Isotopes have the *same* atomic number but *different* mass numbers. For example, the metal lithium (atomic number 3) is a mixture of two isotopes with mass numbers 6 and 7. Lithium-7 is the more common: over 93% of lithium atoms are of this type. On the right, you can see how to represent an atom of lithium-7 using a symbol and numbers. Each different type of atom, lithium-7 for example, is called a **nuclide**.

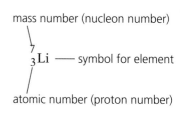

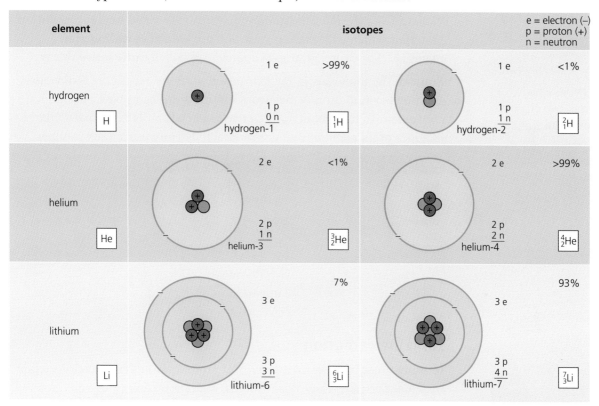

Electron shells*

Electrons orbit the nucleus at certain fixed levels only, called **shells**. There is a limit to how many electrons each shell can hold – for example, no more than 2 in the first shell and 8 in the second. It is an atom's outermost electrons which form the chemical bonds with other atoms, so elements with similar electron arrangements have similar chemical properties.

> The **periodic table** is a chart of all the elements. Elements in the same group have similar electron arrangements and similar chemical properties.

For questions 4 and 5, you will need data from the table of elements on the opposite page.

1 An atoms contains *electrons*, *protons*, and *neutrons*. Which of these particles
   a are outside the nucleus   b are uncharged
   c have a negative charge   d are nucleons
   e are much lighter than the others?

2 An aluminium atom has an atomic number of 13 and a mass number of 27. How many
   a protons   b electrons   c neutrons does it have?

3 Chlorine is a mixture of two isotopes, with mass numbers 35 and 37. What is the difference between the two types of atom?

4 In symbol form, nitrogen-14 can be written $^{14}_{7}N$. How can each of the following be written?
   a carbon-12   b oxygen-16   c radium-226

5 Atom X has 6 electrons and a mass number of 12.
   Atom Y has 6 electrons and a mass number of 14.
   Atom Z has 7 neutrons and a mass number of 14.
   Identify the elements X, Y, and Z.

Related topics: electric charge 8.1–8.2; experimental evidence for nucleus 10.9

10.2 Nuclear radiation (1)

Objectives: – to know the three main types of nuclear radiation and their properties – to describe the effects of a magnetic field on nuclear particles.

Isotope essentials

Different versions of the same element are called isotopes. Their atoms have different numbers of neutrons in the nucleus.

For example, lithium is a mixture of two isotopes: lithium-6 (with 3 protons and 3 neutrons in the nucleus) and lithium-7 (with 3 protons and 4 neutrons).

Some materials contain atoms with unstable nuclei. In time, each unstable nucleus disintegrates (breaks up). As it does so, it shoots out a tiny particle and, in some cases, a burst of wave energy as well. The particles and waves 'radiate' from the nucleus, so they are somtimes called **nuclear radiation**. Materials which emit nuclear radiation are known as **radioactive materials**. The disintegration of a nucleus is called **radioactive decay**.

Some of the materials in nuclear power stations are highly radioactive. But nuclear radiation comes from natural sources as well. Although it is convenient to talk about 'radioactive materials', it is really particular isotopes of an element that are radioactive. Here are some examples:

	isotopes	
stable nuclei	unstable nuclei, radioactive	found in
carbon-12 carbon-13	carbon-14	air, plants, animals
potassium-39	potassium-40	rocks, plants, sea water
	uranium-234	rocks
	uranium-235	
	uranium-238	

▲ If an atom loses (or gains) an electron, it becomes an ion

Ionizing radiation

Ions are charged atoms (or groups of atoms). Atoms become ions when they lose (or gain) electrons. Nuclear radiation can remove electrons from atoms in its path, so it has an **ionizing** effect. Other forms of ionizing radiation include ultraviolet and X-rays.

If a gas becomes ionized, it will conduct an electric current. In living things, ionization can damage or destroy cells (see the next spread).

Alpha, beta, and gamma radiation

Discovering radioactivity

Henri Becquerel discovered radioactivity, by accident, in 1896. When he left some uranium salts next to a wrapped photographic plate, he found that the plate had become 'fogged', and realized that some invisible, penetrating radiation must be coming from the uranium.

There are three main types of nuclear radiation: **alpha particles**, **beta particles**, and **gamma rays**. Gamma rays are the most penetrating and alpha particles the least, as shown below:

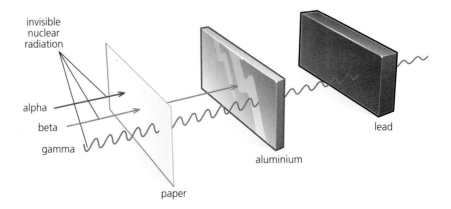

228

ATOMS AND RADIOACTIVITY

type of radiation	alpha particles (α)	beta particles (β)	gamma rays (γ)
	each particle is 2 protons + 2 neutrons (it is identical to a nucleus of helium-4)	each particle is an electron (created when the nucleus decays)	electromagnetic waves similar to X-rays
relative charge compared with charge on proton	+2	−1	0
mass	high, compared with betas	low	–
speed	up to 0.1 × speed of light	up to 0.9 × speed of light	speed of light
ionizing effect	strong	weak	very weak
penetrating effect	not very penetrating: stopped by a thick sheet of paper, or by skin, or by a few centimetres of air	penetrating, but stopped by a few millimetres of aluminium or other metal	very penetrating: never completely stopped, though lead and thick concrete will reduce intensity
effects of fields	deflected by magnetic and electric fields	deflected by magnetic and electric fields	not deflected by magnetic or electric fields

Alpha particles are more ionizing than beta particles. They have a greater charge, so exert more force on electrons. And they are slower, so spend more time close to any electrons they pass. Gamma rays are least ionizing because they are uncharged.

Alpha and beta particles are deflected by a magnetic field (see the diagram on the right). An alpha beam is a flow of positively (+) charged particles, so it is equivalent to an electric current. It is deflected in a direction given by Fleming's left-hand rule (see spread 9.5). Beta particles are much lighter and have a negative (−) charge, so they are deflected more, and in the opposite direction. Being uncharged, the gamma rays are not deflected.

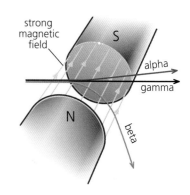

▲ How alpha, beta, and gamma rays are affected by a magnetic field

Alpha and beta particles are also affected by an electric field – in other words, there is a force on them if they pass between oppositely charged plates.

1 Name a radioactive isotope which occurs naturally in living things.
2 *alpha   beta   gamma*
Which of these three types of radiation
  a is a form of electromagnetic radiation
  b carries positive charge
  c is made up of electrons
  d travels at the speed of light
  e is the most ionizing
  f can penetrate a thick sheet of lead
  g is stopped by skin or thick paper
  h has the same properties as X-rays
  i is not deflected by an electric or magnetic field?
3 What is the difference between the atoms of an isotope that is radioactive and the atoms of an isotope that is not?
4 How is an ionized material different from one that is not ionized?

Related topics: electromagnetic waves 7.10–7.11; X-rays 7.11; Fleming's left-hand rule 9.5; isotopes 10.1

10.3 Nuclear radiation (2)

Objectives: – to understand the dangers of nuclear radiation – to know how nuclear radiation can be detected – to know what background radiation is.

Radiation dangers

Nuclear radiation can damage or destroy living cells and stop organs in the body working properly. It can also cause **mutations** (changes in the genetic instructions) in cells, which may then grow abnormally and cause cancer.

(E) The greater the intensity of the radiation and the longer the exposure time, the greater the risk.

Radioactive gas and dust are especially dangerous because they can be taken into the body with air, food, or drink. Once absorbed, they are difficult to remove, and their radiation can cause damage in cells deep in the body. Alpha radiation is the most harmful because it is the most highly ionizing.

Normally, there is much less risk from radioactive sources *outside* the body. Sources in nuclear power stations and laboratories are well shielded, and the intensity of the radiation decreases as you move away from the source. Beta and gamma rays are potentially the most harmful because they can penetrate to internal organs. Alpha particles are stopped by the skin.

▲ Where background radiation comes from (average proportions)

- radon gas from ground
- ground and buildings
- medical (including X-rays)
- food and drink
- cosmic rays from space
  - nuclear test fallout
  - nuclear power stations
  - nuclear waste
  - other

Background radiation

There is a small amount of radiation around us all the time because of radioactive materials in the environment. This is called **background radiation**. It mainly comes from natural sources such as soil, rocks, air, building materials, food and drink – and even space.

In some areas, over a half of the background radiation comes from radioactive radon gas (radon-222) seeping out of rocks – especially some types of granite. In high risk areas, houses may need extra underfloor ventilation to stop the gas collecting or, ideally, a sealed floor to stop it entering in the first place.

Geiger-Müller (GM) tube

This can be used to detect alpha, beta, and gamma radiation. Its structure is shown below. The 'window' at the end is thin enough for alpha particles to pass through. If an alpha particle enters the tube, it ionizes the gas inside. This sets off a high-voltage spark across the gas and a pulse of current in the circuit. A beta particle or burst of gamma radiation has the same effect.

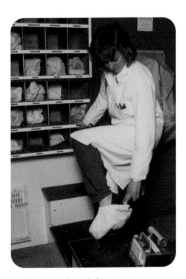

▲ This nuclear laboratory worker is about to use a GM tube and ratemeter to check for any traces of radioactive dust on her clothing

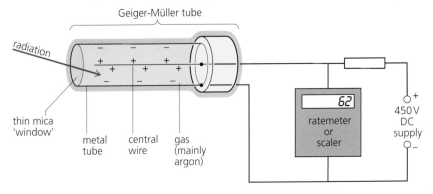

The GM tube can be connected to the following:
- **A ratemeter** This gives a reading in counts per second. For example, if 50 alpha particles were detected by the GM tube every second, the ratemeter would read 50 counts per second.
- **A scaler** This counts the *total* number of particles (or bursts of gamma radiation) detected by the tube.
- **An amplifier and loudspeaker** The loudspeaker makes a 'click' when each particle or burst of gamma radiation is detected.

When the radiation from a radioactive source is measured, the reading always *includes* any background radiation present. So an average reading for the background radiation alone must also be found and subtracted from the total.

Cloud chamber*

This is useful for studying alpha particles because it makes their tracks visible. The chamber has cold alcohol vapour in the air inside it. The alpha particles make the vapour condense, so you see a trail of tiny droplets where each particle passes through. At one time, cloud chambers were widely used in nuclear research, but they have since been replaced by other devices.

> **Safety in the laboratory**
> Experiments with weak radioactive sources are sometimes carried out in school and college laboratories. Such sources are normally sealed so that no radioactive fragments or dust can escape. For safety, a source should be
> - stored in a lead container, in a locked cabinet
> - picked up with tongs, not by hand
> - kept well away from the body, and not pointed at other people
> - left out of its container for as short a time as possible.

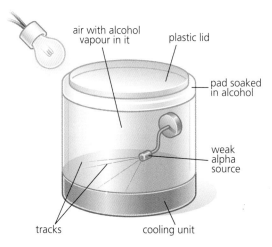
▲ Cloud chamber

▲ Tracks of alpha particles in a cloud chamber. The colours are false and have been added to the picture. The green and yellow lines are the alpha tracks. The red line is the track of a nitrogen nucleus that has been hit by an alpha particle.

1 What, on average, is the biggest single source of background radiation?
2 Radon gas seeps out of rocks underground. Why is it important to stop radon collecting in houses?
3 Which is the most dangerous type of radiation
  a from radioactive sources outside the body
  b from radioactive materials absorbed by the body?
4 In the experiment on the right:
  a What is the count rate due to background radiation?
  b What is the count rate due to the source alone?
  c If the source emits one type of radiation only, what type is it? Give a reason for your answer.

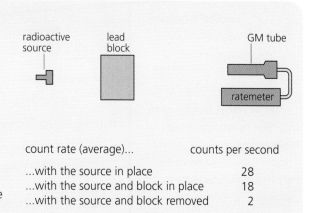

count rate (average)...	counts per second
...with the source in place	28
...with the source and block in place	18
...with the source and block removed	2

Related topics: properties of alpha, beta, and gamma radiation **10.2**

10.4 Radioactive decay (1)

Objectives: – to explain what radioactive decay is – to describe what happens to a nucleus as a result of alpha decay, beta decay, and gamma emission.

The symbol system used for representing atoms can also be used for nuclei and other particles

Nucleus example

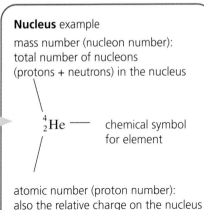

mass number (nucleon number): total number of nucleons (protons + neutrons) in the nucleus

$^{4}_{2}\text{He}$ — chemical symbol for element

atomic number (proton number): also the relative charge on the nucleus compared with +1 for a proton

Alpha particle (helium nucleus)

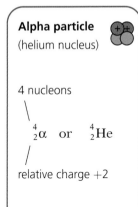

4 nucleons

$^{4}_{2}\alpha$ or $^{4}_{2}\text{He}$

relative charge +2

Beta particle (electron)

mass negligible compared with a proton or a neutron

$^{0}_{-1}\beta$ or $^{0}_{+1}\text{e}$

relative charge equal but opposite to that on a proton

Nuclear essentials

Atoms of any one element all have the same number of protons in their nucleus.

Elements exist in different versions, called isotopes. For example, lithium is a mixture of two isotopes: lithium-6 (with 3 protons and 3 neutrons in the nucleus) and lithium-7 (with 3 protons and 4 neutrons).

Any one particular type of atom, for example lithium-7, is called a **nuclide**. However the word 'isotope' is commonly used instead of nuclide.

Radioactive isotopes have unstable nuclei. In time each nucleus decays (breaks up) by emitting an alpha or beta particle and, in some cases, a burst of gamma radiation as well.

If an isotope is radioactive, it has an unstable arrangement of neutrons and protons in its nuclei. The emission of an alpha or beta particle makes the nucleus more stable, but alters the numbers of protons and neutrons in it. So it becomes the nucleus of a different element. The original nucleus is called the **parent** nucleus. The nucleus formed is the **daughter** nucleus. The daughter nucleus and any emitted particles are the **decay products**.

Ⓔ Alpha decay

Radium-226 (atomic number 88) decays by alpha emission. The loss of the alpha particle leaves the nucleus with 2 protons and 2 neutrons less than before. So the mass number drops to 222 and the atomic number to 86. Radon has an atomic number of 86, so radon is the new element formed:

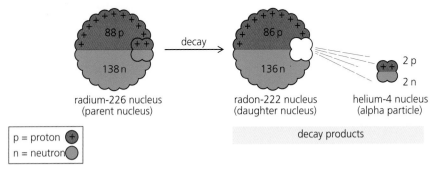

The decay process can be written as a nuclear equation:

$$^{226}_{88}\text{Ra} \rightarrow \,^{222}_{86}\text{Rn} + \,^{4}_{2}\alpha$$

During alpha decay:

- the top numbers balance on both sides of the equation (226 = 222 + 4), so the mass number is conserved (unchanged)
- the bottom numbers balance on both sides of the equation (88 = 86 + 2), so charge is conserved
- a new element is formed, with an atomic number 2 less than before. The mass number is 4 less than before.

232

ATOMS AND RADIOACTIVITY

E) Beta decay

Iodine-131 (atomic number 53) decays by beta emission. When this happens, a neutron changes into a proton, an electron, and an uncharged, almost massless relative of the electron called an antineutrino. The electron and antineutrino leave the nucleus at high speed. As a proton has replaced a neutron in the nucleus, the atomic number rises to 54. This means that a nucleus of xenon-131 has been formed:

Alternative names

atomic number ≡ proton number
mass number ≡ nucleon number

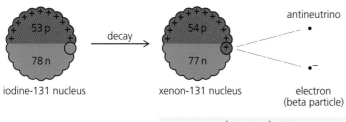

decay products

E) The decay process can be written as a nuclear equation:

$$^{131}_{53}I \rightarrow {}^{131}_{54}Xe + {}^{0}_{-1}\beta + {}^{0}_{0}\bar{\nu}$$ ($\bar{\nu}$ = antineutrino)

During this type of beta decay:

- the top numbers balance on both sides of the equation ($131 = 131 + 0 + 0$), so the mass number is conserved
- the bottom numbers balance on both sides of the equation ($53 = 54 - 1 + 0$), so charge is conserved
- a new element is formed, with an atomic number 1 more than before. The mass number is unchanged.

Beta⁻ and beta⁺

There is a less common form of beta decay, in which the emitted beta particle is a **positron**. This is the **antiparticle** of the electron, with the same mass, but opposite charge (+1). During this type of decay, a proton changes into a neutron, a positron, and a neutrino. The element formed has an atomic number one less than before.

To distinguish the two types of beta decay, they are sometimes called beta⁻ decay (electron emitted) and beta⁺ decay (positron emitted).

Gamma emission

With some isotopes, the emission of an alpha or beta particle from a nucleus leaves the protons and neutrons in an 'excited' arrangement. As the protons and neutrons rearrange to become more stable, they lose energy. This is emitted as a burst of gamma radiation.

- Gamma emission by itself causes no change in mass number or atomic number.

Q

1 The following equation represents the radioactive decay of thorium-232. A, Z, and X are unknown.

$$^{232}_{90}Th \rightarrow {}^{A}_{Z}X + {}^{4}_{2}\alpha$$

a What type of radiation is being emitted?
b What are the values of A and Z?
c Use the table on page 244 to decide what new element is formed by the decay process.
d Rewrite the above equation, replacing A, Z, and X with the numbers and symbols you have found.
e What are the decay products?

2 When radioactive sodium-24 decays, magnesium-24 is formed. The following equation represents the decay process, but the equation is incomplete:

$$^{24}_{11}Na \rightarrow {}^{24}_{12}Mg + \underline{\quad\quad}$$

Assuming that only one charged particle is emitted:
a What is the mass number of this particle?
b What is the relative charge of this particle?
c What type of particle is it?

Related topics: nuclei and isotopes **10.1**; alpha, beta, and gamma radiation **10.2**; more on beta decay **10.10**

10.5 Radioactive decay (2)

Objectives: – to know about the changing rate of decay, what half-life is, and how to find it from a decay curve – to explain why some nuclei are unstable.

Radioactive decay happens spontaneously (all by itself) and at random. There is no way of predicting when a particular nucleus will disintegrate, or in which direction a particle will be emitted. Also, the process is unaffected by pressure, temperature, or chemical change. However, some types of nucleus are more unstable than others and decay at a faster rate.

Rate of decay and half-life

Iodine-131 is a radioactive isotope of iodine. The chart below illustrates the decay of a sample of iodine-131. On average, 1 nucleus disintegrates every second for every 1 000 000 nuclei present.

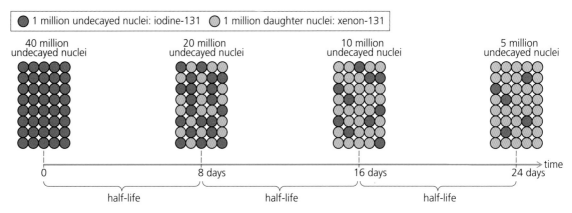

To begin with, there are 40 million undecayed nuclei. 8 days later, half of these have disintegrated. With the number of undecayed nuclei now halved, the number of distintegrations over the next 8 days is also halved. It halves again over the next 8 days... and so on. Iodine-131 has a **half-life** of 8 days.

> The half-life of a radioactive isotope is the time taken for half the nuclei present in any given sample to decay.

radioactive isotope	half-life
boron-12	0.02 seconds
radon-220	52 seconds
iodine-128	25 minutes
radon-222	3.8 days
strontium-90	28 years
radium-226	1602 years
carbon-14	5730 years
plutonium-239	24 400 years
uranium-235	7.1×10^8 years
uranium-238	4.5×10^9 years

The half-lives of some other radioactive isotopes are given on the left. It might seem strange that there should be any short-lived isotopes still remaining. However, some are radioactive daughters of long-lived parents, while others are produced artificially in nuclear reactors.

Activity and half-life

In a radioactive sample, the average number of disintegrations per second is called the **activity**. The SI unit of activity is the **becquerel** (**Bq**). An activity of, say, 100 Bq means that 100 nuclei are disintegrating per second.

The graph at the top of the next page shows how, on average, the activity of a sample of iodine-131 varies with time. As the activity is always proportional to the number of undecayed nuclei, it too halves every 8 days. So 'half-life' has another meaning as well:

> The half-life of a radioactive isotope is the time taken for the activity of any given sample to fall to half its original value.

ATOMS AND RADIOACTIVITY

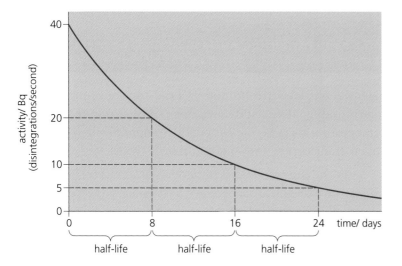

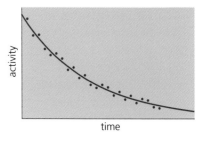

◀ Radioactive decay of iodine-131. Iodine-131 has a half-life of 8 days. From any point on the curve, it always takes 8 (days) along the time axis for the activity to halve.

▲ Radioactive decay is a random process. So, in practice, the curve is a 'best fit' of points which vary irregularly like this.

To obtain a graph like the one above, a GM tube is used to detect the particles emitted by the sample. The number of counts per second recorded by the ratemeter is adjusted to allow for background radiation (see spread 10.3). The adjusted figure is proportional to the activity – though not equal to it, because not all of the emitted particles are detected.

Stability of the nucleus

In a nucleus, some proportions of neutrons to protons are more stable than others. If the number of neutrons is plotted against the number of protons for all the different isotopes of all the elements, the general form of the graph is as shown on the right. It has these features:

- Stable isotopes lie along the stability line.
- Isotopes above the stability line have too many neutrons to be stable. They decay by beta− (electron) emission because this reduces the number of neutrons.
- Isotopes below the stability line have too few neutrons to be stable. They decay by beta+ (positron) emission because this increases the number of neutrons.
- The heaviest isotopes (proton numbers > 83) decay by alpha emission.

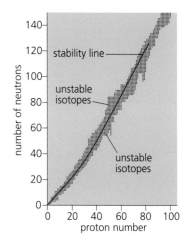

To answer questions 1 and 2, you will need information from the table of half-lives on the opposite page.

1 If samples of strontium-90 and radium-226 both had the same activity today, which would have the lower activity in 10 years' time?

2 If the activity of a sample of iodine-128 is 800 Bq, what would you expect the activity to be after
   a 25 minutes   **b** 50 minutes   **c** 100 minutes?

3 The graph on the right shows how the activity of a small radioactive sample varied with time.
   a Why are the points not on a smooth curve?
   b Estimate the half-life of the sample.

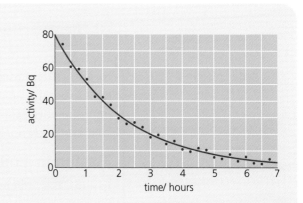

Related topics: nuclei and isotopes 10.1; GM tube 10.3

10.6 Nuclear energy

Objectives: – to describe what happens during nuclear fission, how controlled fission is used in a nuclear power station, and the hazards of nuclear waste.

Nuclear essentials

Atoms of any one element all have the same number of protons in the nucleus. If this number is altered in some way, an atom of a completely different element is formed.

Elements exist in different versions, called isotopes, with different numbers of neutrons in the nucleus. Radioactive isotopes have unstable nuclei. In time, these decay (break up) by emitting one or more particles and, in some cases, gamma radiation as well.

When alpha or beta particles are emitted by a radioactive isotope, they collide with surrounding atoms and make them move faster. In other words, the temperature rises as nuclear energy (potential energy stored in the nucleus) is transformed into thermal energy (heat).

In radioactive decay, the energy released per atom is around a million times greater than that from a chemical change such as burning. However, the rate of decay is usually very slow. Much faster decay can happen if nuclei are made more unstable by bombarding them with neutrons. Whenever a particle penetrates and changes a nucleus, this is called a **nuclear reaction**.

E Fission

Natural uranium is a dense radioactive metal consisting mainly of two isotopes: uranium-238 (over 99%) and uranium-235 (less than 1%). The diagram below shows what can happen if a neutron strikes and penetrates a nucleus of uranium-235. The nucleus becomes highly unstable and splits into two lighter nuclei, shooting out two or three neutrons as it does so. The splitting process is called **fission**, and the fragments are thrown apart as energy is released. If the emitted neutrons go on to split other nuclei... and so on, the result is a **chain reaction**, and a huge and rapid release of energy.

▶ A chain reaction. A neutron causes a uranium-235 nucleus to split, producing more neutrons, which cause more nuclei to split... and so on.

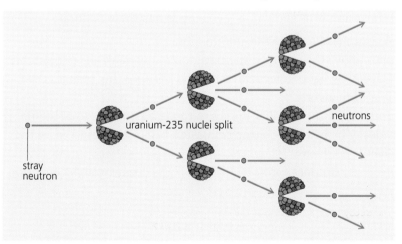

Nuclear safety

Nuclear power stations have safety procedures to

- shield people from direct nuclear radiation
- keep people's time of exposure to radiation as short as possible
- prevent radioactive materials from getting into the body.

Concrete, steel, and lead shielding reduce radiation, and radioactive materials are kept in sealed containers to prevent gas, dust, or liquid escaping.

E For a chain reaction to be maintained, the uranium-235 has to be above a certain **critical mass**, otherwise too many neutrons escape. In the first atomic bombs, an uncontrolled chain reaction was started by bringing two lumps of pure uranium-235 together so that the critical mass was exceeded. In present-day nuclear weapons, plutonium-239 is used for fission.

Fission in a nuclear reactor

In a **nuclear reactor** in a nuclear power station, a controlled chain reaction takes place and thermal energy (heat) is released at a steady rate. The energy is used to make steam for the turbines, as in a conventional power station. In many reactors, the nuclear fuel is uranium dioxide, the natural uranium being enriched with extra uranium-235. The fuel is in sealed cans (or tubes).

ATOMS AND RADIOACTIVITY

Maintaining the reaction* To maintain the chain reaction in a reactor, the neutrons have to be slowed down, otherwise many of them get absorbed by the uranium-238. To slow them, a material called a **moderator** is needed. Graphite is used in some reactors, water in others. The rate of the reaction is controlled by raising or lowering **control rods**. These contain boron or cadmium, materials which absorb neutrons.

Nuclear waste*

After a fuel can has been in a reactor for three or four years, it must be removed and replaced. The amount of uranium-235 in it has fallen and the fission products are building up. Many of these products are themselves radioactive, and far too dangerous to be released into the environment. They include the following isotopes, none of which occur naturally.

- Strontium-90 and iodine-131, which are easily absorbed by the body. Strontium becomes concentrated in the bones; iodine in the thyroid gland.
- Plutonium-239, which is produced when uranium-238 is bombarded by neutrons. It is itself a nuclear fuel and is used in nuclear weapons. It is also highly toxic. Breathed in as dust, the smallest amount can kill.

Spent fuel cans are taken to a reprocessing plant where unused fuel and plutonium are removed. The remaining waste, now a liquid, is sealed off and stored with thick shielding around it. Some of the isotopes have long half-lives, so safe storage will be needed for thousands of years. The problem of finding acceptable sites for long-term storage has still not been solved.

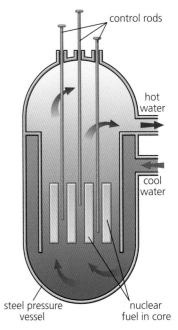

▲ A pressurized water reactor (PWR). For safety, the reactor is housed inside a sealed containment building made of steel and concrete.

Energy and mass

According to Albert Einstein (1905), energy itself has mass. If an object gains energy, its mass increases; if it loses energy, its mass decreases. The mass change m (kg) is linked to the energy change E (joules) by this equation:

$$E = mc^2$$ (where c is the speed of light, 3×10^8 m/s)

The value of c^2 is so high that energy gained or lost by everyday objects has a negligible effect on their mass. However, in nuclear reactions, the energy changes per atom are much larger, and produce detectable mass changes. For example, when the fission products of uranium-235 are slowed down in a nuclear reactor, their total mass is found to be reduced by about 0.1%.

▲ The steel flasks on this train contain waste from a nuclear reactor

Q

1. The high temperatures deep underground are caused by the decay of radioactive isotopes in the rocks. Why does radioactive decay cause a rise in temperature?
2. What is meant by **a** fission **b** a chain reaction?
3. Give one example of
   a a controlled chain reaction
   b an uncontrolled chain reaction.
4. ***a** Where does plutonium-239 come from?
   b Why is plutonium-239 so dangerous?
5. In a typical fission process, uranium-235 absorbs a neutron, creating a nucleus which splits to form barium-141, krypton-92, and three neutrons.

	mass/kg
neutron	1.674×10^{-27}
uranium-235 nucleus	390.250×10^{-27}
barium-141 nucleus	233.964×10^{-27}
krypton-92 nucleus	152.628×10^{-27}

a The reaction can be represented by this equation:
$$^{A}_{92}U + ^{B}_{0}n \rightarrow ^{C}_{56}Ba + ^{D}_{Z}Kr + 3^{B}_{0}n$$
Copy the equation, replacing A, B, C, D, and Z with the correct numbers.

b From the data in the above table, how could you tell that energy is released by the reaction?

Related topics: energy **4.1**; power stations **4.5–4.6**; radiation dangers **10.2–10.3**; radioactive decay **10.4–10.5**; half-life **10.5**; fission reaction **10.7**

10.7 Fusion future

Objectives: – to describe what happens during nuclear fusion – to know that the Sun gets its energy from fusion, and about research into 'clean' fusion reactors.

Nuclear essentials

The nucleus of an atom is made up of protons and (in most cases) neutrons.

Each element has a different number of protons in the nucleus of its atom. The lightest element, hydrogen has just one.

Elements exist in different versions, called isotopes. These have different numbers of neutrons in the nucleus.

Hydrogen

Hydrogen is the most plentiful element in the Universe. The Sun is 75% hydrogen. There is also lots of hydrogen on Earth, though most has combined with oxygen to form water (H_2O).

Most hydrogen is hydrogen-1. Hydrogen-2 (also called deuterium) and hydrogen-3 (tritium) are much rarer forms.

(E) In the nucleus of an atom, the protons and neutrons are held tightly together by a force called, simply, the **strong nuclear force**. However, in some nuclei, they are more tightly held than in others. To get nuclei to release energy, the trick is to make the protons and neutrons regroup into more tightly held arrangements than before. Protons and neutrons in 'middleweight' nuclei tend to be the most tightly held, so splitting very heavy nuclei releases energy: that happens in nuclear fission. However, energy can also be released by fusing (joining) very light nuclei together to make heavier ones. This is called **nuclear fusion**. It is the process that powers the stars. One day, it may drive power stations on Earth.

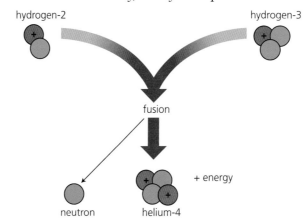

(E) The diagram above shows the fusion of two hydrogen nuclei to form helium. Fusion is difficult to achieve because the nuclei are charged, and repel each other. To beat the repulsion and join up, they must travel very fast – which means that the gas must be much hotter than any temperatures normally achieved on Earth.

Building a fusion reactor

Scientists and engineers are trying to design fusion reactors for use as an energy source on Earth. But there are huge problems to overcome. Hydrogen must be heated to at least 40 million degrees Celsius, and kept hot and compressed, otherwise fusion stops. No ordinary container can hold a superhot gas like this, so scientists are developing reactors that trap the nuclei in a magnetic field.

Fusion reactors will have huge advantages over today's fission reactors. They will produce more energy per kilogram of fuel. Their hydrogen fuel can be extracted from sea water. Their main waste product, helium, is not radioactive. And they have built-in safety: if the system fails, fusion stops.

Fusion in the Sun

The Sun is a star. Like most other stars, it gets its energy from the fusion of hydrogen into helium. Deep in its core, the heat output and huge gravitational pull keep the hydrogen hot and compressed enough to maintain fusion. It has enough hydrogen left to keep it shining for another 6 billion years.

▲ This magnetic containment vessel, called a **tokamak**, is being used to investigate fusion

ATOMS AND RADIOACTIVITY

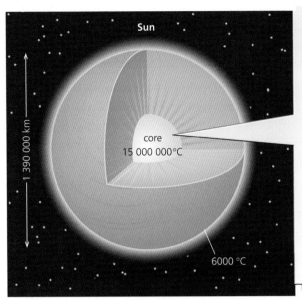

Fusion in the Sun's core
The Sun is 73% hydrogen and 25% helium.
Energy is released as hydrogen is converted into helium.

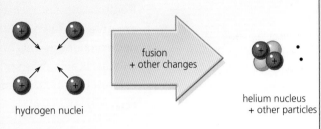

Four hydrogen nuclei fuse together for each helium nucleus formed. This is a multi-stage process which also involves the creation of two neutrons from two protons.

 In the Sun, fusion happens at 'only' 15 million degrees Celsius. But the Sun uses different fusion reactions from those being tried on Earth. If the Sun were scaled down to the size of a nuclear reactor, its power output would be too low to be useful.

Fission and fusion compared

Here is a typical fission reaction of the type that happens in a nuclear reactor:

$$^{235}_{92}\text{U} + ^{1}_{0}\text{n} \rightarrow ^{144}_{56}\text{Ba} + ^{90}_{36}\text{Kr} + 2^{1}_{0}\text{n}$$

A neutron hits a uranium-235 nucleus, which splits to form two lighter nuclei and two neutrons. The energy released per atom is about million times greater than that per atom from a chemical reaction such as burning.

In the fusion process on the opposite page, two rare forms of hydrogen nuclei collide and combine to form a helium nucleus:

$$^{2}_{1}\text{H} + ^{3}_{1}\text{H} \rightarrow ^{4}_{2}\text{He} + ^{1}_{0}\text{n}$$

Although the energy released per fusion is less than 10% of that from a fission reaction, fusion (as in the example above) is a much better energy source if the processes are compared *per kilogram* of material.

▲ Present day nuclear power stations all use fission in their reactors. Fusion is for the future.

1. *Splitting very heavy nuclei to form lighter ones. Joining very light nuclei to form heavier ones.*
   a Which of the above statements describes what happens during nuclear fusion?
   b What process does the other statement describe?
2. What advantages will power stations with fusion reactors have over today's nuclear power stations?
3. Why have fusion reactors have been so difficult to develop?
4. Nuclear reactions are taking place in the Sun's core.
   a What substance does the Sun use as its nuclear fuel?
   b What is the name of the process that supplies the Sun with its energy?
   c What substance is made by this process?
5.* Comparing *burning*, nuclear *fusion*, and nuclear *fission*, which of those processes yields
   a the most energy per kg of fuel
   b the least energy per kg of fuel?

Related topics: gravity **2.9**; power stations **4.5**; energy resources **4.7–4.8**; atoms **10.1**; nuclear fission and energy **10.6**

10.8 Using radioactivity

Objectives: – to describe some industrial uses of radioactive isotopes – to describe their use for medical diagnosis, and for the treatment of cancer.

Nuclear essentials

Elements exist in different versions, called isotopes. For example, lithium is a mixture of two isotopes: lithium-6 (with 3 protons and 3 neutrons in the nucleus of its atoms) and lithium-7 (with 3 protons and 4 neutrons).

Radioactive isotopes have unstable nuclei. In time each nucleus decays (breaks up) by emitting an alpha or beta particle and, in some gases, a burst of gamma radiation as well. In a radioactive sample, the number of nuclei decaying per second is called the activity.

Gamma rays are very penetrating, beta particles less so, and alpha particles least of all. All three types of radiation damage or destroy living cells if absorbed.

Radioactive isotopes are called **radioisotopes** (or **radionuclides**). Some are produced artificially in a nuclear reactor when nuclei absorb neutrons or gamma radiation. For example, all natural cobalt is cobalt-59, which is stable. If cobalt-59 absorbs a neutron, it becomes cobalt-60, which is radioactive.

Here are some of the practical uses of radioisotopes.

Tracers and treatments

Radioisotopes can be detected in very small (and safe) quantities, so they can be used as **tracers** – their movements can be tracked. Examples include:
- Checking the function of body organs. For example, to check thyroid function, a patient drinks a liquid containing iodine-123, a gamma emitter. Over the next 24 hours, a detector measures the activity of the tracer to find out how quickly it becomes concentrated in the thyroid gland.
- Tracking a plant's uptake of fertilizer from roots to leaves by adding a tracer to the soil water.
- Detecting leaks in underground pipes by adding a tracer to the fluid in the pipe.

For tests like those above, artificial radioisotopes with short half-lives are used so that there is no detectable radiation after a few days.

▶ A gamma camera in use. The patient has been injected with a liquid containing weakly radioactive technetium. The camera above her will pick up the gamma rays from the tracer and form an X-ray-type picture of her kidneys.

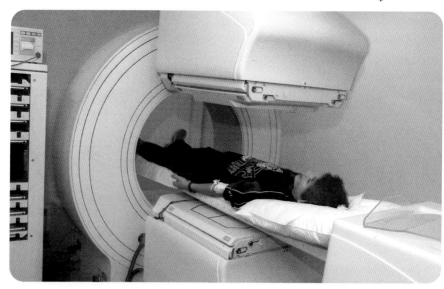

In a hospital, a **gamma camera** like the one in the photograph above may be used to detect the gamma rays coming from a radioactive tracer in a patient's body. The camera forms an image similar that produced by X-rays.

In one form of **radiotherapy**, gamma radiation is used to kill cancer cells. It can penetrate the body to reach a tumour. And by using several beams coming from different directions, their energy can be concentrated at one point.

ATOMS AND RADIOACTIVITY

(E) Thickness monitoring

In some production processes a steady thickness of material has to be maintained. The diagram below shows one way of doing this.

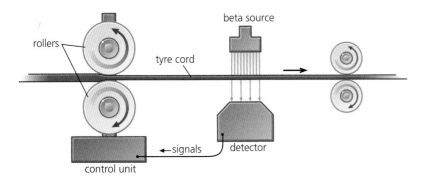

◀ The moving band of tyre cord has a beta source on one side and a detector on the other. If the cord from the rollers becomes too thin, more beta radiation reaches the detector. This sends signals to the control unit, which adjusts the gap between the rollers.

(E) Gamma irradiation

Some food-processing factories use gamma irradiation chambers to kill bacteria (and small insects) in food. The gamma rays penetrate deep into the food. And with the bacteria killed, the food keeps longer without rotting.

Medical instruments can be sterilized using the same method.

Cobalt-60 is used as the gamma ray source. It has a half-life of just over five years and must be replaced before it weakens too much to be effective.

Smoke detectors

A smoke detector can give you a first warning of a fire. It is designed to trigger a very loud alarm whenever smoke particles enter it. It works like this:

In the detector, there are two metal plates with a voltage across them. Between the plates, there is a very weak source of alpha particles (americium-241, half-life 432 years). These ionize the air: the ions move between the plates, so a small current always flows. However, if smoke particles enter the chamber, the ions attach to them. This reduces the current and triggers the alarm.

▲ A smoke detector contains a very weak radioactive source.

Testing for cracks

Gamma rays have the same properties as short-wavelength X-rays, so they can be used to photograph metals to reveal cracks. A cobalt-60 gamma source is compact and does not need electrical power like an X-ray tube.

1 a What are radioisotopes?
  b How are artificial radioisotopes produced?
  c Give *two* medical uses of radioisotopes.

2 In the thickness monitoring system shown above:
  a Why is a beta source used, rather than an alpha or gamma source?
  b What is the effect on the detector if the thickness of the tyre cord increases?

3 a Give *two* uses of radioactive tracers.
  b Why is it important to use radioactive tracers with short half-lives?

4 a In an irradiation chamber, why are gamma rays rather than alpha particles used to kill bacteria in food?
  b Give *one* other use of an irradiation chamber.
  c Give *one* other use of gamma radiation.

Related topics: X-rays **7.11**; alpha, beta, and gamma radiation **10.2–10.3**; radioactive decay **10.4–10.5**; half-life **11.05**

10.9 Atoms and particles (1)

Objectives: – to describe the experiments that have led to the nuclear model of the atom: a tiny nucleus surrounded by orbiting electrons.

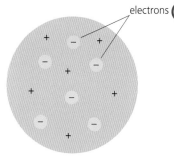

▲ Thomson's 'plum pudding' model of the atom

(E) Atoms are made up of even smaller particles. From experimental evidence collected over the past hundred years, scientists have been able to develop and improve their models (descriptions) of atoms and the particles in them.

Thomson's 'plum pudding' model

The **electron** was the first atomic particle to be discovered. It was identified by J. J. Thomson in 1897. The electron has a negative (−) electric charge, so an atom with electrons in it must also contain positive (+) charge to make it electrically neutral. Thomson suggested that an atom might be a sphere of positive charge with electrons dotted about inside it rather like raisins in a pudding. This became knows as the 'plum pudding' model.

Rutherford's nuclear model

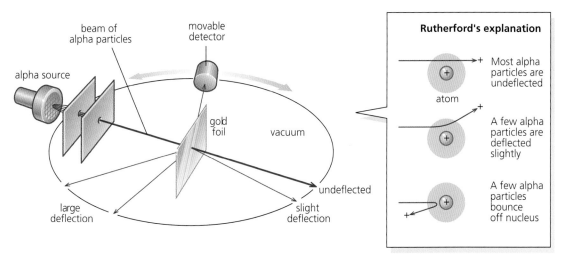

(E) The above experiment was carried out in 1911 by Geiger and Marsden under the supervision of Ernest Rutherford. It produced results which could not be explained by the plum pudding model. Thin gold foil was bombarded with alpha particles, which are positively charged. Most passed straight through the gold atoms, but a few were repelled so strongly that they bounced back or were deflected through large angles. Rutherford concluded that the atom must be largely empty space, with its positive charge and most of its mass concentrated in a tiny **nucleus** at the centre. In his model, the much lighter electrons orbited the nucleus rather like the planets around the Sun.

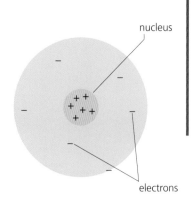

▲ Rutherford's model of the atom: electrons orbit a central nucleus. (If the nucleus were correctly drawn to scale, it would be too small to see.)

Discovering particles in the nucleus*

Rutherford's model said nothing about what was inside the nucleus. However, in 1919, Rutherford bombarded nitrogen gas with fast alpha particles and found that positively charged particles were being knocked out. These were **protons**. In 1932, James Chadwick discovered that the nucleus also contained uncharged particles with a similar mass to protons. He called these **neutrons**.

ATOMS AND RADIOACTIVITY

The problem of spectral lines*

Light comes from atoms. The spectrum of white light is a continuous range of colour from red (the longest wavelength) to violet (the shortest). However, not all spectra are like this. For example, if there is an electric discharge through hydrogen, the glowing gas emits particular wavelengths only, so the spectrum is made up of lines, as shown below. As it stood, Rutherford's model could not explain why spectra like this occurred. To solve this problem, the model had to be modified.

Light essentials
Light is one type of electromagnetic radiation (electromagnetic waves). The colour seen depends on the wavelength of the light.

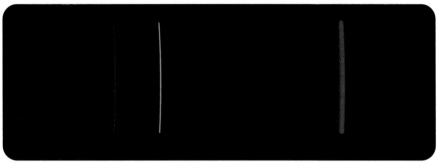

◀ Part of the line spectrum of hydrogen. Each line represents light of a particular wavelength.

shorter................................wavelength..................................longer

The Rutherford-Bohr model*

In 1913, Neils Bohr modified Rutherford's model by applying the **quantum theory** devised by Max Planck in 1900. According to this theory, energy cannot be divided into ever smaller amounts. It is only emitted (or absorbed) in tiny 'packets', each called a **quantum**. Bohr reasoned that electrons in higher orbits have more energy than those in lower ones. So, if only quantum energy changes are possible, only certain electron orbits are allowed. This modified model is known as the **Rutherford-Bohr model**.

Using the model, Bohr was able to explain why atoms emit light of particular wavelengths only (see the next spread). He even predicted the positions of the lines in the spectrum of hydrogen. However, his calculations did not work for substances with a more complicated electron structure. To deal with this problem, scientists have developed a **wave mechanics** model in which allowed orbits are replaced by allowed **energy levels**. However, this is an entirely mathematical approach, and the Rutherford-Bohr model is still used as a way of representing atoms in pictures.

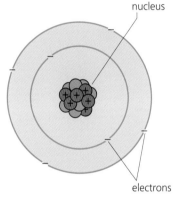

▲ The Rutherford-Bohr model of the atom (with nuclear particles included). In this model, only certain electron orbits are allowed.

Q

1. What is the difference between Rutherford's model of the atom and Thomson's 'plum pudding' model?
2*. What is the difference between the Rutherford–Bohr model of the atom and Rutherford's model?
3. On the right, a beam of alpha particles is being directed at a thin piece of gold foil. How does the Rutherford model of the atom explain why
   a. most of the alpha particles go straight through the foil
   b. some alpha particles are deflected at large angles?
4. Why do the results of the experiment on the right suggest that the nucleus has a positive charge?

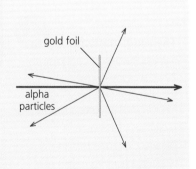

Related topics: light waves **7.1** and **7.10**; spectrum **7.4**; electric charge **8.1–8.2**; particles in the atom **10.1**

10.10 Atoms and particles (2)*

Objectives: – to explain how atoms emit light – to describe how particle accelerators are being used for research into fundamental particles including quarks.

If an electron gains energy...

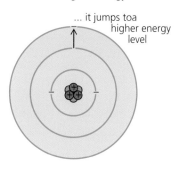

...it jumps to a higher energy level

When the electron drops back to a lower level...

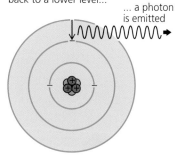

...a photon is emitted

▲ How an atom gives off light

▶ One of the giant detectors surrounding part of the Large Hadron Collider at CERN near Geneva. Hadrons are a family of particles which includes protons and neutrons. In the collider, beams of protons are accelerated by electromagnets round a circular path 27 km long, then made to collide head-on.

The Higgs particle
A major success at CERN, in 2012, was the discovery of the Higgs particle. Long predicted by the standard model, this fundamental particle is required to explain why most particles have mass.

How an atom gives off light

Bohr's explanation of how an atom gives off light was like this.

If an electron gains energy in some way – for example, because its atom collides with another one – it may jump to a higher energy level. But the atom does not stay in this **excited state** for long. Soon, the electron loses energy by dropping back to a lower level. According to the quantum theory, the energy is radiated as a pulse of light called a **photon**. The greater the energy change, the shorter the wavelength of the light.

As a line spectrum contains particular wavelengths only, it provides evidence that only certain energy changes are occurring within the atom – and therefore that only certain energy levels are allowed.

Fundamental particles

A **fundamental particle** is one which is *not* made up of other particles. An atom is not fundamental because it is made up of electrons, protons, and neutrons. But are these fundamental? To answer this and other questions, scientists carry out experiments with **particle accelerators**. They shoot beams of high-energy particles (such as protons) at nuclei, or at other beams, and detect the particles emerging from the collisions. In collider experiments, new particles are created as energy is converted into mass. However, most of these particles do not exist in the atoms of ordinary matter.

The present theory of particles is called the **standard model**. According to this model, electrons are fundamental particles. However, neutrons and protons are made up of other particles called **quarks**, as shown in the chart on the next page. In ordinary matter, there are two types of quark, called the **up quark** and the **down quark** for convenience. Each proton or neutron is made up of three quarks. The quarks have fractional charges compared with the charge on an electron.

ATOMS AND RADIOACTIVITY

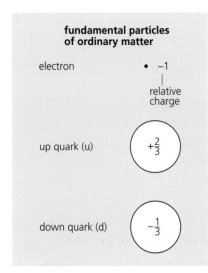
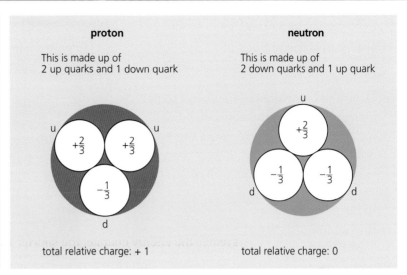

Individual quarks have never been detected. The existence of quarks has only been deduced from the patterns seen in the properties of other particles – for example, how high-energy particles are scattered.

Quark changes in beta decay

In the most common form of beta decay, a neutron decays to form a proton, an electron (the beta particle), and an antineutrino:

neutron → proton + electron + antineutrino

If this is rewritten to show the quarks:

up quark   up quark
down quark → up quark + electron + antineutrino
down quark   down quark

From the above, you can see that this type of beta decay occurs when a down quark changes into an up quark, as follows:

down quark → up quark + electron + antineutrino
$(-1/3)$ $(+2/3)$ (-1) (0)

The relative charges underneath the equation show that there is no change in total charge. In other words, charge is conserved.

In the less common form of beta decay, a proton decays to form a neutron, a positron (the beta particle), and a neutrino. This happens when an up quark in the proton changes into a down quark.

Decay essentials
The break-up of an unstable nucleus is called radioactive decay. During beta decay, a beta particle is shot out. In most cases, this particle is an electron (−). However, more rarely, it a positron (+), an antiparticle with the same mass as an electron, but opposite charge.

1 When an electron drops back to a lower energy in an atom, it loses energy.
  a What happens to this energy?
  b If the difference between the two energy levels was greater, how would this affect the wavelength of the light emitted?
  c Why do atoms emit certain wavelengths only?

2 What is meant by a fundamental particle?

3 Which of the following are thought to be fundamental particles?
   electrons   protons   neutrons   quarks

4 Quarks have a fractional charge. Explain why, if a neutron is made up of three quarks, it is uncharged.

5 In one form of beta decay, an up quark changes into a down quark. Explain why, in this case, the beta particle emitted must be a positron and not an electron.

Related topics: light waves **7.1** and **7.10**; electric charge **8.1–8.2**; particles in the atom **10.1**; beta decay **10.4–10.5**

Check-up on atoms and radioactivity

Further questions

1 | electrons | nuclei | protons | waves |

   a Copy and complete the following sentences using words from the above list. Each word may be used once, more than once or not at all.
      i Radioactive substances have atoms with unstable _____ . [1]
      ii Beta particles are _____ . [1]
      iii Gamma rays are _____ . [1]
   b Name another type of radioactive particle not mentioned in part **a**. [1]

2 The symbol $^{35}_{17}Cl$ represents one atom of chlorine.
   a State the names and numbers of the different types of particle found in one of these chlorine atoms. [3]
   b State where these particles are to be found in the atom. [2]

3

proton number	26
mass number	59
radiation emitted	beta and gamma

The table above shows information about a radioisotope of iron called iron-59.
   a Calculate:
      i the number of neutrons in the nucleus; [1]
      ii the total number of charged particles in a single atom of iron-59. [1]
   b Iron-59 and iron-56 are both isotopes of iron. What are isotopes? [1]
   c Iron-59 emits two types of radiation. Briefly explain how the gamma radiation could be separated from the beta radiation emitted. [1]

4 Phosphorus-32 is a radioactive isotope. It can be used to prove that plants absorb phosphorus from the soil around them.
   a **i** The stable isotope of phosphorus has a mass number of 31. State the structural difference between atoms of phosphorus-31 and phosphorus-32. [2]
      ii* Explain why both isotopes of phosphorus have identical chemical properties. [1]
   b Phosphorus-32 is a **beta-emitter** with a **half-life** of 14 days.
      i What is a beta particle? [1]
      ii (E) The proton number of phosphorus-32 atom is 15. State the new values of the proton number and mass number of the atom just after it has emitted a beta particle. [2]
      iii Explain what is meant by the term **half-life**. [1]
   c A solution of the isotope is watered onto the soil around the plant. Each day for the next week, a leaf is removed from the plan and tested for radioactivity.

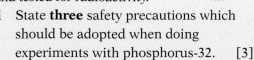

      i State **three** safety precautions which should be adopted when doing experiments with phosphorus-32. [3]
      ii Describe **two** methods which could be used to measure the activity of a leaf. [2]

5 Phyl is in hospital. She is injected with the radioisotope technetium-99m.
This isotope is absorbed by the thyroid gland in her throat. A radiation detector placed outside her body and above her throat detects the radiation.
Technetium-99m has a half-life of 6 hours.
It emits gamma radiation.
   a Why is an emitter of alpha radiation unsuitable? [1]
   b **i** How long will it take for the activity of the technetium-99m to fall to a quarter of its original value? [2]
      ii After 24 hours, how will the activity of the technetium-99m compare with its original value? [2]
   c Eventually the level of radiation from the technetium-99m will fall to less than the level of the background radiation. State **two** naturally occurring sources of background radiation. [2]

6 This question is about an accident at the Chernobyl nuclear power station in which radioactive gas and dust were released into the atmosphere.
The radioactive isotopes in the Chernobyl fallout which caused most concern were iodine-131 and caesium-137. Both are beta and gamma

emitters. Iodine-131, in rainfall, found its way into milk but caesium-137, with a half-life of 30 years, may cause more long-term problems.

a From which part of the atom do the beta and gamma rays come? [1]
b Explain what the number 131 tells you about the iodine atom. [2]
c After the Chernobyl accident, a milk sample containing iodine-131 was found to have an activity of 1600 units per litre. The activity of the sample was measured every 7 days and the results are shown in the table below.

time/days	0	7	14	21	28	35
activity/units per litre	1600	875	470	260	140	77

   i Draw a graph of activity against time, using the grid below as a guide. [2]
   ii Estimate the half-life of iodine-131 and show on the graph how you arrived at your answer. [2]

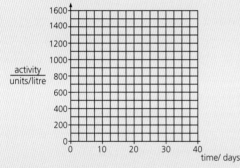

d Give a reason why caesium-137 could cause longer-term problems than iodine-131. [2]

7 a i Explain why some substances are radioactive and some are not. [2]
    ii State the cause of background radiation. [1]
    iii Explain what you understand by the meaning of the **half-life** of a radioactive element. [2]
  b Technetium-99m is a radioactive material with a half-life of 6 hours. It is used to study blood flow around the body. A sample of technetium-99m has an activity of 96 counts per minute when injected into a patient's blood stream. Estimate
    i its activity after 12 hours [1]
    ii how long it will take for the radioactivity from the injection to become undetectable. [1]
  c Technetium-99m is a gamma (γ) emitter and does not produce alpha (α) or beta (β) radiations. Explain why it is safe to inject technetium-99m into the body. [2]
  d Radioactive salt (sodium chloride) is also used in medicine. The radioactive sodium (Na) in the salt decays, according to the equation shown below, to form magnesium (Mg).

$$^{24}_{11}Na \longrightarrow {}^{24}_{12}Mg + X + \gamma \text{ radiation}$$

   i Name the particle **X**. [1]
   ii Use the information given in the equation above to find the
      I total number of charged particles in each sodium atom [1]
      II number of neutrons in the nucleus of a sodium 24 atom. [1]

8 This question is about information in a leaflet.
  a **Extract 1** 'Radon is a naturally occurring radioactive gas. It comes from uranium which occurs in rocks and soils.'
    i Explain the meaning of the word **radioactive**.
    ii Explain the danger of breathing radon gas into the lungs. [4]
  Extract 2 is a diagram showing how radon decays

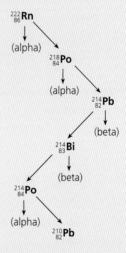

Two of the nuclei shown in the diagram are isotopes of polonium.
  b Explain the meaning of the word **isotope**. [1]

c In the diagram, radon is shown as $^{222}_{86}Rn$. In a neutral radon atom, what is the number of
   i protons ii electrons iii neutrons? [3]
9 A radioactive isotope of gold has the symbol $^{196}_{79}Au$. If this isotope is injected into the bloodstream of a patient, it can be used by doctors as a tracer to monitor the way the patient's heart works. The isotope emits gamma radiation that is detected outside the patient's body.
   a Why would an isotope that emits alpha radiation be unsuitable as a tracer to monitor the working of the heart? [1]
   b Give one non-medical use for a radioactive tracer. [1]

10 Isotopes of the radioactive element uranium occur naturally in small proportions in some rocks. The table gives information about **one** uranium isotope.

nucleon (mass) number	238
proton (atomic) number	92
radiation emitted	alpha particle

   a How many neutrons are there in an atom of this uranium isotope? [1]
   b From which part of the uranium atom does the alpha particle come? [1]
   c What does an alpha particle consist of? [2]

Use the list below when you revise for your IGCSE examination. The spread number, in brackets, tells you where to find more information.

Revision checklist
Core Level
- [ ] The particles in an atom and the charges on them. (10.1)
- [ ] The meanings of atomic number (proton number) and mass number (nucleon number). (10.1)
- [ ] What isotopes are. (10.1)
- [ ] What a nuclide is. (10.1)
- [ ] Representing nuclides in symbol form. For example: $^{7}_{3}Li$ (10.1)
- [ ] What radioactive materials are. (10.2)
- [ ] What radioactive decay means. (10.2)
- [ ] How atoms become ions. (10.2)
- [ ] Alpha and beta particles, their properties and detection. (10.2)
- [ ] Gamma rays, their properties and detection. (10.2)
- [ ] The ionizing and penetrating effects of alpha, beta, and gamma radiation. (10.2)
- [ ] The dangers of nuclear radiation. (10.3)
- [ ] What background radiation is, and where it comes from. (10.3)
- [ ] Detecting and measuring nuclear radiation. (10.3)
- [ ] Handling and storing radioactive materials safely. (10.3 and 10.6)
- [ ] How the emission of an alpha or beta particle changes an atom into one of a different element. (10.4)
- [ ] The random nature of radioactive decay. (10.5)
- [ ] How the rate of radioactive decay changes with time. (10.5)
- [ ] The meaning of half-life. (10.5)
- [ ] Working out a half-life from a radioactive decay curve or other data. (10.5)

Extended Level
As for Core Level, plus the following:
- [ ] Why alpha particles, beta particles, and gamma rays have different ionizing effects. (10.2)
- [ ] How alpha and beta particles are deflected by electric and magnetic fields. (10.2)
- [ ] Why gamma rays are not deflected by electric and magnetic fields. (10.2)
- [ ] Identifying which types of radiation are coming from a radioactive source. (10.2 and 10.3)
- [ ] Allowing for background radiation when dealing with data about radioactive decay. (10.3 and 10.5)
- [ ] Using symbol equations to represent the changes that happen during alpha decay and beta decay. (10.4)
- [ ] How the stability of a nucleus is affected by radioactive decay. (10.5)
- [ ] What happens during nuclear fission. (10.6 and 10.7)
- [ ] What happens during nuclear fusion. (10.7)
- [ ] The practical applications of alpha, beta, and gamma emissions. (10.8)
- [ ] How the scattering of alpha particles by metal foil provides evidence for a nucleus in an atom. (10.9)

11

The Earth in space

- SUN, EARTH, AND MOON
- DAY AND NIGHT
- SEASONS
- THE SOLAR SYSTEM
- ORBITS
- STARS AND GALAXIES
- BIRTH AND DEATH OF STARS
- INSIDE THE SUN
- THE BIG BANG
- THE EXPANDING UNIVERSE

Glowing gas surrounds a black hole at the heart of the Messier 87 galaxy, 55 million light years from Earth. The black hole is huge – about the size of our Solar System – but is so far away that a network of eight radio telescopes around the world was needed to detect and create this image. The 'shadow' area is caused by the gravitational bending and capture of light. It is about three times larger than the black hole itself, which is invisible.

11.1 Sun, Earth, and Moon

Objectives: – to explain why day, night, and seasons occur - to explain why the Moon appears in different phases during a month - to understand orbital speed.

Diameters
Sun   1 390 000 km
Earth   12 800 km
Moon   3500 km

Distances (average)
Earth to Sun 149 600 000 km
Earth to Moon 384 000 km

The Sun

The Sun is a huge, glowing ball of gas called a star. It is extremely hot, with a temperature of 6000 °C on the surface rising to 15 million °C in its centre. Its energy is released by nuclear reactions in its core, and it radiates mostly in the infrared, visible, and ultraviolet regions of the electromagnetic spectrum.

The Earth: day, night, and seasons

Earth turns slowly on its axis once a day. Half of the Earth is in sunlight, and half in shadow (in darkness). As each place moves from sunlight into shadow, it passes from daytime into night.

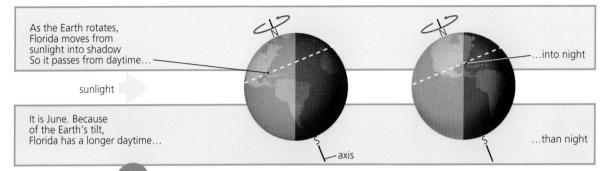

As the Earth rotates, Florida moves from sunlight into shadow So it passes from daytime…

sunlight

It is June. Because of the Earth's tilt, Florida has a longer daytime…

…into night

…than night

axis

The Earth moves round the Sun in a near-circular path. The path is called an **orbit**. Each orbit takes just over 365 days, or one year. The Earth's axis is tilted by 23.5° as shown in the diagrams. Because of this, most regions get varying hours of daylight through the year and varying climatic conditions. In other words, they have different seasons. When it is summer, there are two reasons for the higher temperatures. First, because of the tilt, the surface is more 'square on' to the Sun, so the Sun's radiation is less spread out, as explained in the panel on the left. Second, there are more hours of sunlight.

Solar heating

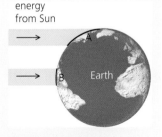

energy from Sun

The Earth is heated by energy transferred from the Sun. Zones A and B receive equal amounts of energy per second. But zone A has a larger surface area, so the energy is more spread out. This is one reason why zone A is colder than zone B.

▶ Many parts of the world get four seasons: spring, summer, autumn, and winter. However in equatorial regions other seasons apply.

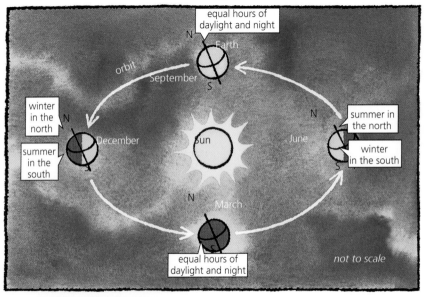

equal hours of daylight and night

orbit

September

winter in the north

December

summer in the south

Sun

June

summer in the north

winter in the south

March

equal hours of daylight and night

not to scale

250

THE EARTH IN SPACE

The Moon

The Moon moves around the Earth in a near-circular orbit. Each orbit takes about a month – or, more accurately, 27.3 days. The Moon also takes 27.3 days to turn once on its axis, which is why it always keeps the same face toward us.

The Moon is smaller than the Earth and has a rocky, cratered surface. The craters were mainly caused by the impact of large meteorites over a billion years ago. The 'seas' are not seas at all, but fairly flat areas of basalt rock.

▲ When we look at the Moon, only the sunlit part is visible.

We see the Moon because its surface reflects sunlight. Once a month, when there is a 'full Moon', the whole of the sunlit side is facing us. But most of the time, we can only see part of the sunlit side. The rest is in shadow.

New Moon | waxing crescent | first quarter | waxing gibbous | Full Moon | waning gibbous | third quarter | waning crescent

▲ The phases of the Moon Half of the Moon is always in sunlight, but we see an increasing then decreasing proportion of that during the course of a month.

The views above are from well north of the equator. For the view well south, in Australia for example, rotate each image of the Moon by 180°.

Orbital speed

Knowing the radius r of the Moon's orbit around the Earth and the **orbital period** (time for one orbit) T, you can calculate the Moon's orbital speed.

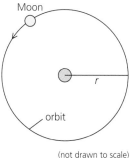

(not drawn to scale)

Assuming the orbit is circular, as in the diagram on the right:

distance travelled during one orbit = $2\pi r$

So: orbital speed = $\dfrac{\text{distance}}{\text{time}} = \dfrac{2\pi r}{T}$

For the Moon, $r = 3.84 \times 10^5$ km, $T = 27.3$ days = 655.2 hours

Using these values in the equation: orbital speed = 3.68×10^3 km/h

That's about the same speed as the world's fastest ever jet aircraft.

1 Give the time taken (to the nearest day) for
   a one rotation of the Earth on its axis
   b one rotation of the Moon on its axis
   c one orbit of the Earth about the Sun
   d one orbit of the Moon about the Earth.
2 Explain the following.
   a From the Earth, we always see the same side of the Moon.
   b The Moon sometimes appears as a crescent.
   c Well north of the equator, there are more hours of daylight in June than in December.
   d Well north of the equator, average temperatures are higher in June than in December.
3 Using information you can find on the opposite page, calculate the average orbital speed of the Earth around the Sun in km/h.

Related topics: shadows 7.1; the Sun 11.5–11.6

11.2 The Solar System (1)

Objectives: – to interpret data about the planets orbiting the Sun – to describe the main features of planetary orbits – to know that other stars have planets.

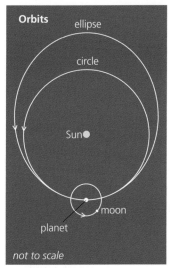

Orbits — *not to scale*

The Earth is one of many **planets** in orbit around the Sun. The Sun, planets, and other objects in orbit, are together known as the **Solar System**. The Solar System is so large that it is almost impossible to show the planets' sizes and their distances from the Sun on the same scale diagram. That is why *two* diagrams have been used here – one below and one on the next page.

Planets are not hot enough to give off their own light. We can only see them because they reflect light from the Sun. From Earth, they look like tiny dots in the night sky. Without a telescope, it is difficult to tell whether you are looking at a star or a planet.

(E) The planets are kept in orbit by the gravitational pull of the Sun. Most are in an almost circular orbit with the Sun at the centre. However, for Mercury and Mars, the orbit is more of an **ellipse** ('stretched out circle'). The planets all travel round the Sun in the same direction. They also travel in approximately the same plane. Most of the planets have smaller **moons** in orbit around them.

The table at the top of the next page gives some data about the planets. Here are two features shown by the data:
- The further a planet is from the Sun, the slower it travels, and the more time it takes to complete an orbit.
- In general, the further the planet is from the Sun, the lower its average surface temperature. This is because the intensity of the Sun's radiation weakens with distance. (Doubling the distance means that the energy per second reaching each square metre of a surface is reduced to a quarter.)

On the next spread, there is more information about the planets, and about other objects orbiting the Sun.

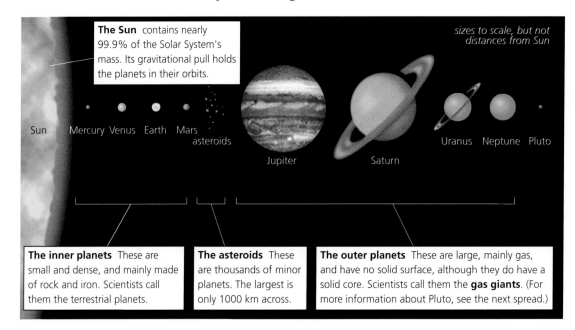

The Sun contains nearly 99.9% of the Solar System's mass. Its gravitational pull holds the planets in their orbits.

sizes to scale, but not distances from Sun

Sun — Mercury — Venus — Earth — Mars — asteroids — Jupiter — Saturn — Uranus — Neptune — Pluto

The inner planets These are small and dense, and mainly made of rock and iron. Scientists call them the terrestrial planets.

The asteroids These are thousands of minor planets. The largest is only 1000 km across.

The outer planets These are large, mainly gas, and have no solid surface, although they do have a solid core. Scientists call them the **gas giants**. (For more information about Pluto, see the next spread.)

THE EARTH IN SPACE

	Mercury	Venus	Earth	Mars	Jupiter	Saturn	Uranus	Neptune
average distance from Sun/ million km	58	108	150	228	778	1420	2870	4490
time for one orbit/ years	0.24	0.62	1.00	1.88	11.86	29.46	84.01	164.8
diameter (at equator)/ km	4880	12 100	12 800	6790	142 980	120 540	51 120	49 530
mass compared with Earth (Earth = 1)	0.06	0.82	1.00	0.11	318	95.2	14.5	17.2
gravitational field strength at surface/ N/kg	3.8	8.8	9.8	3.8	25	10.4	10.4	13.8
average density/ g/cm^3 *	5.4	5.2	5.5	3.9	1.3	0.7	1.3	1.6
average surface temperature	170 °C	460 °C	15 °C	−23 °C	−110 °C	−140 °C	−210 °C	−200 °C
number of moons	0	0	1	2	79	82	27	4

* For comparison: water has a density of 1.0 g/cm^3 (1 cubic centimetre has a mass of 1 gram); iron has a density of 7.9 g/cm^3

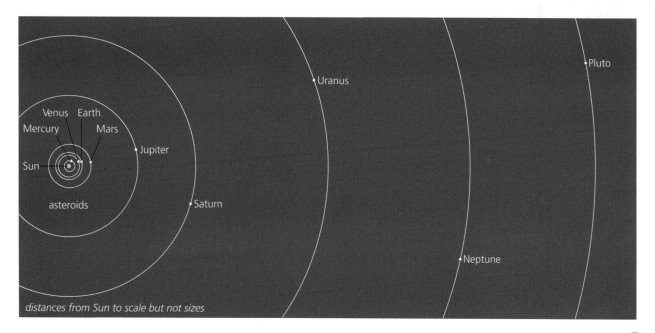

distances from Sun to scale but not sizes

The Sun is just one of billions of stars. Many planets have been detected around other stars, but they are much too far away to be viewed directly through a telescope. To fit the nearest star on the diagram above, the page would have to be more than a kilometre wide!

** In 2006, Pluto was reclassified as a dwarf planet (see the next spread).

To answer the following, you will need to refer to data in the table at the top of the page.
1 Which is the largest of the gas giants?
2 Which planet has the highest gravitational field strength at its surface?
3 Which planet orbits the Sun at the highest speed?
4 Ceres, a dwarf planet, takes 4.6 years to orbit the Sun. Between the orbits of which two planets does Ceres lie? Give a reason for your answer.

5 a Why does Mars have a lower average surface temperature than Earth?
  b Extreme global warming has given one of the planets a much higher surface temperature than expected. Which planet do you think this is? Give a reason for your answer.
6 If the speed of light is 300 000 km/second, use the equation time = distance/speed to work out how long it takes the Sun's light to reach us.

Related topics: Sun, Earth, and Moon **11.1**; gravity and orbits **2.9**, **2.14**, and **11.4**; formation of the Solar System **11.5**

253

11.3 The Solar System (2)

Objectives: – to describe the main characteristics of the inner planets, asteroids, uter planets, and comets – to know that there are also dwarf planets.

The inner planets

Mercury is the closest planet to the Sun. In many ways, it is a larger version of our own Moon. It has a cratered surface and no atmosphere.

Venus is the brightest object in the night sky (apart from the Moon). It is almost the same size as the Earth, but conditions there are very different. The planet is covered by thick clouds of sulfuric acid, and its atmosphere (97% carbon dioxide) causes a severe greenhouse effect. Down on the surface, the temperature can reach nearly 500 °C.

Earth is the only planet in the Solar System known to support life.

Mars is sometimes called the red planet because of its surface colour. It has a thin atmosphere (mainly carbon dioxide), a dusty surface, and polar caps. From the shapes of some of its valleys, scientists think that water may once have flowed there. It also has volcanoes, although none are active.

The asteroids

The asteroids have dimensions ranging from a few kilometres up to 1000 km. Most have orbits between those of Mars and Jupiter. But some have much more elliptical orbits that cross the paths of other planets. There is some evidence that, around 65 million years ago, an asteroid about 10 km across struck the Earth. The effects of this may have caused the extinction of the dinosaurs.

▲ This asteroid, photographed by the *Galileo* spacecraft, is nearly 60 km long.

The outer planets

Jupiter is more massive than all the other planets put together. It is mainly gas (its atmosphere is 90% hydrogen) and has no solid surface. The **Great Red Spot** is a huge storm that has raged for centuries. One of Jupiter's moons, Io, has active volcanoes on it – the first to be found beyond the Earth.

▲ Jupiter's Great Red Spot is a storm so big that the Earth would fit inside it!

▲ Saturn's rings are millions of pieces of ice, each in its own orbit about the planet.

THE EARTH IN SPACE

Saturn too is mainly gas. It is surrounded by very thin rings. These are not a solid mass, but millions of pieces of ice (mostly), ranging in size from grains to boulders. Each is a tiny 'moonlet' in its own orbit.

Uranus is another gas giant. It is unusual in that its axis of rotation is tilted at more than 90°. It too has rings, but much fainter ones than Saturn's.

Neptune is the outermost of the gas giants. It also has a faint ring system.

Pluto was, for many years, commonly known as the outermost planet. However, because of its small size and other factors, most astronomers no longer consider it to be one of the main planets (see right).

Comets, meteors, and meteorites

Comets have highly elliptical orbits which, in some cases, can take them out beyond Pluto and then close in to the Sun. In the 'head' of a comet, there is an icy lump, typically several kilometres across. Heated by the Sun, particles of dust and gas stream off it into space, forming a huge 'tail' millions of kilometres long. This is visible because it reflects the sunlight.

*As the Earth moves through space, it runs into tiny grains of material which hit the atmosphere so fast that they burn up. Each one causes a streak of light called a **meteor**. Rarely, a larger chunk of material reaches the ground without completely burning up. The chunk is called a **meteorite**.

Ice in space
To astronomers, 'ice' does not necessarily mean frozen water. It can also mean frozen carbon dioxide, methane, or ammonia.

Pluto reclassified
In 2006, an international committee of astronomers carried out a reclassification of the planets. As a result, Pluto lost its status as a planet and was reclassified as a dwarf planet, along with Ceres (in the asteroid belt) and Eris (further out than Pluto). There are at least six other dwarf planets beyond Pluto.

▲ Halley's comet can be seen from Earth every 76 years. The last time was in 1986.

▲ A large meteorite caused this impact crater in Arizona, USA, around 50 000 years ago. The crater is 800 metres across.

To answer the following, you may need to refer to Spread 11.2 and data in the table at the top of this page.
1. **a** List the *inner planets* and the *outer planets* in order.
   b Give the *two* main ways in which these are different from each other.
2. Name one dwarf planet.
3. Expain why it would be very difficult for astronauts to land on **a** Venus **b** Jupiter.
4. **a** What is the difference between the orbit of a comet and that of most planets?
   b* How is the 'tail' of a comet formed?

Related topics: planetary data **11.2**; comets in orbit **11.4**; formation of the Solar System **11.5**

11.4 Objects in orbit

Objectives: – to know that orbits are the result of speed and gravitational attraction – to describe uses of artificial satellites and geostationary orbits.

Gravity and orbits

With no force acting on it, a planet, a moon, or any other object, would travel through space in a straight line. To move in the curved path of an orbit, there must be an inward force acting on it. That force is provided by gravity.

There is gravitational attraction between *all* masses. The force between everyday objects is far too weak to detect, but with objects as massive as moons and planets, the force is considerable, and controls their motion.

Gravitational force weakens with distance. Isaac Newton found that the force obeys an inverse square law: *doubling* the distance between two masses reduces the gravitational force between them to a *quarter*... and so on.

Comets in orbit

Comets have highly elliptical orbits. Here is an example:

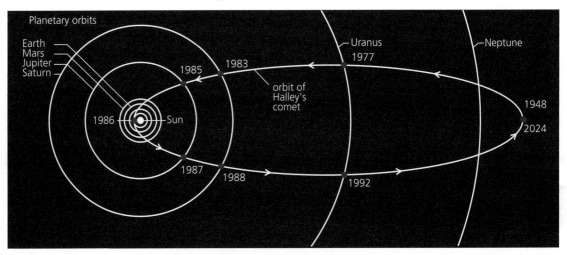

▲ The orbit of Halley's comet. From the positions and dates, you can see that the comet speeds up as it approaches the Sun and slows down as it moves away from it.

A comet has least speed when it is furthest from the Sun. That is also when the gravitational pull on it is weakest. As the comet moves closer to the Sun, the force of gravity increases. Also, the comet speeds up as it 'falls' towards the Sun – its gravitational potential energy is converted to kinetic energy.

Satellites in orbit

Any object in orbit around a more massive one is called a **satellite**. So the Moon is a natural satellite of the Earth. However, when people talk about 'satellites', they usually mean artificial satellites launched from Earth.

There are hundreds of artificial satellites in orbit around the Earth. Most are in circular orbits. A satellite in a low orbit needs the highest speed. For example, a satellite at a height of 300 km, just above the Earth's atmosphere, must travel at 29 100 km/hour to maintain a circular orbit. At this speed, it takes 86 minutes to orbit the Earth. This is the **period** of its orbit. A higher orbit requires a lower speed, and the period is longer.

THE EARTH IN SPACE

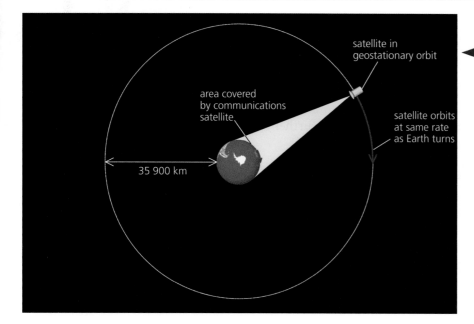

▲ A satellite in a geostationary orbit always appears in the same position relative to the ground because the period of its orbit, 24 hours, matches the period of the Earth's rotation.

To be geostationary, a satellite must be put in a circular orbit 35 900 km above the equator. The required speed is 11 100 km/hour.

Communications satellites beam radio, TV, and other signals from one part of the Earth to another. Satellites like this are normally put into a **geostationary orbit**: their motion exactly matches the Earth's rotation so that they appear stationary relative to the ground. As a result, dish aerials on the ground do not have to track them, but can point in a fixed direction.

Navigation satellites are used by cars, boats, and planes to locate their position. The **Global Positioning System (GPS)** has a network of satellites transmitting synchronized time signals. Down on the ground, a receiver picks up signals from different satellites, compares their arrival times, and uses the data to calculate its position to within a few metres.

Monitoring satellites, such as weather satellites, contain cameras or other detectors for scanning and surveying the Earth. Some are in low orbits which pass over the North and South Poles. As the Earth rotates beneath them, they can scan the whole of its surface.

Astronomical satellites, such as the Hubble Space Telescope, contain equipment for observing distant stars and galaxies. Unlike observatories on Earth, the light or other radiation they receive is not disrupted and weakened by the presence of the atmosphere.

▲ Using a GPS receiver. The readings on the display give the position.

1. The diagram on the right shows the orbit of a comet. At which point
   a is the Sun's gravitational pull on the comet greatest
   b is the Sun's gravitational pull on the comet least
   c does the comet have the greatest speed
   d does the comet have the least speed?
2. a What is a *geostationary* orbit?
   b What is the period (orbit time) of a satellite in a geostationary orbit?
   c Why are communications satellites normally put into a geostationary orbit?

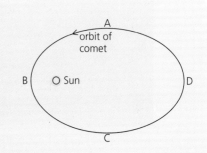

Related topics: force, motion, and gravity **2.7** and **2.9**; circular motion and orbits **2.14**; sending signals **7.12**

11.5 Sun, stars, and galaxies (1)

Objectives: – to know what a light year is – to explain what galaxies are and that there are billions of them in the Universe – to describe how stars are formed.

The Sun is a star. As stars go, it is rather average. There are much bigger and much brighter stars. However all other stars look like tiny dots to us because they are much further away.

Millions and billions
1 million = 1 000 000
= 10^6
1 billion = 1000 million
= 1 000 000 000
= 10^9

Light year

Distances between stars are so vast that astronomers have special units for measuring them. For example:

One **light year** (**ly**) is the distance travelled by light in one year. The speed of light is nearly 300 000 kilometres per second, so light travels more than 9 million million kilometres in one year. Expressed more accurately:

$$1 \text{ light year} = 9.5 \times 10^{12} \text{ km}$$

The nearest star to the Sun, *Proxima Centauri*, is 4.2 light years away. In other words, its light takes 4.2 years to reach us: we see it as it was 4.2 years ago. For comparison, light from the Sun takes 8 minutes to reach us.

Galaxies

The Sun is a member of a huge star system called a **galaxy**. This contains at least 100 billion (10^{11}) stars, and is more than 100 000 light years across. Between the stars, there is thinly spread gas and dust called **interstellar matter**. The gas is mainly hydrogen. The galaxy is slowly rotating, and is held together by gravitational attraction.

Our galaxy is called the **Milky Way**. You can see the edge of its disc as a bright band of stars across the night sky. It is just one of many billions of galaxies in the known **Universe**.

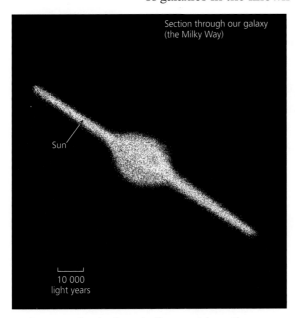

▲ Our own Sun is about halfway out from the centre of our galaxy.

▲ The Andromeda Galaxy is 2 million light years away. It is very similar in structure to our own galaxy.

THE EARTH IN SPACE

The birth of a star

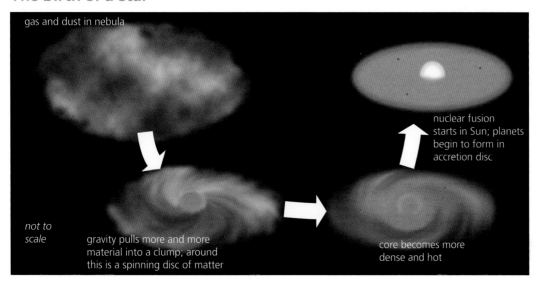
▲ The formation of the Solar System

Scientists think that the Sun and the rest of the Solar System formed about 4500 million years ago in a huge, rotating cloud of gas (mainly hydrogen) and dust called a **nebula**. Because of gravity, the nebula slowly collapsed inwards, rotating faster as it went. As more and more material was pulled in, a massive clump started to form at its centre. And as the gravitational potential energy of the incoming material was converted into thermal energy this **protostar** heated up. Deep inside it, the gas became hotter and more compressed. Eventually, the temperature and pressure were high enough to trigger **nuclear fusion** with hydrogen as its fuel (see next page). The protostar had become a star.

When the outward pressure from its radiation balanced the pull of gravity, this new star became stable. Around it, the remaining gas and dust formed a huge, rotating **accretion disc** ('accretion' means gradual growth by the addition of material). Here, grains of material were slowed by collisions and pulled into clumps by gravity. These clumps would become planets and moons.

The Sun's radiation drove off most of the gas from the inner planets, so they were left small and rocky. However, further out, where it was cooler, the planets retained gas: they became **gas giants**.

▲ The Great Nebula in the constellation of Orion. Stars are forming in this huge cloud of gas and dust.

1 Explain what is meant by the terms
   a nebula b galaxy c Milky Way d Universe.
2 a What is an *accretion disc*?
   b Why does the material in an accretion disc start to collect in clumps?
   c When the Sun was a protostar, what made the material in it heat up?
3 a What is the name of the process that supplies the Sun with its energy?
   b What element does the Sun use as its nuclear fuel?
4 *Proxima Centauri*, a star, is 4.2 light years away.
   a What is meant by a light year?
   b How far away is *Proxima Centauri* in km?
   c Today, the fastest rockets can reach speeds of up to 50 000 km/hour. If a spacecraft travelled to *Proxima Centauri* at this average speed, about how long would its journey take?

Related topics: potential and thermal energy 4.1; Solar System 11.2–11.3; gravity 11.4

259

11.6 Sun, stars, and galaxies (2)

Objectives: – to know what powers stars – to describe what eventually happens to stars, and how super novae, neutron stars and black holes are formed.

Atomic essentials

Everything is made from about a hundred basic substances called elements. The smallest bit of an element is an atom. The nucleus of an atom is made up of protons and (in most cases) neutrons. Each element has a different number of protons in the nucleus.

hydrogen atom (1 proton)

helium atom (2 protons)

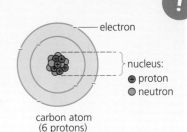
carbon atom (6 protons)

Fusion power in the Sun

If hydrogen nuclei can be made to fuse (join) together to form helium nuclei, energy is released. But nuclei do not readily join because they are electrically charged, and repel each other. To make them fuse, they have to collide at extremely high speeds. In practice, this means maintaining a gas at an extremely high temperature: for example, 15 million °C in the Sun's core.

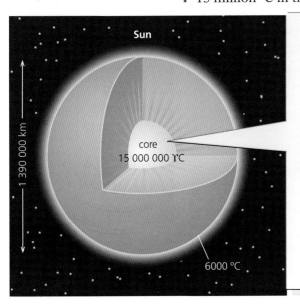

Fusion in the Sun's core

The Sun is 73% hydrogen and 25% helium. Energy is released as hydrogen is converted into helium.

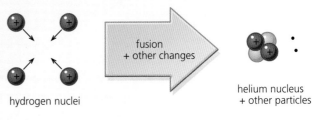

Four hydrogen nuclei fuse together for each helium nucleus formed. This is a multi-stage process which also involves the creation of two neutrons from two protons.

Death of a star

In the Sun's core, thermal activity prevents gravity from pulling the material further inwards. However, in about 6 billion years time, all of the hydrogen in the core will have been converted into helium, hydrogen fusion will cease, and the core will collapse. At the same time, the Sun's outer layer will expand to about 100 times its present diameter and cool to a red glow. The Sun will then be a **red giant**. Eventually, its outer layer will drift into space, exposing a hot, extremely dense core called a **white dwarf**. This tiny star will use helium as its nuclear fuel, converting it into carbon by fusion. When the helium runs out, the star will cool and fade for ever.

THE EARTH IN SPACE

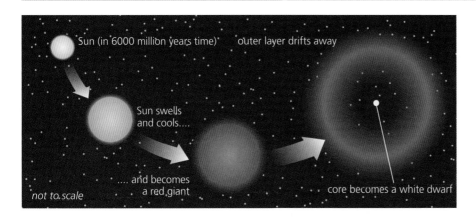

◀ In about 6 billion years time, the Sun will become a red giant. Later, its core will become a white dwarf, before it cools and fades for ever.

E Supernovae...

In every galaxy, new stars are forming and old ones are dying. But more massive stars have a different fate from that of our Sun. Eventually, they become **red supergiants**, and blow up in a gigantic nuclear explosion called a **supernova**. This leaves a core in which matter is so compressed that electrons and protons react to form neutrons. The result is a **neutron star**.

...and black holes

When the most massive stars of all explode, the core cannot resist the pull of gravity and goes on collapsing. The result is a **black hole**. Nothing can escape from it, not even light, so it cannot be seen. However the presence of a black hole can be detected by its effect on matter near it or light going past it.

Scientists think that there is a massive black hole at the centre of most galaxies.

Made from stardust

In stars, fusion reactions change lighter elements into heavier ones. However, to make very heavy elements (gold and uranium for example), the extreme conditions that create a supernova are needed. That is because, to make elements heavier than iron, energy must be *supplied* for fusion, and is not released by it. The Sun and inner planets contain very heavy elements. This suggests that the nebula in which they formed included 'stardust' from an earlier supernova. In other words, the Sun is a **second-generation star**.

Astronomical torches
Type 1 a supernovae all reach the same brightness at their peak, and can even outshine their galaxy. This makes them very useful for estimating the galactic distances.
Imagine two identical small torches. One is close to you and looks bright, the other is much further away and looks dim. Type 1a supernovae are like astronomical torches. By comparing their apparent brightness, their relative distances can be calculated.

1. What element will the Sun use as its nuclear fuel when its core runs out of hydrogen?
2. Describe in stages what will happen to the Sun when its core runs out of hydrogen.
3. Stars and planets contain many elements, but the nebulae in which they form are mainly hydrogen. What process produces all the other elements?
4. What is a supernova?
5. Why are some supernovae useful in estimating the distances of galaxies?
6. How are *neutron stars* and *black holes* formed?
7. What evidence is there that the nebula in which the Solar System formed contained remnants from an earlier supernova?

Related topics: atoms and elements **10.1**; gravity **10.4**; nuclear reactors **10.6–10.7**; proton-neutron conversion **10.9**

11.7 The expanding Universe

Objectives: – to describe the evidence that makes scientists think that the Universe is expanding – to know how its age can be estimated from this.

Stars, galaxies, and gravity
The Sun is one star in a huge star system called a galaxy. This contains over 100 billion stars. Stars and galaxies are shaped by gravity. There is gravitational attraction between *all* masses: the greater the masses and the closer they are, the stronger the force.

The light year
One light year is the distance travelled by light in one year. It is equal to 9.5×10^{12} km. Typically, neighbouring stars are a few light years apart.

Electromagnetic waves
These include light and microwaves. Here are some typical wavelengths:
Visible spectrum from 0.000 4 mm (violet light) to 0.000 7 mm (red light)
Microwaves from 0.001 mm to 300 mm

wavelength

▶ With radio telescope arrays like this, scientists have been able to detect radio waves from the most distant parts of the Universe, including microwaves which may be the remnants of radiation from the Big Bang.

There are billions of galaxies in the Universe. Neighbouring galaxies are, typically, a few million light years apart. The most distant galaxies detected from Earth are more than 13 billion light years away – their light has taken over 13 billion (13×10^9) years to reach us.

The expanding Universe

When objects move away from Earth at high speed, the light waves from them become 'stretched out'. This is known as the **Doppler effect**. It means that the wavelengths are shifted towards the red (longer wavelength) end of the visible spectrum. This is called **red shift**, and it can be used to calculate the speed.

In the 1920s, Edwin Hubble observed that light from distant galaxies is red shifted, and that, in general, the red shift increases with the distance of the galaxy. This implies that the more distant galaxies are receding (moving away) from us at high speed. We are living in an expanding Universe.

The Big Bang theory

According to this theory, the Universe (and time) began many billions of years ago when a single, hot 'superatom' erupted in a burst of energy called the **Big Bang**. All the matter in the Universe came from this. Here are two pieces of evidence to support the theory:
- As the galaxies appear to be moving apart, they may once have been together in the same space.
- Radio telescopes have picked up microwave radiation of a particular frequency coming from every direction in space. This may be the heavily red-shifted remnants of radiation from the Big Bang. It is called **cosmic microwave background radiation**, or **CMBR** for short.

The Big Bang was not an explosion into existing space. Space itself started to expand: the galaxies are separating because the space between them is increasing. To think about this, it helps to use a simplified model of an expanding Universe, such as the one shown on the next page.

THE EARTH IN SPACE

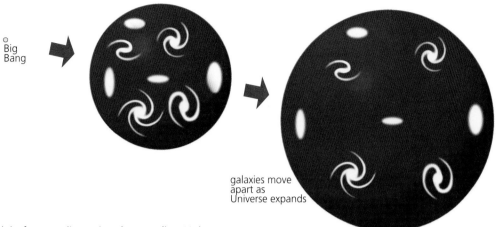

▲ A model of a two-dimensional expanding Universe

Above, the Universe is represented by the surface of a balloon. Imagine that you are on one of the galaxies as the balloon inflates. All the others appear to move away from you. The more distant they are, the faster they recede. This applies wherever you are. No single galaxy is at the centre of the expansion.

Ⓔ Estimating the age of the Universe

Using red shift, scientists have measured the rate at which the galaxies appear to be moving apart. They represent this using the **Hubble constant**, H_0

$$H_0 = \frac{v}{d}$$
v = speed at which a galaxy is moving away from Earth
d = distance of the galaxy from Earth

Different methods of measuring H_0 have produced different results, but many scientists now agree on a value of 2.3×10^{-18} per second. Assuming that at the Big Bang, the Earth and galaxy were at the same point and have separated at a constant rate, and using time = distance/speed:

$$\text{age of universe} = \text{time to separate} = \frac{d}{v} = \frac{1}{H_0}$$

Using the value of H_0 above gives the age as 4.35×10^{17} s, or 13.8 billion years.

Expanding faster

There is evidence that the Universe's rate of expansion is increasing - an unexpected result because it was assumed that gravitational attraction would slow it down. The most likely cause is thought to be a repulsive force produced by **dark energy** (see spread 12.4). This can't be detected directly, but it is needed to explain some gravitational effects.

Q

1 4   4 million   13 billion   14 billion
         (4×10^6) (1.3×10^{10}) (1.4×10^{10})

Which of the above numbers could represent
a the separation between two neighbouring galaxies, in light years
b the separation between two neighbouring stars, in light years
c the distance from Earth to the most distant galaxies observed, in light years
d the age of the Universe, in years.

2 Light from distant galaxies shows red shift.
a What is meant by red shift?
b What is thought to be the cause of the red shift?

3 What evidence is there that the Universe may have started with a Big Bang?

4 a What is the connection between the Hubble constant and the age of the Universe?
b What assumption does this connection make?
c If H_0 is found to be 2.5×10^{-18} /s, what value does this give for the age of the Universe?

Related topics: light, radio waves, microwaves **7.10–7.11**; gravitational attraction **11.4**; distances of galaxies **11.6**; dark matter and dark energy **12.4**; Edwin Hubble **12.4**

Check-up on the Earth in space

Further questions

1

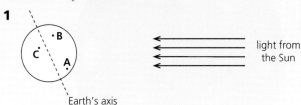

The diagram shows the Earth and three cities **A**, **B**, **C** on the Earth's surface.

a State:
  (i) which cities are in daylight; [1]
  (ii) which city receives the most amount of light during the day. [1]
b State how long it would take city **C** to return to the same place again as the Earth spins on its axis. [1]

WJEC

2 The diagram shows the Earth in four different positions in its orbit around the Sun. Assume that the Earth is always the same distance from the Sun.

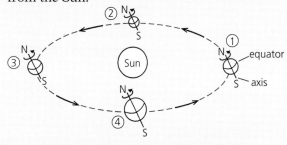

a i When the Earth is in position 1, it is summer in the northern hemisphere. Why is this? [2]
  ii What is the season in the northern hemisphere when the Earth is in position 2? position 3? [2]
  iii What is the season in the southern hemisphere when the Earth is in position 3? [1]
b The order of some planets outwards from the Sun is as follows:
Mercury   Venus   Earth   Mars
Jupiter   Saturn   Uranus
How would you expect the average day-time temperature on Uranus to compare with that on Mars? Explain your answer. [2]

MEG

3 a | *Earth*   *Sun*   *Moon* |

Copy the sentences below and select **one** or **two** words from the boxes above to complete the sentences. Each word can be used once, more than once or not at all.
  i The _____ takes 24 hours to spin on its own axis. [1]
  ii The _____ takes 365 days to orbit the _____ [1]
  iii The _____ takes 27 days to orbit the _____ [1]
b Explain why
  i in summer, daytime lasts longer than night, [1]
  ii we can see the Moon, [1]
  iii the shape of the Moon appears to change over a four week period. [1]

WJEC

4 a The box contains the names of eight of the nine planets in the Solar System.

| Earth | Jupiter | Mars | Mercury |
| Neptune | Saturn | Uranus | |

  i Name the planet which has **not** got its name in the box. [1]
  ii Which planet takes about 365 days to complete its orbit? [1]
  iii Which planet has the shortest orbit? [1]
  iv Which star do **all** the planets orbit? [1]
  v Which planet has the longest orbit? [1]
  vi What is the shape of each orbit? [1]
b i We can see the Moon. Explain why. [2]
  ii We can see stars. Explain why. [2]
  iii On some nights we cannot see any stars. Explain why. [2]
c Copy and complete each sentence by choosing the correct words from the box. You may use the words once, more than once or not at all.

| Big Bang | black hole | galaxy | Milky Way |
| Solar | System | star | Universe |

The Sun is at the centre of the _____ which is part of a _____ called the _____.

This is all part of a much larger system called the _____ which scientists think began with the _____.

E When a supermassive star eventually collapses, a _____ is formed at its centre. [6]

5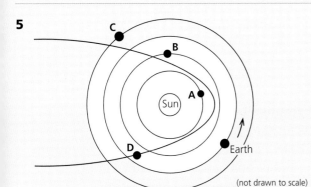

(not drawn to scale)

The diagram shows the orbits of some bodies around the Sun. The arrow shows the direction of the Earth's orbit.
- **a** Choosing from **A**, **B**, **C** and **D**, state which body is
  - **i** a comet [1]
  - **ii** Venus. [1]
- **b** Copy the diagram and mark on the orbit of **A** an arrow to show the direction in which it moves. [1]
- **c i** Describe the shape of the orbit of **D**. [1]
  - **ii** Name the force which keeps **D** in its orbit. [1]

WJEC

6 The sun is orbited by eight planets as well as dwarf planets and other objects.
- **a** Give the **two** main differences (apart from temperature) between the four inner planets and the four outer planets. [2]
- **b** Apart from a dwarf planet, Ceres, what else orbits the Sun between the inner and outer planets? [1]
- **c** Ceres takes 4.04×10^4 hours to orbit the Sun. Calculate its orbital speed in km/hr, assuming that it has a circular orbit of radius 4.13×10^8 km. [2]

7 Copy and complete the passage below, using words from the box (you do not need to use them all).

accretion disc	gravity	supernova
nebula	Big Bang	protostar
planets	star	galaxy

The Solar System formed in a huge cloud of gas and dust called a _____. This contained remnants from the explosion (called a _____) of a much older star. In the cloud, _____ started to pull material together into clumps. At the centre, the biggest clump (called a _____) got hotter and hotter until fusion started and it became a _____. Around it a huge, rotating _____ had formed. In this, material was starting to clumped together to form the _____. [7]

8 a Astronomers think that the Universe is expanding. Give one piece of evidence supporting this idea. [2]
E b Explain what is meant by Cosmic Microwave Background Radiation. [2]
c The Sun is powered by nuclear fusion. Explain what is meant by nuclear fusion. [2]

9 It is a very clear night. Luke is looking up and sees the Milky Way, a hazy band of light across the night sky.

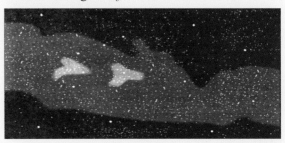

- **a** What in the Milky Way produces this hazy band of light? [1]
- **b** The planet Venus is also seen as a bright speck of light. Explain why Venus shines brightly. [2]

not to scale

- **c** The diagram shows the orbits of Venus and the Earth about the Sun.
  - **i** What causes these planets to orbit the Sun? [1]
  - **ii** Suggest why the orbit time for Venus is less than for the Earth. [2]
- **d** A year later Luke looks up into the sky from the same place and at the same time of night.
  - **i** Copy the table and put ticks in the boxes that would fit in with his observations. [1]

	in the same position	in a new position
Venus		
Milky Way		

  - **ii** Explain Luke's observations. [2]

MEG

Revision checklist

Core Level
- ☐ The Earth is a planet in orbit around a star (the Sun). (11.1)
- ☐ The types of radiation emitted by the Sun. (11.1)
- ☐ How the Earth's rotation causes day and night. (11.1)
- ☐ The tilt of the Earth's axis: why a year has different seasons. (11.1)
- ☐ How the Moon orbits the Earth. (11.1)
- ☐ Why we see different phases of the Moon during the course of a month. (11.1)
- ☐ What the Solar System is. (11.2)
- ☐ How the Sun contains most of the mass of the Solar System. (11.2)
- ☐ The order of the planets in the Solar System. (11.2)
- ☐ The four inner planets are small and rocky; the four outer planets are gas giants. (11.2 and 11.3)
- ☐ Moons, asteroids, dwarf planets, and comets. (11.2 and 11.3)
- ☐ How the Sun's gravitational pull keeps the planets in orbit. (11.4)
- ☐ How gravitational force weakens with distance. (11.4)
- ☐ How the planets and moons were formed. (11.5)
- ☐ What an accretion disc is. (11.5)
- ☐ The Sun is a medium-sized star in a galaxy of billions of stars called the Milky Way. (11.5)
- ☐ Measuring astronomical distances in light years. (11.5)
- ☐ How the distances between stars compares with the distances between galaxies. (11.5)
- ☐ The Universe is made up of billions of galaxies. (11.5)
- ☐ The evidence that the galaxies are rushing away from each other. (11.7)
- ☐ How the expansion of the Universe supports the theory that it started with a Big Bang. (11.7)

Extended Level
As for Core Level, plus the following:
- ☐ How to calculate orbital speed, knowing the radius and period (time) of an orbit. (11.1)
- ☐ The link between a planet's distance from the Sun and the period of its orbit. (11.2)
- ☐ The link between a planet's distance from the Sun and its surface temperature. (11.2)
- ☐ Interpreting other data about the planets. (11.2)
- ☐ How some objects in space, including comets, have elliptical orbits. (11.4)
- ☐ How the speed of an object in an elliptical orbit changes. (11.4)
- ☐ How stars are formed in a nebula. (11.5)
- ☐ What a protostar is. (11.5)
- ☐ How stars are powered by nuclear fusion reactions. (11.5 and 11.6)
- ☐ What happens to a star at the end of its life. (11.6)
- ☐ What red giants, supernovae, neutron stars, and black holes are. (11.6)
- ☐ How the brightness of some supernovae can be used to estimate how far away other galaxies are. (11.6)
- ☐ What Cosmic Microwave Background Radiation is, and why it is important to scientists. (11.7)
- ☐ How redshifted light can be used to estimate the speed at which a galaxy is moving away from Earth. (11.7)
- ☐ What the Hubble constant is. (11.7)
- ☐ How the Hubble constant can be used to estimate the age of the Universe. (11.7)

12

History of key ideas

- CHANGING IDEAS ON FORCE, MOTION, ENERGY, AND HEAT
- CHANGING IDEAS ON LIGHT, RADIATION, AND ATOMS
- CHANGING IDEAS ON MAGNETISM AND ELECTRICITY
- CHANGING IDEAS ABOUT THE EARTH IN THE UNIVERSE

This ancient stone circle at Stonehenge in Wiltshire, England, was built before 1500 BCE. Its builders left no written records to explain its purpose. It may have been a centre for ceremonies associated with death or healing, but the alignments of the stones also suggest that it could have been used to observe the movements of the Sun and the Moon and for identifying the seasons.

12.1 Force, motion, and energy*

Forces and motion

On Earth, unless there is a force to overcome friction, moving things eventually come to rest. Over 2300 years ago, this led Aristotle and other Greek philosophers to believe that a force was always needed for motion. The more speed something had, the more force it needed. But in the heavens, the Sun, Moon, and stars obeyed different rules. They moved in circles for ever and ever. These ideas were generally accepted until the early 1600s, when Galileo Galilei started to come up with new ideas about motion. From his observations, Galileo deduced that, without friction, sliding objects would keep their speed. Also, all falling objects, light or heavy, would gain speed at the same steady rate.

Our present-day ideas about forces and motion mainly come from Isaac Newton, who put forward his three **laws of motion** in 1687. Our definition of force is based on his second law: force = mass × acceleration. Newton also realized that 'heavenly bodies' did not obey different rules from everything else. The motion of the Moon around the Earth was controlled by the same force – gravity – that made objects fall downwards on Earth. Newton is supposed to have had this idea while watching an apple fall from a tree, although his mathematical treatment of gravity was much more complicated than this simple experience suggests.

In 1905, Albert Einstein put forward his **special theory of relativity**. From this, we now know that, near the speed of light, Newton's second law is no longer valid. However, at the speeds we normally measure on Earth, the law is quite accurate enough.

▲ It is said that Galileo investigated the laws of motion by dropping cannon balls from the top of Pisa's famous tower. There is no evidence to support this story. However, Galileo was born in Pisa, Italy (in 1564), and studied and lectured there.

Energy and heat

The modern, scientific meaning of energy arose in the early 1800s when scientists and engineers were developing ways of measuring the performance of steam engines. Steam engines used forces to move things. They did **work**. To do this, they had to spend **energy**. So it made sense to measure energy and work in the same units (we now use the joule).

▶ A modern replica of Stephenson's *Locomotion*. The original, built in 1825, was used on the world's first public steam railway. The development of steam engines like this led to advances in scientists' understanding of the relationship between work, energy, and heat.

HISTORY OF KEY IDEAS

With the idea of energy established, people soon realized that energy could exist in different forms – electrical, potential, kinetic, and so on. However, the law of conservation of energy was not developed until 1847.

Today, we link heat with energy. However, scientists once thought that heat was an invisible, weightless fluid called 'caloric' which flowed out of hot things and was squeezed from solids when they were rubbed. In the 1790s, Count Rumford did some experiments which suggested that the caloric theory was wrong. While boring cannon barrels, he found that he could get an endless supply of heat by keeping the borer turning. If heat was a fluid, then the supply should run out. Instead, the amount of heat seemed to be directly linked with the amount of work being done. The link between work and heat was firmly established in 1849 by James Joule. He found that it always took 4.2 joules of work to produce 1 calorie of heat (an old unit, equivalent to the heat required to increase the temperature of 1 gram of water by 1 °C). However, Joule's work did not explain what heat *was*.

▲ Boring out cannon barrels made them hot. Count Rumford discovered that the amount of heat produced was related to the amount of work done during the boring process.

We now know that materials are made up of particles (atoms or molecules) which are in a state of random motion, and that heat is associated with that motion. In a solid or liquid, the particles vibrate. In a gas they move about freely at high speed. The higher the temperature, the faster the particles move. The random motion of the particles is called **thermal activity**, and the energy which an object has because of it is called **internal energy** (the sum of the kinetic and potential energies of all the particles).

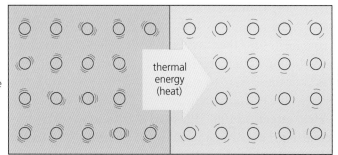

If a hot object is put in contact with a colder one, as above, energy is transferred from one to the other because of the temperature difference. This energy is called **heat**. So internal energy is the *total* amount of energy due to thermal activity, while heat represents an amount of energy *transferred*. However, for simplicity, both can be called **thermal energy**.

◀ In engines, releasing thermal energy by burning fuel is one stage in the process of producing motion.

12.2 Rays, waves, and particles*

Light and radiation

Over 2300 years ago, the Ancient Greeks knew that light travelled in straight lines. The Romans used water-filled glass spheres to magnify things. But it was not until the 1200s that glass lenses were first made, for spectacles. The telescope was invented in the early 1600s. In the same century, Snell discovered the law of refraction, Huyghens suggested that light was a form of wave motion, and Newton demonstrated that white light was a mixture of colours. Newton also tried to explain the nature of light. He thought that light was made up of millions of tiny 'corpuscles' (particles).

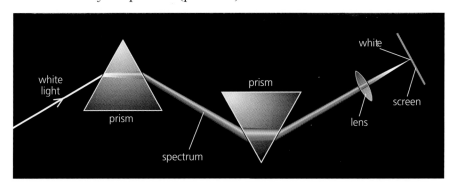

▶ One of Newton's experiments. Newton passed white sunlight into a glass prism, and produced a spectrum. When he recombined the colours with a second prism, he obtained white light again. He concluded that the colours must be from the white light and not produced by the glass.

In the early 1800s, Thomas Young investigated the interference and diffraction of light and successfully used the wave theory to explain these effects. From his results, he was also able to calculate a value for the wavelength of light. Young's work seemed to put an end to Newton's 'corpuscles', but these were to appear later in another form. Some materials give off electrons when they absorb light. This is called the **photoelectric effect**. In 1905, Einstein was able to explain it by assuming that light consisted of particle-like bursts of wave energy, called **photons**. Light, it seemed, could behave like waves *and* particles.

James Clerk Maxwell was the first to put forward the idea that light was a type of **electromagnetic radiation**. He did this in 1864. From his theoretical work on electric and magnetic fields, he predicted the existence of electromagnetic waves, calculated what their speed should be, and found that it matched the speed of light. His equations also predicted the existence of radio waves, although 'real' radio waves were not detected until the 1880s. X-rays were discovered by Wilhelm Röntgen in 1895, but their electromagnetic nature was not established until 1912.

In 1896, Henri Becquerel detected a penetrating radiation coming from uranium salts. He had discovered **radioactivity**. Later, Marie Curie showed that the radiation came from within the atom and was not due to reactions with other materials. In 1899, Ernest Rutherford investigated radioactivity and identified two types of radiation, which he called alpha and beta. The following year, he discovered gamma rays. Today, we know that waves, such as gamma radiation, can behave like particles, and that particles can also behave like waves. Scientists call this **wave–particle duality**.

▲ Marie Curie in her laboratory

HISTORY OF KEY IDEAS

Atoms and electrons

The word 'atom' comes from the Greek *atomos*, meaning indivisible. The first modern use of the word was by John Dalton, who put forward his atomic theory in 1803, in order to explain the rules governing the proportions in which different elements combined chemically. Dalton suggested that all matter consisted of tiny particles called atoms. Each element had its own type of atom, and atoms of the same element were identical. Atoms could not be created or destroyed, nor could they be broken into smaller bits.

But what were atoms made of? The first clues came in the 1890s, when scientists were studying the conduction of electricity through gases. They found that atoms could give out invisible, negatively charged rays. J. J. Thomson investigated the rays and deduced that they were particles much lighter than atoms. This was in 1897. Thomson had discovered the **electron**.

If an atom contained electrons, it must also contain positive charge to make it electrically neutral. But where was this charge? In 1911, a team led by Ernest Rutherford directed alpha particles at thin gold foil. They found that most passed straight through but a few were deflected at huge angles. To explain this, Rutherford suggested that each atom must be largely empty space, with its positive charge and most of its mass concentrated in a tiny **nucleus**.

▲ Ernest Rutherford (right) and his assistant Hans Geiger, in 1912. They are standing next to the apparatus which they used for detecting alpha particles.

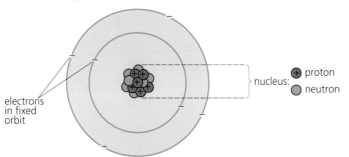

◀ The Rutherford–Bohr model of the atom (with nuclear particles included). The picture is not to scale. With an atom of the size shown, the nucleus would be far too small to see.

In Rutherford's model (picture) of the atom, electrons orbited the nucleus like planets around the Sun. Unfortunately, the model had a serious flaw: according to classical theory, an orbiting electron ought to radiate energy continuously and spiral into the nucleus, so its orbit could not be stable. In 1913, Neils Bohr used the **quantum theory** to solve this problem. According to Bohr, electrons were in fixed orbits and could not radiate continuously. They could only lose energy by jumping to a lower orbit and emitting a quantum ('packet') of electromagnetic energy – in other words, a photon. Using this model, Bohr was able to predict the positions of the lines in the hydrogen spectrum. However, his calculations did not work for elements with a more complicated electron structure. To deal with this problem, scientists later developed a mathematical, **wave-mechanics** model of the atom.

The Rutherford–Bohr model of the atom said nothing about what was inside the nucleus. However, in 1919, Rutherford used alpha particles to knock positively charged particles out of the nucleus. These were **protons**. In 1932, James Chadwick discovered that the nucleus also contained **neutrons**. In recent years, experiments with particle accelerators have suggested that protons and neutrons are made from particles called **quarks**.

▲ Richard Feynmann giving a lecture at the California Institute of Technology. Advances made by Professor Feynmann in quantum electrodynamics won him the 1965 Nobel Prize for Physics.

12.3 Magnetism and electricity*

> **Year dates**
> It is becoming more common to give dates in the form 300 BCE and 1600 CE (or just 1600) rather than 300 BC and 1600 AD. The letters CE stand for 'common era' and BCE for 'before common era'.

Magnetism

Around 2600 years ago, the Ancient Greeks knew that a certain type of iron ore, now known as magnetite or lodestone, could attract small pieces of iron. They found the ore in a place called Magnesia, which is how magnetism got its name. The Chinese had also come across the mysterious ore and, by 200 BCE, knew that a piece of lodestone, if free to turn, would always point in the same direction.

By around 800 CE, the Chinese had discovered how to make magnetic needles by stroking small pieces or iron with lodestone. The first compasses probably consisted of a magnetized needle supported by a straw floating in a bowl of water. However, they were not really suitable for use on ships. Compasses with pivoted needles did not appear until the 1200s.

▶ No signs of a mountain at the North Pole. At one time, sailors thought that a huge magnetic mountain here might be the source of the Earth's magnetism.

At this time, no one really understood why a compass needle points north. Some sailors believed that there was a huge mountain of lodestone at the North Pole, whose force was so strong that it would pull the iron nails out of a ship's hull. Then, in 1600, William Gilbert published the results of his experiments with magnets. He introduced the term **magnetic pole** and suggested that the Earth itself might behave like a bar magnet. But what caused magnetism? The answer to that would come from an understanding of electricity.

Electricity

The Ancient Greeks also knew of the strange properties of a solidified resin called amber. When rubbed, it attracted dust and other small things. The Greek word for amber is *elektron*, from which the word electricity comes.

Our modern knowledge of electricity really began in the 1600s, when experimenters started to investigate amber and other rubbed materials more closely. They found that it was possible to produce repulsion as well as attraction, and that there were two different kinds of **electric charge**. In 1752, Benjamin Franklin carried out a famous – and extremely dangerous – experiment in which he flew a kite in a thunderstorm and got sparks to jump from a key attached to the line. The sparks were just like those produced by rubbing amber. Here was evidence that lightning and electricity were the same thing.

▲ Amber

HISTORY OF KEY IDEAS

At this time, electrical experiments were with 'static electricity' – charges on insulators that could be transferred in sudden jumps. However, in 1800, Alessandro Volta discovered that two metals with salt water between them could cause a continuous flow of charge – in other words, an **electric current**. He had made the first battery. Within 50 years, the electric motor, generator, and lamp had all been invented. However, no one had any evidence to explain what electricity really was until J. J. Thomson's discovery of the **electron** in 1897. From this, we now know that the current in a circuit is a flow of electrons.

Electromagnetism...

Until the early 1800s, electricity and magnetism were regarded as two different phenomena. Then in 1820, in Denmark, Hans Oersted demonstrated that a compass needle could be deflected by an electric current. The following year, Michael Faraday succeeded in using the force from a magnet on the current in a wire to produce rotation. He had made a very simple form of **electric motor**. Later, in the 1830s, he discovered **electromagnetic induction**, the effect in which a voltage is generated in a conductor by moving or varying a magnetic field around it. Today's generators and transformers make use of this idea.

▲ Faraday's first transformer, made in 1831

In the 1860s, James Clerk Maxwell linked electricity and magnetism mathematically. Later, following the discovery of the electron, the cause of magnetism became clear. As an electron orbits in an atom, it produces a magnetic field, rather as the current in a coil produces a field. In most materials, the various fields are in random directions and cancel each other out, but in a magnetized material, some of the fields line up and reinforce each other.

◀ An aurora, another example of the link between magnetism and electricity. Charged particles (mainly electrons) streaming out from the Sun are directed towards polar regions by the Earth's magnetic field. When the particles hit atoms or molecules in the upper atmosphere, light is emitted.

...and beyond

Electric and magnetic forces are so closely related that they are classed as **electromagnetic forces**. The forces that hold atoms together are of this type. But three other kinds of force also operate in the natural world. Gravity is the most familiar. The others are the **weak nuclear force**, responsible for radioactivity, and the **strong nuclear force** which binds the particles in the nucleus of an atom. Today, physicists are developing models to link these forces, but gravity remains a problem, and the search is still on for a satisfactory unified-field theory that combines all the known forces of nature.

12.4 The Earth and beyond*

The centre of the Universe?

The ancient view of the Earth was that it was flat. However, by about 600 BCE, Greek mariners had observed how the positions of the stars altered as they sailed north or south, and realized that the Earth's surface must be curved. Around 350 BCE, Aristotle believed that the Earth was a stationary sphere at the centre of the Universe. The Sun, Moon, planets, and stars lay on transparent, crystal spheres which rotated about the Earth, so they moved in perfect circles.

The idea that the 'heavenly bodies' must move in perfect circles around the Earth was to cause difficulties for many centuries. When viewed from Earth, the planets do not move across the sky at a steady rate, and sometimes appear to move backwards and forwards. Around 150 CE, Ptolemy came up with an elaborate explanation for this. The planets *did* move in perfect circles, but sometimes they followed small circles superimposed on a larger one.

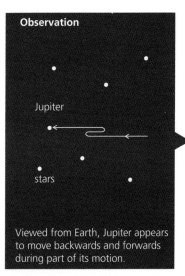

Viewed from Earth, Jupiter appears to move backwards and forwards during part of its motion.

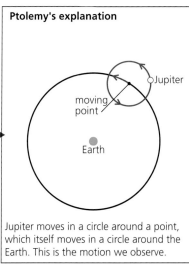

Jupiter moves in a circle around a point, which itself moves in a circle around the Earth. This is the motion we observe.

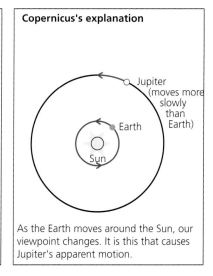

As the Earth moves around the Sun, our viewpoint changes. It is this that causes Jupiter's apparent motion.

It was not until the 1500s that the views of Aristotle and Ptolemy were seriously questioned. The person responsible was Nicolaus Copernicus, who decided to take a fresh look at the problem of the observed motion of the planets. In 1543, he published his theory that the Sun must be at the centre of the Universe, with the Earth and planets moving around it. Over the following years, this idea was strongly opposed by the Church, which insisted that the Earth must be central. Later, Galileo supported Copernicus's ideas, but was forced to renounce them or risk torture and execution. In 1610, he had observed tiny moons moving around Jupiter – evidence that the Earth was not central to all objects in the heavens.

During the late 1500s, observations made by Tycho Brahe greatly increased the amount of accurate data on the positions of the planets. During the 1600s, the evidence for the Copernican model became overwhelming. Kepler established the laws of planetary orbits, Newton published his theory of gravitation, and put Kepler's laws on a firm mathematical basis.

▲ Galileo using a telescope he designed and built himself.

HISTORY OF KEY IDEAS

◀ When William Herschel built this reflecting telescope in 1789, it was the largest in the world. It could be raised or lowered by pulleys, and there were rollers under the platform so that it could be turned.

Caroline Herschel, William's sister, was also an expert astronomer. She discovered comets and nebulae, and made a huge catalogue of her brother's observations.

Dark matter and dark energy

Less than 5% of the Universe seems to be made up of ordinary matter as we know it. The rest is **dark matter** and **dark energy**. Neither can be detected directly, but their existence has been suggested in order to explain gravitational effects seen in galaxies, and how the Universe is expanding. Observations and mathematical analysis suggests that the rate of expansion is increasing and that dark energy is the most likely cause.

Sun, stars, and galaxies

By the late 1600s, it was clear that the stars were similar to the Sun, but much further away. In the late 1700s, William Herschel used a large telescope to study how the stars were distributed. He concluded that the Sun was near the centre of a huge, lens-shaped system of stars, which he called the **Galaxy**.

By the early 1800s, astronomers were making increasingly accurate estimates of the distances to the stars. These were based on the following principle. During the course of a year, as the Earth moves round the Sun, our viewpoint in space changes, so nearby stars appear to move against the background of very distant stars. This apparent movement is called **parallax**. By measuring it, the distance to nearby stars can be calculated using trigonometry.

In 1918, Harlow Shapley mapped the relative distances of star clusters and found that the Sun was not at the centre of our Galaxy after all. And in the 1920s, Edwin Hubble discovered that our Galaxy was not alone. There were millions of other galaxies in the Universe. Hubble also made another significant discovery about galaxies. From the altered wavelengths of their light, he concluded that they must be rushing away from each other. This discovery led to the development of the **Big Bang theory** – the idea that, billions of years ago, the whole of space and everything in it started to expand from a single, atom-sized concentration of matter and energy. Evidence suggests that the Big Bang occurred 13.8 billion years ago.

▲ The Hubble Space Telescope is in orbit around the Earth. It transmits pictures back to the ground which enable astronomers to see distant stars and galaxies without the distorting effects of the Earth's atmosphere. A larger replacement is due to be launched in 2021.

275

Key developments in physics

c. 400 BCE	Democritus suggests that there might be a limit to the divisibility of matter. (*Atomos* is the Greek word for indivisible.)
c. 350 BCE	Aristotle suggests that the Earth is at the centre of the Universe, with the Sun, Moon, and planets on crystal spheres around it.
c. 240 BCE	Eratosthenes estimates the diameter of the Earth by comparing shadow angles in different places.
CE	
c. 60	Hero makes a small turbine driven by jets of steam.
c. 150	Ptolemy suggests that the Earth is at the centre of the Universe, and that the Sun, Moon, and planets are moving in perfect circles.
c. 1000	Magnetic compass used in China.
1543	Copernicus suggests that the Sun is at the centre of the Universe, with the Earth and planets moving around it.
1600	Gilbert suggests that the Earth acts like a giant bar magnet.
1604	Galileo shows that all falling objects should have the same, steady acceleration.
1621	Snell states his law of refraction.
1644	Torricelli makes the first mercury barometer.
1654	Guericke demonstrates atmospheric pressure.
1662	Boyle states his law for gases.
1678	Huygens puts forward his wave theory of light.
1679	Hooke states his law for elastic materials.
1687	Newton publishes his theory of gravity and laws of motion.
1714	Fahrenheit makes the first mercury thermometer.
1752	Franklin performs a hazardous experiment with a kite to show that lightning is electricity.
c. 1790	Herschel discovers the shape of our galaxy.
1800	Volta makes the first battery.
1803	Dalton suggests that matter is made up of atoms.
	Young demonstrates the wave nature of light.
1821	Faraday makes a simple form of electric motor.
1825	Ampère works out a law for the force between current-carrying conductors.
1827	Ohm states his law for metal conductors.
1832	Faraday demonstrates electromagnetic induction.
1832	Sturgeon makes the first moving-coil meter.
1840	First use of the words 'physicist' and 'scientist'.
1849	Fizeau measures the speed of light.
	Joule establishes the link between heat and work.
1852	Kelvin states the law of conservation of energy.
1864	Maxwell predicts the existence of radio waves and other electromagnetic waves.
1879	Swan and Edison make the first electric light bulbs.
1888	Hertz demonstrates the existence of radio waves.
1894	Marconi transmits the first radio signals.
1895	Röntgen discovers X-rays.
1896	Becquerel discovers radioactivity.
1897	Thomson discovers the electron.
1898	M. Curie discovers radium and polonium.
1899	Rutherford identifies alpha and beta rays.
1900	Planck proposes the quantum theory.
1905	Einstein uses the quantum theory to explain the photoelectric effect, and publishes his special theory of relativity.
1911	Rutherford proposes a nuclear model of the atom.
1913	Bohr uses the quantum theory to modify Rutherford's model of the atom.
1916	Einstein publishes his general theory of relativity.
1919	Rutherford splits the atom and discovers the proton.
1924	De Broglie suggests that particles can behave as waves.
1925	Schrödinger wave-mechanics model of the atom.
1927	Lemaitre suggests the possibility of the Big Bang.
1928	Geiger and Müller invent their radiation detector.
1929	Hubble discovers that the Universe is expanding.
1932	Chadwick discovers the neutron.
	Cockroft and Walton produce the first nuclear change using a particle accelerator.
1938	Hahn discovers nuclear fission.
1942	Fermi builds the first nuclear reactor.
1947	Bardeen, Brattain, and Shockley make the first transistor.
1957	First artificial satellite, *Sputnik I*, put into orbit.
1958	St Clair Kilby makes the first integrated circuit.
1960	Maiman builds the first laser.
1963	First geostationary communications satellite.
1969	First manned landing on the Moon.
1971	Intel Corporation makes the first microprocessor.
1977	First experimental evidence of quarks.
1990	Hubble Space Telescope launched.
2012	Higgs particle discovered.
2019	New definitions for kilogram and other units.

Source: *the Biographical Encyclopedia of Scientists*, published by the Institute of Physics   c. = *circa* (about)

13
Practical physics

- WORKING SAFELY
- PLANNING AND PREPARING
- MEASURING AND RECORDING
- DEALING WITH DATA
- EVALUATING AND IMPROVING
- INVESTIGATIONS TO TRY
- PRACTICAL TESTS

The worker inside the cage is quite safe, despite the 2.5 million volt sparks from the huge Van de Graaff generator. The electric discharges strike the metal bars, rather than pass between them, so the cage has a shielding effect. In fact, if safety procedures were ignored, some of the experiments done in a school laboratory would be much more dangerous than this one.

13.1 Working safely

When carrying out physics experiments, you need to be able to do the following:
- Handle equipment and materials safely.
- Follow instructions carefully.
- Change how you carry out each step of an experiment, depending on what happened the time before.

Here are some reminders about how to work safely with different types of equipment:

Bunsens and tripods

- If a bunsen burner is alight, but not in use, always leave it on the yellow flame setting so that the flame can be seen.
- Make sure that bunsens and tripods have a heatproof mat underneath.
- Give a hot tripod plenty of time to cool down before attempting to move it.
- Don't attempt to move a tripod when there is a beaker resting on it.

▲ A yellow bunsen flame is easier to see than a blue one.

Glass thermometers

- Don't put glass thermometers where they can roll off the bench.
- Keep glass thermometers away from bunsen flames.
- Support thermometers safely: see **Safe support** below.
- Mercury, used in some thermometers, is toxic. If a thermometer breaks and mercury runs out, don't handle it.

Glass tubing

- Never attempt to push glass tubing (or glass thermometers) through a hole in a bung. The laboratory technician has a special tool for doing this.
- Always handle hot glass tubing with tongs. Rest it on a heatproof mat; don't put it straight on the bench.
- Hot glass tubing can stay hot for a long time. Give it plenty of time to cool down before you attempt to pick it up.

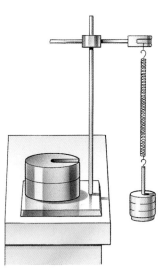
▲ In experiments like this, make sure that the apparatus is stable enough to support the heaviest load.

Safe support

- When clamping a test-tube, don't overtighten the clamp. And make sure that the clamp has soft pads to touch against the glass. This also applies when clamping a glass thermometer.
- In experiments where you have to suspend a load, make sure that the supporting clampstand is stable enough to take the heaviest load you will be using. You may need to weigh it down for this, as shown in the diagram on the left.

PRACTICAL PHYSICS

Electricity

- Before making any changes to the wiring in your circuits, always switch off the power or disconnect the battery.
- Remember: low voltage circuits may not give you a shock, but they can cause burns if the current is too high and a wire overheats.
- Never make a direct connection across the terminals of a battery. Don't put wires or tools where they might connect across the terminals.
- If a mains appliance is faulty, switch off the power and pull out the plug. Don't change the fuse. Ask the laboratory technician to deal with the fault.
- Electrical fires: see **Fire** below.

In many modern laboratories, the mains circuits are protected by RCDs (residual current devices), so the risk of shocks is reduced. But...

- If someone has been electrocuted, and is still touching the faulty appliance, don't touch the person. Switch off the power and pull out the plug.

▲ Emergency! But the first job is to switch off the power and pull out the plug.

Eye protection

- Always wear eye protection (e.g. safety goggles) when:
  – stretching metal wires or plastic cords
  – breaking or grinding solids (e.g. rock samples)
  – heating liquids
  – dealing with acids, alkalis, or any other liquid chemicals that might splash.

Light

- Don't look directly into a laser beam or other source of bright light. Don't stand where laser light might be reflected into your eyes.
- If you need to study the Sun's image, project it onto a card. Never look through a telescope or binoculars pointing straight at the Sun – even if there is a filter in front.

Radioactive sources

- The radioactive sources used in school laboratories should always be sealed.
- Radioactive sources should be kept well away from the body, and never placed where they are pointing at people.

Fire

- Don't heat flammable liquids (e.g. methylated spirits) over a bunsen. If heating is required, a water bath should be used – with hot water heated well away from the experiment.
- Don't throw water on burning liquids (e.g. methylated spirits). Smother the fire with a fire blanket or use a carbon dioxide extinguisher.
- Don't throw water on electrical fires. Switch off the supply and use a carbon dioxide extinguisher.

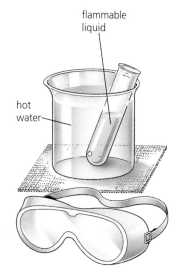

▲ The only safe way to heat a flammable liquid is to use a water bath

13.2 Planning and preparing

This spread should help you plan an experimental procedure. The handwritten notes show part of one student's commentary on her procedure.

Presenting the problem

Start by describing the problem you are going to investigate, and the main features of the method you will use to tackle it.

> I am going to investigate how the resistance of nichrome wire depends on its length.
>
> I know that resistance can be calculated with this equation:
>
> $$\text{resistance (in } \Omega\text{)} = \frac{\text{voltage (in V)}}{\text{current (in A)}}$$
>
> So to find the resistance of a length of nichrome wire, I need to put the wire in a circuit, then measure the voltage across it and the current in it. I will do this for different lengths of nichrome.

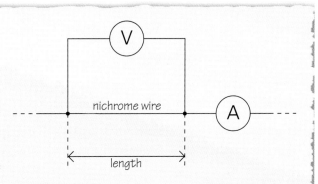

> I think I can predict how the resistance will vary with length. If the length of wire is doubled, the current (flow of electrons) has to be pushed between twice as many atoms. So I would expect the resistance to double as well.

Making a prediction

You may have an idea of what you expect to happen in your enquiry. This prediction is called your **hypothesis**. You should write it down. It may not be right! It is just an idea. The aim of your procedure is to test it.

Dealing with variables

Quantities like length, current, and voltage are called **variables**. They can *change* from one situation to another.

Key variables These are the variables that can affect what happens in an experiment. You must decide what they are. For example, in the nichrome wire experiment, length is one of the key variables because changing the length of wire changes the resistance.

You must also decide how to measure the variables, and over what range. For example, in planning the nichrome wire experiment, you would have to:
- decide what the highest voltage and current values should be (safety must be considered here)
- decide what lengths of wire to use.

> In my experiment, three of the key variables are:
>
> <u>length</u> of nichrome wire - to be measured with a ruler marked in mm
>
> <u>voltage</u> – to be measured with a voltmeter
>
> <u>current</u> – to be measured with an ammeter
>
> I shall start with 50 cm of thin nichrome wire, put a voltage of 6 V across it, and measure the current in it. From the voltmeter and ammeter readings, I can calculate the resistance.
>
> I will take more sets of readings, shortening the wire by 5 cm each time until it is only 10 cm long.
>
> For convenience, I will probably keep the voltage fixed at 6 V throughout the experiment.

Controlling variables Some variables don't have to be measured, but they do need to be controlled. For example, in the nichrome wire experiment, you might want to keep the wire at a steady temperature, in case the temperature affects the resistance.

Some variables can be difficult to control. In your experiment, you may want to use the same thickness of nichrome wire each time, but this depends on how accurately the wire was manufactured. You must take factors like this into account when deciding how reliable your results are.

A fair test When doing an experiment, you should change just one variable at a time and find out how it affects one other. If lots of variables change at once, it will not be a fair test. For example, if you want to find out how the length of a wire affects its resistance, it wouldn't be fair to compare a long, thick wire with short, thin one.

Final preparations

Decide what equipment you need, how you will arrange it, and how you will use it.

To help your planning, you may need to carry out a trial run of the experiment. Before you do this, make sure that all your procedures are safe.

Prepare tables for your readings *before* you start your experiments. Look at the next spread on **getting the evidence** before doing this.

> Equipment needed:
> voltmeter (0–6 V), ammeter (0–3 A), 50 cm of 0.28 mm diameter nichrome wire, ...

nichrome:			
length	voltage	current	resistance
cm	V	A	Ω
50			
45			
40			
35			
30			
25			
20			

> There are two more variables I need to control:
>
> <u>temperature</u> – I know from reference books that the resistance of nichrome changes with temperature. So I will use a large beaker of cold water to keep the temperature of the nichrome steady.
>
> <u>diameter</u> (thickness) of nichrome wire – this could affect the resistance. To make sure that I have the same diameter all the time, I will use lengths of wire taken from the same reel, and check each piece with a gauge before using it.

> I will set up this circuit:
>
>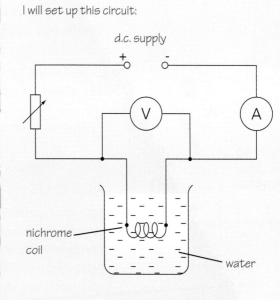
>
> I am not sure how big the maximum current will be, so I will do a trial run of the experiment first. I will start with an ammeter that can measure several amperes, but may be able to change to a more sensitive meter for the main experiment.
>
> Safety:
> I must make sure that the power supply is switched off before I remove the nichrome wire to change its length.

13.3 Measuring and recording

This spread should help you take and record measurements correctly.

Units

When you write down a measurement, remember to include the unit. For example:

voltage = 2.3 V

If you just write down '2.3', you may not be able to remember whether this was supposed to be a voltage of 2.3 V or 2.3 mV.

When writing measurements in a table, you don't need to put the unit after each number. But be sure to include the unit in the heading at the top of each column. You can see an example on the left.

▲ When recording readings in a table (see spread 13.2), remember to include a unit in the heading at the top of each column.

Uncertainties

No measurement is exact. There is always some **uncertainty** about it. For example, you may only be able to read a voltmeter to the nearest 0.1 V.

Say that you measure a voltage of 2.3 V and a current of 1.2 A. To work out the resistance in ohms (Ω), you divide the voltage by the current on a calculator and get...

1.916 666 7

This should be recorded as 1.9 Ω. Uncertainties in your voltage and current readings mean that you cannot justify including any more figures. In this case, you are giving the result to two **significant figures**.

Take enough readings

For a graph, you should have at least five sets of readings.

Not all experiments give you readings for a graph. Sometimes, you have to measure quantities that don't change – the diameter of a wire for example. In cases like this, you should repeat the measurement at least three times and find an average. Repeating a measurement helps you spot mistakes. It also gives you some idea of the uncertainty. Look at this example.

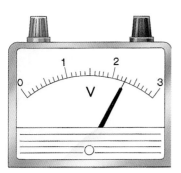

▲ You can only read this voltmeter to the nearest 0.1 V.

The diameter of a wire was measured four times:

1.41 mm   1.34 mm   1.19 mm   1.30 mm

You can work out the average like this:

$$\text{average} = \frac{(1.41 + 1.34 + 1.19 + 1.30)}{4} = 1.31 \text{ mm}$$

The original four numbers ranged from less than 1.2 to more than 1.4. So, the last figure, 1, in the average of 1.31, is completely uncertain. Therefore, you should write down the average diameter as 1.3 mm.

PRACTICAL PHYSICS

Reading scales

On many instruments, you have to judge the position of a pointer or level on a scale and work out the measurement from that. Here are some ways of making sure that you take the correct reading:

A **Using a glass thermometer** to measure the temperature of a liquid: keep the liquid well stirred, give the thermometer time to reach the temperature and keep the lamp in the liquid while you take the reading.
B **Using a ruler**: be sure that the scale is right alongside the point you are trying to measure. (Errors due to an incorrect line of sight are called **parallax** errors.)
C **Measuring a liquid level** on a scale: look at the level of the liquid's flat surface, not its curved meniscus.
D **Reading a meter**: look at the pointer and scale 'square on'. (The pointer may have a flat end like that shown here, so that you can look at it edge on.)

Can you read the instruments below correctly? The answers are on page 331.

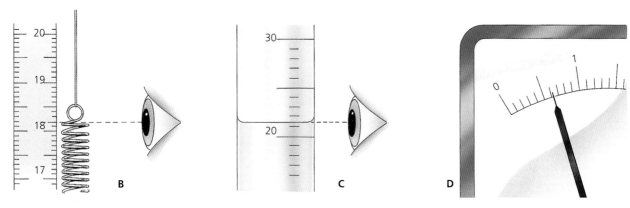

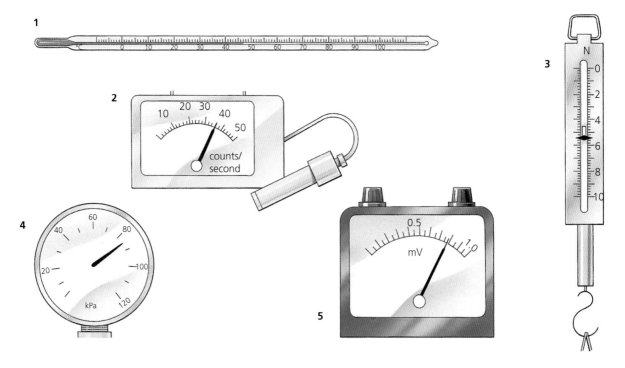

13.4 Dealing with data

I have used my voltage and current readings to calculate the resistance of each length of nichrome wire. Now I shall use these values to plot a graph of resistance against length.

Length is the <u>independent</u> variable (the one I chose to change), so it goes along the bottom axis. Resistance goes up the side.

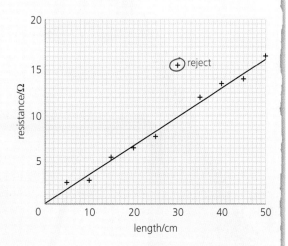

The points on my graph are a little scattered, but I think that the line of best fit is a straight line.

The line ought to go through the origin. If the wire has zero length, there is no metal to resist the current, so the resistance should also be zero.

I have rejected one point on my graph. In my table, the current reading for that point seems far too low. I probably misread the ammeter.

As the graph is a straight line through the origin, the resistance of the nichrome wire is in direct proportion to its length. This agrees with my original hypothesis that doubling the length of wire ought to make it twice as difficult to push electrons through.

This page should help you to analyse your data and draw conclusions from it. The handwritten notes show part of one student's commentary on her enquiry.

Drawing a graph

A graph can help you see trends in your data.

Choosing axes Decide which variable to put along the bottom axis. Usually, it is the one you chose to vary by set amounts – the length of nichrome wire, for example. This is the **independent variable**. The resistance would be the **dependent variable** because its value depends on the length you chose. It goes up the side axis.

Choosing scales Check your highest readings, then choose the largest scales you can for your axes.

Labelling axes Along each axis, write in what is being measured and the units being used.

Drawing the best line Because of uncertainties, the points on a graph will be uneven. So don't join up the points! Instead, draw the straight line or smooth curve that goes closest to most of them. This is called a **line of best fit**. Before you draw it:
- Decide whether the line should go through the origin.
- Decide whether any readings should be rejected. Some may be so far out that they are probably due to mistakes rather than uncertainties. See if you can find out why they occurred.

From the way points scatter about a line of best fit, you can see how reliable your readings are. But for this, you need plenty of points.

Trends and conclusions

From the shape of your graph, you can draw conclusions about the data.

The simplest form of graph is a straight line through the origin. A graph of resistance against length of wire might be like this. If so, it means that if the length doubles, the resistance doubles… and so on. In this case, resistance and length are in **direct proportion**.

If you think that your graph supports your original prediction, then say so and explain your reasons.

13.5 Evaluating and improving

This page should help you decide how reliable your conclusions are, and how your procedure could be improved or extended.

> The points on the graph are uneven. But as they zig-zag at random, I am fairly sure that, without uncertainties, they would lie on a straight line.
>
> There are several reasons why the points may have been so scattered...

> To get a more reliable graph, I need to find a more accurate method of measuring resistance...

> To extend my enquiry, I could find out how the resistance of the nichrome wire depends on the diameter...

Reliability

In reaching your conclusions, remember that there are uncertainties in your measurements, and variables that you may not have allowed for. So your results can never *prove* your original prediction. You must decide how far they *support* it.

If you think that your results are unreliable in any way, see if you can explain why.

You may have some results which do not agree with the others and look like mistakes. These are called **anomalous results**. Try to explain what caused them.

Suggesting improvements

Having completed your procedure, suggest ways of improving it so that your conclusions are more reliable.

Looking further

Suggest some further work which might produce extra evidence or take your procedure further.

Writing your report

The student's commentary was designed to help you understand the different stages of an procedure. It includes far more detail than you would normally put in a report. When producing your own report, these are the things you *should* include:

Planning   **1**
- A description of what the procedure is about.
- A prediction of what you think will happen, and why.
- A list of key variables, and a description of how you will measure or control each one.
- A list of the equipment needed.
- Diagrams showing how the equipment will be set up.
- A description of what you plan to do.

Getting evidence   **2**
- A description of what you did, including comments about any difficulties and how you overcame them.
- Tables showing all measurements, including units.

Analysing and concluding   **3**
- Graphs and charts.
- Calculations based on your data.
- A conclusion, including details of:
  – what you found out
  – whether your findings matched your prediction.

Evaluating   **4**
- Comments about:
  – how reliable you think your results were
  – any anomalous results, and their possible causes
  – how your procedure could be improved
  – further work that could be done.

13.6 Some experimental investigations

Here are some suggestions for practical work. Some are full investigations. Others are shorter exercises to help you develop your experimental skills.

Measuring newspaper

Plan and carry out experiments to measure:

a the thickness of one sheet of newspaper **b** the mass of one sheet of newspaper **c** the density of the paper used.

Start by thinking about the following:
If a single sheet is too thin to measure accurately, how can you improve the accuracy?

Wet or dry?

The makers of a well-known brand of soft tissue paper claim that their tissues are just as strong wet as dry. Are they right? Plan and carry out an enquiry to test their claim.

Start by thinking about the following:
What is meant by the 'strength' of a tissue? Do you need use a whole tissue? When comparing tissues, how can you make sure that your test is fair?

Fine or coarse?

Coarse glasspaper ('sandpaper') rubs through a wooden surface more quickly than fine glasspaper. But does it produce more friction? Plan and carry out experiments to find out.

Start by thinking about the following:
How can you measure the frictional force when glasspaper is rubbed on wood? How can you keep the glasspaper pressed against the wood? Will the force used to press the glasspaper against the wood affect the result? How can you make sure that your test is fair?

Pendulum

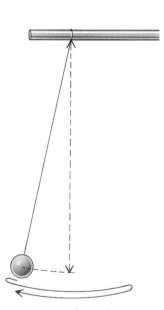

one complete swing

The time of one complete swing of a pendulum is called its period.

The period of swing *might* be affected by these factors: the mass of the bob, the amplitude (size) of the swing, the length of the pendulum.

Plan and carry out an enquiry to find out which factors affect the period.

Start by thinking about the following:
The period of your pendulum will probably be a couple of seconds at most. How are you going to find the time of one swing accurately? How are you going to measure the size of the swing?

Note: make sure that the top of the pendulum string is firmly held so that there is no movement at that point.

Further work:
Find out how the period of one pendulum compares with another of four times the length. Is there a simple connection between the length and the period? Does the connection work for other lengths as well?

PRACTICAL PHYSICS

Stretching rubber

A company wants to market a cheap spring balance for weighing letters. Their designer suggests that, to save money, they could use a rubber band instead of a spring. Their technician says that this would be unsatisfactory because rubber bands change length and 'springiness' once they have been stretched. Who is correct? Plan and carry out an enquiry to find out.

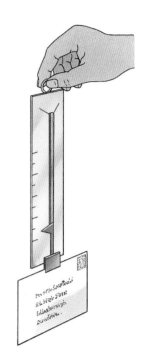

Find the mass

Plan and carry out an experiment to find the mass of a lump of Plasticine (or some other solid). You are not allowed to use a balance with a mass scale already marked on it. And you are not allowed to use slotted masses of less than 50 g.

Further work:
Take the problem a stage further. Plan and carry out experiments to measure a much smaller mass – such as the mass of a pen or pencil.

This time, you can use a selection of standard masses down to 5 g.

Start by thinking about the following:
Your original design will probably not be sensitive enough to measure a small mass. Can it be modified in some way to make it more sensitive?

Bouncing ball

Some table tennis balls have more 'bounce' than others. Plan and carry out an enquiry to compare the bounce of two table tennis balls.

Start by thinking about the following:
What is meant by 'bounce'? What do you need to measure? When comparing the balls, how can you make sure that your test is fair?

Parachute design

The diagram on the right shows a simple model parachute. Plan and carry out an enquiry to find out if there is a link between the design of the parachute and the speed at which it falls.

Start by thinking about the following:
Shape and area are two possible features of the design. Will you investigate both? How will you make sure that your tests are fair? How will you work out the speed of fall?

Double-glazing

In cooler countries, people fit double-glazing in their houses because two layers of glass, with air between, are supposed to lose thermal energy (heat) more slowly than a single layer. But does double-glazing cut down thermal energy loss? Plan and carry out an enquiry to find out.

Start by thinking about the following:
How are you going to set up a double layer of glass with air between? What will you use as a source of thermal energy? How will you tell whether the flow of thermal energy is reduced when the extra layer of glass is added? Will your test be fair?

PRACTICAL PHYSICS

Salt on ice

During winter, salt is often sprayed on the roads to melt the ice. Pure ice has a melting point of 0 °C. Adding salt to ice affects the melting point.

Plan and carry out experiments to find out how the melting point of ice changes when salt is mixed in. Find out if there is a connection between the melting point and the concentration of salt in the ice. (The concentration can be measured in grams of salt per cm^3 of ice.)

Start by thinking about the following:
How will you make sure that the salt and ice are properly mixed?
How are you going to measure the melting point?

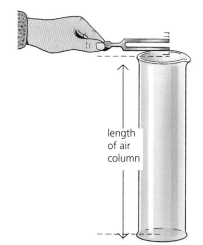

The speed of sound

In the diagram on the left, someone is holding a vibrating tuning fork above a measuring cylinder. Sound waves travel down the cylinder and back, and make the air inside vibrate. If the length of the air column is exactly a quarter of the wavelength of the sound, the air vibrations are strongest and the air gives out its loudest note. The effect is called resonance.

The speed of sound is linked to its frequency and wavelength by this equation:

speed (m/s) = frequency (Hz) × wavelength (m)

Using the information above, plan and carry out an enquiry to find the speed of sound in air.

Start by thinking about the following:
As a measuring cylinder has a fixed length, how will you vary the length of the air column inside?

Apparent depth

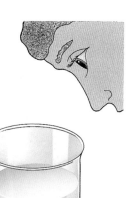

The person in the diagram on the left is looking at a pin on the bottom of a beaker of water. Light from the pin is refracted (bent) when it leaves the water. As a result, the water looks less deep than it really is and the pin appears closer to the surface than it really is.

Plan and carry out an experiment to find the apparent depth of some water in a beaker.

Start by thinking about the following:
If you look at a pin in some water, it is an image of the pin which you are seeing. How can you locate the position of this image? Could you use a similar method to that used to find the position of an image in a mirror?

Two pairs or one?

People claim that two pairs of socks are warmer than one. But does an extra pair cut down the loss of thermal energy (heat)? Plan and carry out an enquiry to find out. (You do not have to use warm feet as your source of thermal energy!)

Start by thinking about the following:
How are you going to tell that one object is losing heat more rapidly than another? How will you make sure that your test is fair?

PRACTICAL PHYSICS

Image size and distance

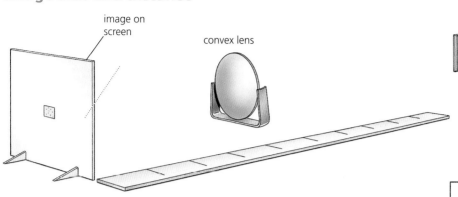

Place a bright object well away from a convex lens as in the diagram, and you can get a clear image on a screen. If you move the object closer, the size and the position of the image both change, and you need to move the screen to get a clear image again.

Is there a connection between the size of the image and its distance from the lens? Plan and carry out an enquiry to find out.

Current–voltage investigations

Plan and carry out experiments to find out how the current in each of the following depends on the voltage across it:
1) nichrome wire, kept at constant temperature
2) the filament of a lamp
3) a semiconductor diode.

Start by thinking about the following:
How will you vary the voltage across each component and measure the current in it? What checks must you do to make sure that the current in each component is safe, and does not cause damage?

Making a resistor

Resistors are used for keeping voltages and currents at correct levels in electronic circuits.

Using nichrome wire, make a resistor with a resistance of 5 Ω.

Start by thinking about the following:
How does the length of wire affect its resistance? How is resistance calculated? What circuit will you use to test the nichrome? From your measurements, how can you work out how much wire you need?

Thermistor investigation

Thermistors have a resistance that varies considerably with temperature. They can be used as temperature sensors. Plan and carry out an experiment to find out how the resistance of a thermistor varies between 0 °C and 100 °C.

Start by thinking about the following:
How will you change and control the temperature of the thermistor? How will you measure the resistance of the thermistor? How will you make sure that your circuit doesn't heat up the thermistor?

13.7 Taking a practical test

This spread should help you if you have to take a practical test in physics. The test is the same at both Core and Extended Level. You will not need any knowledge of physics beyond Core Level. There are two typical questions on the opposite page.

Instead of doing a practical test, you may have to sit an alternative-to-practical examination paper. Your teacher will be able to tell you which form of assessment applies to you. There are some sample alternative-to-practical questions in Section 15 (IGCSE practice questions).

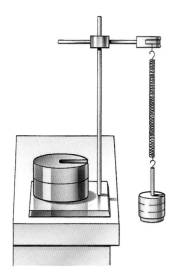

Apparatus used in the test

In your test, you could be asked to carry out experiments involving the following:

- Measuring physical quantities such as length, volume, or force.
- Cooling and heating (for example, Question 1 on the next page).
- Springs and balances (for example, Question 2 on the next page).
- Timing motion or oscillations.
- Electric circuits.
- Optics equipment such as mirrors, prisms, and lenses.

Preparing for the test

Before taking a practical test, there are certain things you need to be familiar with. These are listed in **Practical preparation** on page 292. Most also apply if you are taking the alternative-to-practical paper.

Go through the list and check them one by one.

During the test

Make sure that you can do the following:

- Take plenty of readings.
- When you record your readings, remember to include the correct units. If you are putting your readings in a table, the column headings should also include the correct units.
- Record readings or results with a suitable degree of accuracy.
- Identify any anomalous results.
- Justify your conclusions by referring to your data.
- Identify any possible causes of uncertainty.

For more information about any of the above, see page 292.

Check-up on practical physics

Further questions

1 In this experiment, you will investigate the effect of insulation on the cooling of water.

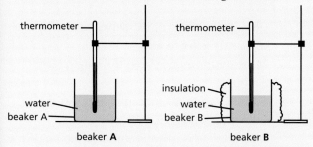

a i Pour 200 cm³ of hot water into beaker **A**.
  ii Measure the starting temperature of the hot water. Record this temperature in **Table 1.1** for time t = 0.0 s
  iii Immediately start the stopwatch. In **Table 1.1**, record the temperature of the water θ every 30 s until you have eight sets of readings.
  iv Repeat steps **i** to **iii** using beaker **B**. Beaker **B** has insulation around it but no lid.

	beaker **A** (no insulation)	beaker **B** (with insulation)
t/s	θ/	θ/
0.0		

Table 1.1 [3]

b Complete the missing units in the column headings in **Table 1.1** [1]

c i Plot a graph of θ (y axis) against t (x axis) for beaker **A**. Draw the line of best fit. [4]
  ii Plot a graph of θ against t for beaker **B** on the same axes you used for beaker **A**. Draw the line of best fit for beaker **B**. [2]

d Describe what your results show about the effect of insulation on the cooling of water. Use data from your table and graph in your answer. [2]

e Suggest two conditions that you had to keep the same so that the comparison between beaker **A** and beaker **B** is fair. [2]

f Describe one precaution that you took to make sure your temperature readings were as accurate as possible. [1]

2 In this experiment, you will investigate the refraction of light as it passes through a rectangular transparent block.

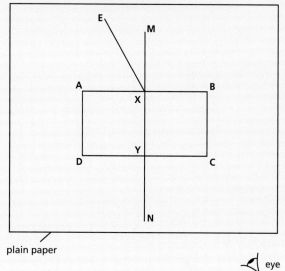

Fig. 2.1

a i Place the rectangular block, largest face down, in the middle of a piece of plain paper.
  ii Draw around the block and then remove it. Label **ABCD**.
  iii Draw the normal **MN** at the centre of side **AB**. Label the point **X** where the normal **MN** crosses side **AB**.
  iv Continue **MN** so that it crosses side **CD**. Label the point **Y** where **MN** crosses side **CD**.
  v Draw a line **EX** at an angle of incidence $i = 30°$ to the left of the normal, as shown in **Fig. 2.1**.
  vi Place the paper on a pin board. Place one pin P_1 on the line **EX** close to point **E**.
  vii Place another pin P_2 at a suitable distance from P_1 on the line **EX**.

viii Place the rectangular block back in the same position on the plain paper. Observe the images of the two pins through side **CD** from the position of the eye in **Fig. 2.1**, so that P_1 and P_2 appear behind one another.

ix Place two more pins, P_3 and P_4, between your eye and the rectangular block. Make sure that pins P_3 and P_4, and the images of P_1 and P_2 appear behind one another.

x Label the positions of all the pins P_1, P_2, P_3 and P_4. Remove the block.

xi Draw a line that passes through positions P_3 and P_4 and continue the line until the point where it meets the normal **MN**. Label this point **Z**. [3]

b i Measure the angle ϑ between the line joining positions P_3 and P_4 and the line **ZN**.

ii Measure the length l between **Y** and **Z**. [2]

c Change the angle of incidence to $i = 50°$ and repeat steps **a v** to **a xi**.
Measure the angle ϑ and the length l when $i = 50°$. [2]

d State whether your results support the following suggestion:
'The angle ϑ is always equal to the angle of incidence i.
Use your results to justify your answer. {1}

e Describe one precaution that you should take with this experiment to make sure that your results are as accurate and reliable as possible. {1}

Use the list below to help you prepare for your practical test. The page number, in brackets, tells you where to find more information.

Practical preparation
Core and Extended Level
Check to make sure that you know how to do each of the following. Most of these also apply if you are taking an alternative-to-practical examination paper.

☐ Identify key variables. (page 280)

☐ Explain why certain variables should be controlled. (page 281)

☐ Measure lengths to the nearest half millimetre using a rule. (page 296)

☐ Measure angles to the nearest half degree using a protractor. (page 296)

☐ Measure time using a stopwatch or stopclock. (page 17)

☐ Measure mass using a balance. (pages 14 and 22)

☐ If measuring the mass of a liquid, allow for the mass of its container. (page 20)

☐ Measure the volume of a liquid using a measuring cylinder. (page 20)

☐ Measure a force, such as weight, using a spring balance. (page 38)

☐ When using meters or other instruments with scales on them, be able take readings that lie between the divisions on the scale. (page 283)

☐ Calculate simple areas and volumes: for example, the area of a rectangle or triangle, or the volume of a rectangular block. (pages 20 and 296)

☐ Allow for zero errors when making measurements. (page 17)

☐ Record readings, or do calculations, with a suitable level of accuracy – and not include too many significant figures. (page 282)

☐ Include the correct units with your readings. (page 282)

☐ Find the average value of several similar readings. (pages 282 and 295)

☐ Draw a line of best fit on a graph. (page 284)

☐ Find the gradient of a graph. (page 30)

☐ Read off new values from a graph line. (page 295)

☐ Understand direct and inverse proportion. (page 295)

☐ Draw circuit diagrams using symbols. (pages 178 and 321)

14 Mathematics for physics

- THE ESSENTIAL MATHEMATICS

A summary of the mathematical concepts and skills required for IGCSE examinations.

The essential mathematics

For IGCSE examinations, you will need some basic skills in mathematics. The following are typical of what is required.

Adding, subtracting, multiplying, and dividing

You should be familiar with the symbols +, −, ×, and ÷ and the processes they represent. This may sound obvious, but it includes understanding the link between division and fractions, as described next.

Using fractions and decimals

A half can of course be written as $\frac{1}{2}$ (sometimes printed as ½ or 1/2). However, you should also understand that it means 1 ÷ 2

Improper fractions are those with a bigger number on the top than the bottom: for example, $\frac{48}{12}$, which means 48 ÷ 12.

You should be able to write fractions using decimals. So, one half is 0.5. Decimals may have several numbers after the point. However, you should understand that 0.489, for example, is smaller than 0.5.

Using percentages

You should understand that percentages are fractions of one hundred. So, for example, one half is $\frac{50}{100}$, which is 50%. The percent symbol % really means 'divided by 100'.

Using ratios

Ratios are another way of expressing fractions. If some apples are being shared between two people in the ratio 2 : 3, there are 5 'parts' to divide up, so one person gets $\frac{2}{5}$ of the total, and the other gets $\frac{3}{5}$ of the total.

If there were 10 apples to share:

one person would get $\frac{2}{5} \times 10$, which is 4 apples;

the other person would get $\frac{3}{5} \times 10$, which is 6 apples.

Working out reciprocals

$\frac{1}{\text{number}}$ is called the **reciprocal** of the number. For example:

The reciprocal of 2 is $\frac{1}{2}$, or 0.5.

The reciprocal of 10 is $\frac{1}{10}$, or 0.1.

You may have to work out reciprocals when preparing readings for a graph. Many calculators have a special key for doing this.

MATHEMATICS FOR PHYSICS

Drawing and interpreting graphs and charts

Graphs are a way of presenting sets of scientific data so that trends or laws can more easily be seen. For more information on how to prepare a graph and draw a line of best fit, see spread 13.4.

You should be able to read off new values from a graph line. Finding values between existing points is called **interpolation**. Extending a graph line to estimate new values beyond the measured points is called **extrapolation**.

Data can also be presented in charts. A **table** is one form of chart. A **pie chart** is another: it shows the different proportions or percentages making up the whole. From the example on the right, you should be able to deduce that 25% of the people at a football match supported the away team.

▲ Proportions of home and away supporters attending a football match

Understanding direct and inverse proportions

Look at the sets of values for X and Y in the table on the right. If X doubles, Y doubles; if X triples, Y triples. Also, dividing Y by X always gives the same number (4), and a graph of Y against X is a straight line through the origin. These all indicate that X and Y are in **direct proportion**. This can be expressed in the form $Y \propto X$, where $\propto$ is the symbol for 'directly proportional to'.

Now look at the sets of values for X and Z on the right. In this case, if X doubles, Z halves, and so on. And *multiplying Z by X* always gives the same number (12). Here, Z and X are in **inverse proportion**.

The table also includes a column for the reciprocals of X. Note that Z and $1/X$ are in *direct* proportion. So, another way of expressing the inverse proportion between Z and X is to write: $Z \propto 1/X$

X	Y
1	4
2	8
3	12
4	16

X	Z	$1/X$
1	12	1
2	6	0.5
3	4	0.33
4	3	0.25

Understanding indices

You should know that 2^3 means $2 \times 2 \times 2$, and that 10^4 means $10 \times 10 \times 10 \times 10$. The tiny numbers ^{3} and ^{4} are called **indices**.

For more advanced work, it is useful to know about negative indices. For example, 10^{-4} means $\frac{1}{10^4}$

Understanding numbers in standard notation

In standard notation (also called standard form, or scientific notation), numbers are expressed using powers of 10. For example, 1500 is written as 1.5×10^3. Using standard notation, you can indicate how accurately a value is known. This is explained in spread 1.1.

Using a calculator

You need to able to use a calculator correctly. For example, to work out a value for $\frac{6 \times 7}{11 \times 3}$, you would key in this: $6 \times 7 \div 11 \div 3 =$
As a result, the calculator display will show this: 1.2727273

Working out an average

If you have, say, five similar readings and need to find the average, you add up the five readings and divide the total by 5. (See also spread 13.3.)

295

MATHEMATICS FOR PHYSICS

If a calculator display reads 1.2727273, you must be able to interpret this correctly. If the original numbers came from experimental data, you could not justify giving the result so accurately. 1.27, or 1.3 if you round it up, would be more appropriate.

You should also be able to interpret high numbers on a calculator. For example, 2.5×10^9 will probably be displayed as 2.5 E 09, or just 2.5 09.

Making approximations and estimations

You should be able to check whether a result is reasonable by doing a rough estimate without a calculator. For example, if you divide 12 by 2.95, you should realise that the answer will be just over 4. So, if the calculator displays 40.67, you have made a mistake in keying in the numbers.

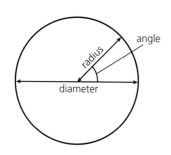

Understanding units

Most measurements have units as well as numbers: for example, a speed of 10 m/s. When giving a result, you must always include the unit. For more about units, see spread 1.1.

Understanding number accuracy

You should understand the significance of whole numbers. You may count 12 students in a room as an exact number, but if a length measurement is given as 12 mm, this only indicates that the value lies somewhere between 11.5 mm and 12.5 mm.

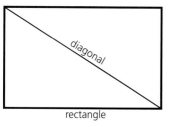

Manipulating equations

If you are given an equation like this: $Z = \dfrac{Y}{X}$

you should be able to rearrange it to give $Y = Z \times X$, and $X = \dfrac{Y}{Z}$

Understanding the terms for shapes, lines, and angles

As well as the circle, sphere, triangle, square, and cube, you need to know the terms shown in the first three diagrams above left.

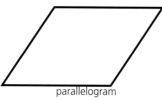

Using the links between length, area, and volume

You should be able to calculate the area of a rectangle, the area of a right-angled triangle (see left), and the volume of a rectangular block (see spread 1.5).

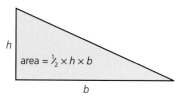

Using mathematical instruments

The basic instruments are a ruler for measuring length, a protractor for measuring angles in degrees (°), compasses for drawing circles, and a set square for use in drawing right angles.

Knowing the points of the compass

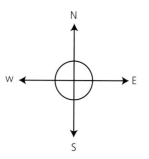

The directions north, south, east, and west are called the points of the compass. To make sure you know which is which, look at the diagram on the left.

296

15 IGCSE practice questions

- MULTICHOICE QUESTIONS (CORE)
- MULTICHOICE QUESTIONS (EXTENDED)
- IGCSE THEORY QUESTIONS
- ALTERNATIVE-TO-PRACTICAL QUESTIONS

IGCSE PRACTICE QUESTIONS

Multichoice Questions (Core)

1 The diagram shows the mass of a measuring cylinder before and after a liquid is poured into it.

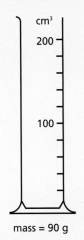
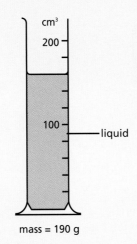

What is the density of the liquid?

A $\dfrac{100}{160}$ g/cm³
B $\dfrac{130}{190}$ g/cm³
C $\dfrac{100}{130}$ g/cm³
D $\dfrac{190}{160}$ g/cm³

2 A motorcycle accelerates from rest. The graph shows how its speed changes with time.

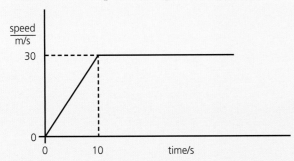

What distance does the motorcycle travel before it reaches a steady speed?

A 3 m
B 30 m
C 150 m
D 300 m

3 Which type of power station does **not** use steam from boiling water to turn the generators?

A coal-fired
B geothermal
C hydroelectric
D nuclear

4 An electric motor is used to lift boxes onto a lorry. Which of the following values would you use to calculate the power of the motor?

A the current used and the work done only
B the force used and the distance moved only
C the mass of the boxes and the current used only
D the work done and the time taken only

5 The diagram shows two forces acting on a small object.

What is the resultant force on the object?

A 2 N downwards
B 2 N upwards
C 8 N downwards
D 8 N upwards

6 How long, approximately, does the Earth take to rotate once on its own axis?

A 12 hours
B 24 hours
C 1 month
D 365 days

7 When a car is travelling along a road, the temperature of its tyres increases.
Why does the air pressure in the tyres also increase?

A The molecules in the air expand.
B The molecules in the air hit the sides of the tyres less often.
C The molecules in the air increase in number.
D The molecules in the air move at a higher speed.

8 A vacuum flask has two walls of glass with a vacuum between the two walls.
Which types of heat transfer are reduced by the vacuum?

A conduction and convection only
B conduction, convection, and radiation only
C convection and radiation only
D radiation only

IGCSE PRACTICE QUESTIONS

9 A fire alarm is too quiet when it rings. The fire alarm is adjusted so that it produces a louder sound at the same pitch.
What effect does this have on the amplitude and the frequency of the sound produced?

	amplitude	frequency
A	larger	same
B	larger	larger
C	same	larger
D	same	same

10 Waves in a ripple tank spread out when they pass through a gap.

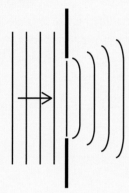

What is the name of this effect?
A diffraction   C refraction
B reflection   D radiation.

11 Some of the waves in the electromagnetic spectrum are shown:

long - - - - - - - - - - wavelength - - - - - - - - - short

radio waves	infrared	light	P	X-rays	Q

What are the names of wave **P** and wave **Q**?

	wave P	wave Q
A	gamma	ultraviolet
B	microwave	gamma
C	microwave	ultraviolet
D	ultraviolet	gamma

12 Which component has a resistance which decreases when the temperature increases?
A filament lamp
B relay
C thermistor
D transformer

13 The p.d. and current readings of four electric heaters are shown in the table below. Which of the heaters has the highest resistance?

	p.d. / V	current / A
A	110	4.0
B	110	8.0
C	230	4.0
D	230	8.0

14 Which of the following shows the correct order of planets from the Sun?
A Mars   Venus   Saturn   Uranus
B Mercury   Earth   Jupiter   Saturn
C Saturn   Jupiter Mercury   Mars
D Venus   Earth   Neptune   Saturn

15 A transformer has 200 turns on its primary coil and 400 turns on its secondary coil. An alternating voltage of 50 V is applied across the primary coil.

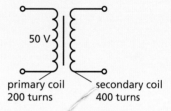

primary coil   secondary coil
200 turns   400 turns

What is the voltage across the secondary coil?
A 25 V   B 50 V   C 100 V   D 200 V

16 A sample contains 800 mg of a radioactive isotope. The isotope has a half-life of 6 days and decays into a stable isotope by emitting alpha particles.
What mass of the isotope is still radioactive after 18 days?
A 0 mg   B 100 mg   C 200 mg   D 400 mg

17 A loudspeaker emits sound waves of frequency 640 Hz. They travel through cold air at a speed of 320 m/s. What is their wavelength?
A 0.050 m
B 0.5 m
C 2.0 m
D 20 m

299

Multichoice Questions (Extended)

1 The graph shows how the speed of a motorcycle changes with time.

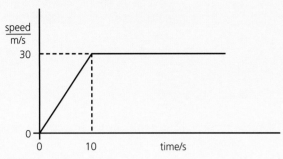

What is the acceleration of the motorcycle during the first 10 seconds?
A 0.33 m/s^2
B 3.0 m/s^2
C 150 m/s^2
D 300 m/s^2

2 A car of mass 1000 kg accelerates at 2 m/s^2 along a straight, flat road.
What is the resultant force acting on the car?
A 0.0020 N
B 500 N
C 2000 N
D 5000 N

3 The table gives information about the length of a spring when different loads are applied.

load / N	0	1	2	3
length / cm	22	32	42	52

What is the spring constant of the spring?
A 0.031 N/m
B 0.10 N/m
C 3.1 N/m
D 10 N/m

4 Which of the following energy resources does **not** have the Sun as its main source of energy?
A geothermal
B hydroelectric
C oil
D wind

5 Which of these units is the same as the watt (W)?
A J   B J/m^2 C J/s   D J/s^2

6 The table shows the performance of four different electric motors.

	A	B	C	D
input power / W	200	300	200	100
output power / W	160	210	150	50

Which motor has the highest efficiency?
A motor A
B motor B
C motor C
D motor D

7 Air is trapped in a sealed cylinder at a pressure of 1200 kPa. The piston is pulled out slowly so that the air expands to three times its volume. Assume that there is no change in the temperature of the air.

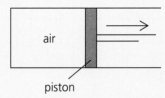

What is the new pressure of the air?
A 3600 kPa
B 600 kPa
C 400 kPa
D 200 kPa

8 Copper is a better thermal conductor than glass. Which statement explains why?
A In glass, the atoms vibrate more quickly than in copper.
B In glass, there are free protons that transfer the thermal energy.
C In copper, the atoms are further apart than in glass.
D In copper, there are free electrons that transfer the thermal energy.

9 A lorry with a mass of 3500 kg has 1100 kJ of energy in its kinetic energy store. Calculate the speed of the lorry.
A 2.5 m/s
B 6.4 m/s
C 25 m/s
D 630 m/s

10 An object is viewed through a convex lens which is being used as a magnifying glass.

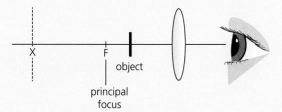

Which statement describes the image?
A The image is real and at position X.
B The image is real and between the lens and the eye.
C The image is virtual and at position X.
D The image is virtual and between the lens and the eye.

11 Two wires, X and Y are made of the same metal and are at the same temperature. Y is twice as long as X and has twice the cross-sectional area.

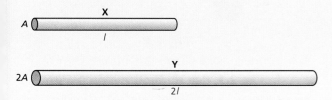

Which statement about resistance is correct?
A X and Y have the same resistance.
B X has double the resistance of Y.
C X has four times the resistance of Y.
D X has half the resistance of Y.

12 Which component, used in electronic circuits, allows current to flow through in one direction only?
A diode
B resistor
C thermistor
D transformer

13 Three resistors are arranged in this combination in a circuit.

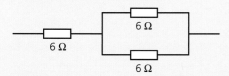

What is the total resistance of the three resistors?
A 6 Ω   B 9 Ω   C 12 Ω   D 18 Ω

14 The table shows data about the Earth's orbit around the Sun.

average radius of orbit / km	1.5×10^8
orbital period / s	3.2×10^7

What is the average orbital speed of the Earth?
A 1.3×10^{-3} m/s
B 1.3 m/s
C 29 m/s
D 29 000 m/s

15 A transformer has 200 turns on its primary coil and 400 turns on its secondary coil. The alternating current in the primary coil is 4.0 A and the voltage across the primary coil is 50 V.

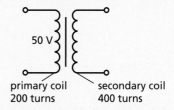

primary coil   secondary coil
200 turns   400 turns

What is the current in the secondary coil? Assume that the transformer is 100% efficient.
A 0.5 A   B 1.0 A   C 2.0 A   D 8.0 A

16 An unstable nucleus has 137 neutrons and 88 protons. It decays by emitting a β-particle. How many neutrons and protons does the nucleus have after emitting the β-particle?

	neutrons	protons
A	136	88
B	136	89
C	137	87
D	137	89

17 A low mass star, such as the Sun, expands to form a red giant when most of its hydrogen has been converted into helium.
Which of the following will the Sun become near the end of its life?
A black hole
B neutron star
C supernova
D white dwarf

IGCSE Theory questions

1. Drops of water fall from the corner of the roof of a bungalow.

Fig. 1.1

A student uses a stopwatch to time how long it takes for a drop of water to fall from the roof to the ground.

The student:
- sets the stopwatch to zero
- starts the stopwatch when a drop of water falls from the roof
- stops the stopwatch when the drop of water lands on the ground
- repeats this method for another four drops of water.

a The student records his result in **Table 1.1**.

drop number	time for one drop / s
1	0.70
2	0.85
3	0.91
4	0.79
5	

Table 1.1

Fig. 1.2 shows the reading on the student's stopwatch for the final drop of water.

Fig. 1.2

Complete **Table 1.1** with the time shown on the stopwatch in **Fig. 1.2**. [1]

b Calculate the average time it takes for 1 drop of water to fall from the roof to the ground. [2]

c Calculate the average speed of a drop of water. [3]

d Explain why it is better to calculate the average time for 5 drops of water to fall rather than just 1 drop. [1]

2. a Describe the difference between mass and weight. [2]

b A student wants to measure the density of a small, irregularly shaped piece of iron. She is given the equipment in **Fig. 2.1**.

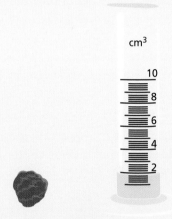

Fig. 2.1

Name a piece of equipment, not shown in **Fig. 2.1**, that the student will use to measure the mass of the piece of iron. [1]

c Describe how the student determines the volume of the piece of iron, using the equipment in **Fig. 2.1**. [3]

d These are the student's results:
- mass of iron = 23.6 g
- volume of iron = 3.0 cm³

Calculate the density of iron. [3]

3. **Fig 3.1** shows the distance-time graph of a cyclist travelling to his friend's house at position S.

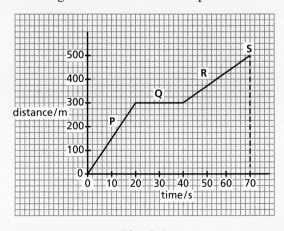

Fig. 3.1

a Describe the motion of the cyclist during:
  i section **P** [1]
  ii section **Q** [1]
b State the total distance travelled by the cyclist to his friend's house. [1]
c State the total time it takes the cyclist to travel to his friend's house. [1]
d Calculate the average speed of the cyclist over the whole journey. [1]
e In which section of the journey does the cyclist travel the fastest? Explain your answer. [1]

4 Fig 4.1 shows the speed–time graph for a car.

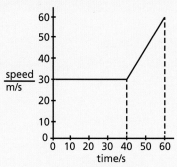

Fig. 4.1

a Describe the motion of the car during the first 40 s. [1]
b (Supplement) Calculate the acceleration of the car between $t = 40$ s and $t = 60$ s. [3]
c Calculate the total distance travelled by the car. [4]
d Calculate the average speed of the car over the whole journey. [3]

5 A rectangular brick has a mass of 2900 g.

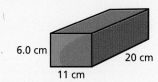

a Calculate the volume of the brick. [2]
b Calculate the density of the brick in kg/m³. [3]

6 (Supplement) Fig 6.1 shows two trolleys colliding.
Trolley **A** has a mass of 200 g. Trolley **B** has a mass of 300 g.

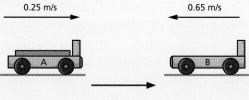

Fig. 6.1

a Calculate the momentum of trolley **A** before the collision. [1]
b Calculate the momentum of trolley **B** before the collision. [1]
The trolleys stick together when they collide.
c Calculate the velocity of the two trolleys after the collision. [4]

7 (Supplement) Fig. 7.1 shows the speed–time graph of a solid ball falling towards the Earth.

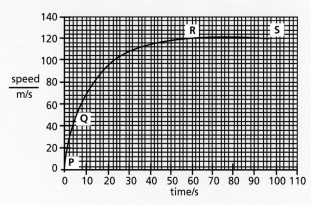

Fig. 7.1

a Describe the motion of the ball between position **R** and position **S**. [1]
b The ball in **Fig.7.2** is drawn when it is at position R.
Add two arrows to the diagram to show the forces acting on the ball.
Label each arrow with the name of the force.

Fig. 7.2 [3]

c Explain the motion of the ball between position **P** and position **R**. Use ideas about forces in your answer. [4]

8 a State two conditions needed for an object to be in equilibrium. [2]

Fig. 8.1 shows two students, **A** and **B**, sitting on a see-saw.

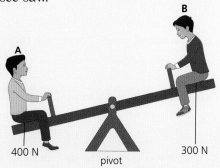

Fig. 8.1

Each student sits 1.5 m from the pivot.
b Calculate the moment of student **A** about the pivot. [2]
c Calculate the moment of student **B** about the pivot. [2]
d **(Supplement)** Calculate where another student of weight 250 N will sit to balance the seesaw. [3]

9 A bucket of water has a mass of 4.5 kg. Gravitational field strength = 9.8 N/kg.
a Calculate the weight of the bucket of water. [2]
b **(Supplement)** The bucket of water hangs from a spring. The spring does not exceed its limit of proportionality.
   i Define the term 'limit of proportionality'. [1]
   ii The unstretched length of the spring is 15 cm. Its length increases to 35 cm when the bucket of water hangs from it. Calculate the spring constant of the spring. [2]

10 An aeroplane of mass 3.1×10^5 kg accelerates uniformly from rest along a runway.
a i Name one energy store that increases as the aeroplane accelerates along the runway. [1]
  ii Name one energy store that decreases as the aeroplane accelerates along the runway. [1]
b **(Supplement)** The aeroplane reaches a speed of 70 m/s on the runway after 25 s.
   i Calculate the average acceleration of the aeroplane. [2]
   ii Calculate the resultant force acting on the aeroplane while it is accelerating. [2]
c **(Supplement)** After the aeroplane takes off, it gains height and then flies at a constant height in a circular path around the Earth.
   i What is the direction of the resultant force acting on the aeroplane when it is flying at a constant height? [1]
   ii The speed of the aeroplane increases but its height above the Earth remains constant. State what happens to the size of the resultant force acting on the aeroplane. [1]

11 **(Supplement)** **Fig. 11.1** shows a cylinder 20 m below the surface of the water.

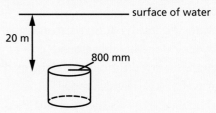

a The density of water is 1000 kg/m³ and the acceleration of free fall is 10 m/s². Calculate the pressure that the water exerts on the top of the cylinder. [3]
b The cylinder has a radius of 800 mm. Calculate the force the water exerts on the top of the cylinder. [4]

12 A student sets up the simple pendulum in **Fig. 12.1**. The pendulum swings backwards and forwards between **A** and **C**.

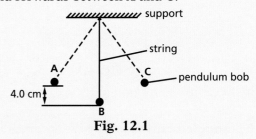

Fig. 12.1

a The student measures the time for the pendulum to swing from **A** to **C**. Describe how the student determines this time as accurately as possible. [2]
b What are the names of the two vertical forces acting on the pendulum bob when it is at position **B**? [2]
c **(Supplement)** The mass of the pendulum bob is 150 g. When the pendulum swings, it is 4.0 cm higher at position **A** than at position **B**. Gravitational field strength = 10 N/kg Calculate the change in gravitational potential energy of the pendulum bob between **A** and **B**. [4]

13 a Describe the difference between renewable and non-renewable energy. [2]
b i State two advantages of using nuclear fission as an energy resource. [2]
  ii State two disadvantages of using nuclear fission as an energy resource. [2]
c Other than nuclear and solar, state one renewable energy resource and one non-renewable energy resource. [2]

d (Supplement) Fig 13.1 shows solar cells arranged in a rectangle.

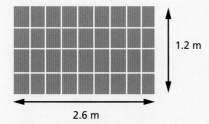

Fig. 13.1

The solar cells receive 250 W of energy from the Sun per square metre.
The cells produce a current of 2.4 A and a potential difference of 80 V.
Calculate the efficiency of the solar cells. [5]

14 a State the name of the galaxy that the Earth is in. [1]
b Describe what is meant by redshift of electromagnetic radiation. [2]
c The speed of light is 3.0×10^8 m/s. Calculate the time it takes light to reach Earth from the Sun if the distance between the Earth and the Sun is 1.5×10^8 km. [2]

15 Fig 15.1 A man pushes a cupboard towards a van at a constant speed.

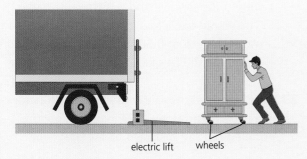

Fig. 15.1

a The frictional force acting on the wheels of the cupboard is 50 N.
What is the size of the force that the man pushes the cupboard with? [1]
b An electric lift moves the cupboard a vertical distance of 1.7 m from the ground to the van's floor level. The weight of the cupboard is 450 N.
Calculate the work done to lift the cupboard. [2]
c The power of the electric lift is 0.20 kW.
Calculate the time it takes for the electric lift to move the cupboard to the van's floor level. [3]

d A 12 V battery powers the electric lift. Calculate the current flowing in the lift. [2]

16 A group of students use the equipment in **Fig. 16.1** to compare how well rods of different materials conduct heat.

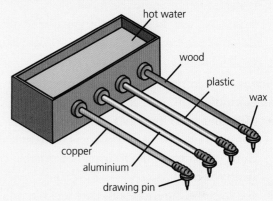

Fig. 16.1

a Explain how the students will use this equipment to determine which material is the best conductor of heat. [1]
b (Supplement) Describe the difference between thermal conduction in copper and in plastic. [4]
c (Supplement) Describe why thermal conduction is poor in gases. [1]

17 Complete the table by writing **T** if the statement is **true** and **F** if the statement is **false**. [2]

statement	(T) or (F)
The Earth orbits the Sun once in approximately 365 days.	
The Moon orbits the Earth once every 24 hours.	
The strength of the gravitational field around a planet increases as the distance from the planet increases.	

18 Fig. 18.1 shows a fixed volume of gas in a sealed container.

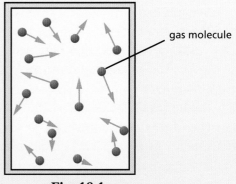

Fig. 18.1

a Explain how the gas molecules exert pressure on the walls of the container. [1]
b The gas is heated. Explain what happens to the pressure of the gas. [2]
c (Supplement) A fixed mass of gas at a constant temperature has a volume of 1200 cm³. The pressure of the gas is 4000 kPa. Calculate the volume of the gas at a pressure of 2500 kPa. [3]

19 The planets in the Solar System differ in size and composition.
a Describe the differences in the size and composition of the four planets nearest the Sun compared to the four planets furthest from the Sun. [2]
b Explain the differences in part **a** using an accretion model for Solar System formation. [4]

20 (Supplement) A student sets up the experiment in **Fig. 20.1**.

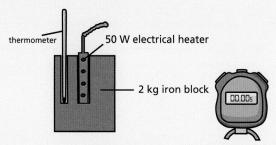

Fig. 20.1

a Describe how the student uses the equipment to determine the specific heat capacity of iron. In your answer, include any equations the student will need to use. [4]
b The student calculates the value of the specific heat capacity of iron to be 620 J/kg°C. The true value of the specific heat capacity of iron is 450 J/kg°C. Explain why the value calculated by the student is higher than the true value. [1]
c Suggest how the student could improve the accuracy of the value they obtained for the specific heat capacity of iron. [1]

21 (Supplement) A student pours a small amount of water onto a table outside on a warm, sunny day. The water slowly evaporates.
a Explain why the table in contact with the water cools down. Use ideas about molecules in your answer. [3]

b The student heats 500 g of water in a beaker until it starts to boils. The starting temperature of the water is 15°C. The specific heat capacity of water is 4200 J/kg°C. Calculate the amount of heat energy supplied to the water to increase its temperature to 100°C. [4]
c State two ways in which the evaporation of a liquid is different from when a liquid boils. [2]

22 A sound wave is produced at **Q**. The sound wave travels towards the wall.

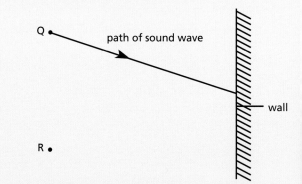

a Explain why a person standing at **R** hears an echo. [1]
b A sound wave is a longitudinal wave. Describe the difference between a longitudinal wave and a transverse wave. [2]
c The sound wave has a speed of 330 m/s and a wavelength of 0.80 m. Calculate the frequency of the sound wave. [3]

23 The diagram shows a ray of white light incident on a glass prism in air. The path of only the refracted red ray is shown inside the prism.

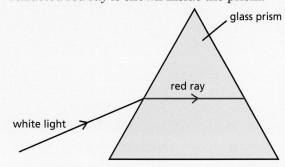

a Draw a ray to show the path of the red light when it emerges from the prism into the air. [1]
b On the same diagram, draw a ray to show the path of violet light in the prism and when it emerges from the prism into the air. [2]

c **(Supplement)** The refractive index of glass for red light is 1.52. The angle of incidence of the white light is 50°.
Calculate the angle of refraction of the red light in the glass. [3]

24 (Supplement) The table shows data about the planets in the Solar System.

Planet	Orbital distance / million km	Orbital duration / years	Average surface temperature / °C	Gravitational field strength at surface N/kg
Earth	150	1.0	15	9.8
Jupiter	779	12	–110	25
Mars	228	1.9	–23	4.0
Mercury	58	0.20	170	4.0
Neptune	4490	165	–200	14
Saturn	1420	30	–140	10.4
Uranus	2870	84	–210	10.4
Venus	108	0.60	460	9.0

a Give one reason why the average surface temperature of Uranus is less than the average surface temperature of the Earth. [1]
b An object has a mass of 10 kg. Explain on which planet the object will have the greatest weight. [2]
c Calculate the circumference of Jupiter's orbit around the Sun. [2]
d Calculate how many times bigger the Earth's average orbital speed is compared to Jupiter's. [3]

25 A student shines a ray of light from a ray box at **X** towards a mirror. The ray of light is reflected by the mirror towards **Y**.

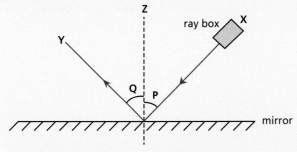

a State the name of the dotted line **Z**. [1]
b State the name of angle **Q**. [1]
c Describe the relationship between angle **P** and angle **Q**. [1]

d A ray of light enters an optical fibre at end **M**. The light undergoes total internal reflection as it travels through the optical fibre.

Complete the path of the light ray until it emerges from end **N**. [2]
e **(Supplement)** The optical fibre material has a refractive index of 1.5. Calculate the critical angle. [3]

26 a Student **A** and student **B** stand 250 m from a flat wall. Student **A** has a stopwatch. Student **B** claps her hands.
Describe how the students determine the speed of sound in air. Include an equation in your answer. [3]
b The students look at traces of two sound waves, **X** and **Y**, on an oscilloscope.

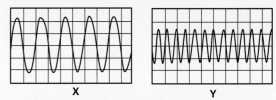

Student **A** says that wave **X** has a higher pitch and is louder than wave **Y**.
Is student **A** correct? Explain your answer. [2]

27 There are 7 regions in the electromagnetic spectrum.

radio waves
P
infrared
visible light
ultraviolet
X-rays
Q

a State the names of regions **P** and **Q** in the electromagnetic spectrum. [2]
b State which region is used in electric grills and television remote controllers. [1]
c State which region is used in security markings and sterilising water. [1]

307

28 (Supplement) The diagram shows the position of a convex lens and an object **QR**.

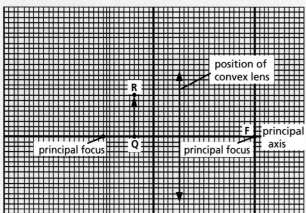

a Draw two rays from **R** that pass through the lens. Use the two rays to determine the position of the image.
On the diagram, mark the position of the image and label it **I**. [2]
b Describe two characteristics of the image formed. [2]
c Describe how a diverging lens can be used to correct short-sightedness. [3]

29 The diagram shows two rays of light, **X** and **Y**, from an object **O**. The two rays are incident on a plane mirror.

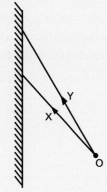

a Draw the normal to the mirror at the point where ray **X** reaches the mirror. [1]
b Draw the path of ray **X** after it reaches the mirror. [1]
c Describe two characteristics of the image formed by the plane mirror. [1]
d **(Supplement)** Draw the path of ray **Y** after it reaches the mirror.
Use the paths of ray **X** and ray **Y** after they reach the mirror to locate the image of the object **O**. Label the image **I**. [3]

30 a A student sets up a circuit using a voltmeter, an ammeter, a cell, and a fixed resistor. Draw a circuit diagram of the circuit the student will set up to determine the resistance of the fixed resistor. [3]
b The potential difference across the resistor is 10 V. The current through the resistor is 0.10 A. Calculate the resistance of the resistor. [3]
c **(Supplement)** Draw an arrow on your circuit to show the direction of the conventional current. [1]
d **(Supplement)** Calculate the time it takes for a current of 0.10 A to transfer a charge of 1.4 C. [3]

31 A student sets up a circuit to investigate the resistance of an unknown component, **X**.

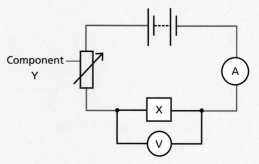

a State the name of component **Y**. [1]
b **(Supplement)** The student uses the circuit to take measurements of the current through **X** and the potential difference across **X**. She plots a current–potential difference graph of her results for component **X**.

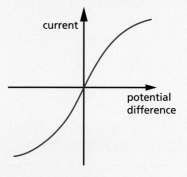

i State the name of component **X**. [1]
ii Explain the shape of the current–potential difference graph for component **X**. Use ideas about temperature and resistance in your answer. [3]

IGCSE PRACTICE QUESTIONS

32 (Supplement) The diagram shows two parallel conducting plates connected to a high-voltage supply.

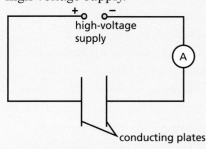

An electric field is present between the conducting plates.

a Describe what is meant by an electric field. [1]
b Draw lines on the diagram to show the electric field pattern between the two conducting plates.
Draw an arrow on one line to show the direction of the electric field. [2]
c A piece of resistance wire is used to join the two conducting plates. A charge of 0.050 C passes through the wire in 25 s.
Calculate the reading on the ammeter. [2]
d The potential difference of the high-voltage supply is changed to 2000 V.
Calculate the new reading on the ammeter if 300 J of energy is supplied in 15 s. [3]

33 Fig. 33.1 shows an experiment a student sets up to investigate the magnetic field due to a current in a straight wire.

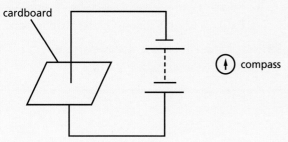

a Describe how the student uses this equipment to determine the magnetic field pattern produced by the current in the wire. [3]
b Draw the pattern and direction of the magnetic field around the wire in **Fig. 33.1**. [2]
c **(Supplement)** The student reverses the direction of the current in the wire. Describe the effect this has on the magnetic field around the wire. [1]

34 The diagram shows a transformer.

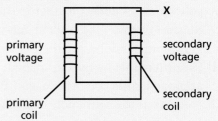

a State the name of the part of the transformer labelled **X**. [1]
b **(Supplement)** Explain how the transformer works. [3]
c A student fills in a table about transformers:

	input voltage / V	output voltage / V	step-up or step-down
A	20	60	step-up
B	100	40	step-up
C	230	1500	step-down
D	70	20	step-down

Which two rows in the table has the student filled in incorrectly? [2]
d The student finds out information about the high-voltage transmission of electricity.
voltage across primary coil = 10 000 V
number of turns on primary coil = 4000
number of turns on secondary coil = 200
current through secondary coil = 250 A
  i Calculate the voltage across the secondary coil. [3]
  ii **(Supplement)** Calculate the current through the primary coil. Assume that the transformer is 100% efficient. [3]
  iii State the advantage of using high-voltage transmission of electricity. [2]

35 The diagram shows a bar magnet next to a circuit containing a coil of wire and a galvanometer.

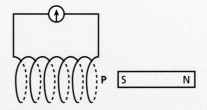

a A student moves the bar magnet into the coil. Explain why the galvanometer briefly deflects to the left. [2]

309

b Suggest one way the student can make the galvanometer deflect further to the left. [1]
c Suggest one way the student can make the galvanometer deflect to the right. [1]
d (Supplement) As the bar magnet approaches the coil of wire, end **P** of the coil acts as the south pole of a magnet. Explain why end **P** has this polarity. [1]

36 (Supplement) The Hubble constant describes how fast the universe is expanding and can be used to estimate the age of the Universe.
  a State the definition of the Hubble constant and an estimate of its current value. [2]
  b A galaxy is moving away from the Earth at a speed of 1.6×10^5 m/s. Calculate the distance from the Earth to this galaxy. [2]
  c The graph shows the relationship between the speed of galaxies that are moving away from the Earth and their distance from the Earth. The value of the Hubble constant can also be expressed in km/s/Mpc.
  Use the graph to show that the value of the Hubble constant is approximately 67 km/s/Mpc. [2]

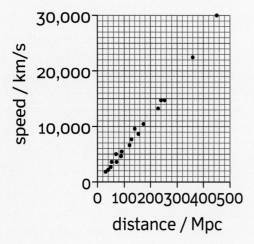

37 Two fixed resistors are connected in a circuit with identical batteries.

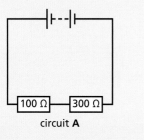

circuit **A**

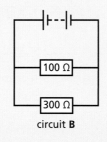

circuit **B**

a i In which circuit is the current the same though both resistors? [1]
  ii (Supplement) In which circuit is the potential difference the same across both resistors? [1]
  iii Calculate the combined resistance of the resistors in circuit **A**. [1]
  iv (Supplement) Calculate the combined resistance of the resistors in circuit **B**. [3]
b (Supplement) Circuit **C** contains a thermistor and a 1.5 kΩ resistor in series. There is no current through the voltmeter.

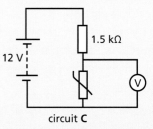

circuit **C**

Calculate the reading on the voltmeter when the thermistor has a resistance of 4.5 kΩ. [3]

38 Radium-226 is the most stable isotope of radium. Radium-226 has a proton number of 88 and a nucleon number of 226.
In nuclide notation, radium is written as:
$$^{A}_{Z}\text{Ra}$$
  a i State the value of A. [1]
    ii State the value of Z. [1]
    iii Calculate the number of neutrons in the nucleus of radium-226. [2]
  b Radium-228 is another isotope of radium. Explain what is meant by the term isotope. [2]

39 The graph shows the decay curve of radioactive isotope **A**.

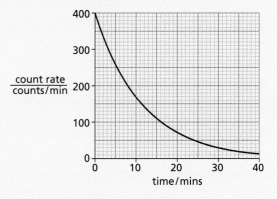

  a i Define the term half-life. [1]
    ii Determine the half-life of radioisotope **A** from the graph. [2]

b Another radioisotope, **B**, starts with the same count rate of 400 counts/min. Radioisotope **B** has a longer half-life than radioisotope **A**. Draw a line on the graph to show the decay curve of radioisotope **B**. [2]

c **(Supplement)** A scientist wants to determine the half-life of radioisotope **C**. The background radiation measured by a detector when no radioactive isotopes are present is 30 counts/min.
The scientist moves radioisotope **C** closer to the detector. The reading on the detector increases to 350 counts/min. After 20 days, the count rate is 50 counts/min when radioisotope **C** is at the same distance from the detector. Calculate the half-life of radioisotope **C**. [4]

40 Three types of nuclear emissions are:
alpha radiation
beta radiation
gamma radiation

a **i** State which of the three nuclear emissions has the greatest ionizing effect. [1]
ii State which of the three nuclear emissions has the lowest penetrating ability. [1]

b **i** Describe what is meant by the term background radiation. [1]
ii State two sources of background radiation. [1]

c **(Supplement)** A beam of β particles enter a magnetic field between the poles of a strong magnet.

Describe the path of the β particles in the magnetic field. [3]

41 (Supplement) Americium-241 is radioactive. An americium-241 nucleus decays to a neptunium nucleus by emitting an α-particle.

a Complete the equation to show the decay of americium-241. [2]

$$^{241}_{95}\text{Am} \rightarrow \text{Np} + \alpha$$

b The diagram shows how americium-241 is used in a household smoke alarm. Use the diagrams to describe how the smoke alarm works and explain why the smoke alarm rings when there is a fire. [2]

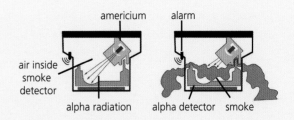

c A beam of α-particles enter an electric field between two charged plates.

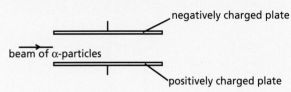

Sketch the path of the beam of α-particles between the charged plates. [1]

42 (Supplement)

a State the time, in years, that it takes light to travel 10 million light years in space. [1]

b Calculate the time, in years, it takes for light to travel 1.9×10^{13} km in space. [2]

c Complete the diagram to show the life cycle of a star. [6]

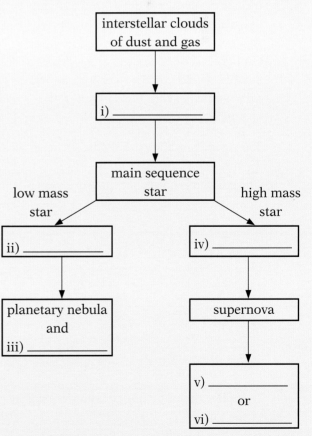

IGCSE PRACTICE QUESTIONS

If you do not take a practical examination, you will sit an alternative-to-practical paper instead. Here are some examples of typical questions. For some of them, you will require graph paper.

IGCSE Alternative-to-Practical Questions

1 A student is investigating springs. The student measures the unstretched length l_0 of the spring in mm. She attaches the spring to a clamp, as shown in **Fig. 1.1**.

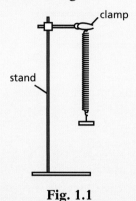

Fig. 1.1

The student:
- hangs a load L on the spring
- measures the new length l of the spring
- calculates the extension e of the spring
- repeats the method with different loads.

a On **Fig. 1.1**, draw the position where the student would place a ruler in order to accurately measure the length l of the spring. [1]

The unstretched length l_0 of the spring is 45 mm.
The student records her measurement in a table.

load L / N		
0.0	45	0.0
1.0	48	
2.0	50	
3.0	54	
4.0	57	
5.0	60	

b Complete the two missing column headings in the table. [2]

c Calculate the extension of the spring for each load added and write your results in the table. [3]

d Plot a graph of extension e / mm (x-axis) against load L / N (y-axis). Draw a line of best fit. [5]

e A student suggests that the length l of the spring is directly proportional to the load L. Is the student correct? Explain your answer. [1]

f Use your graph to determine the extension of the spring when a load of 2.3 N is added. [1]

2 A student wants to determine a value for the acceleration of free fall g.
Fig. 2.1 and **Fig. 2.2** show the equipment the student sets up.

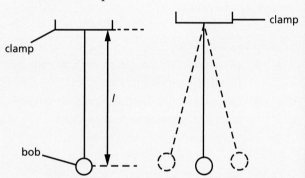

Fig. 2.1   **Fig. 2.2**

a The student sets up the equipment so that the length l of the pendulum is actually 44 cm. Describe how the student prevents a parallax error when measuring the length l. [1]

b The student moves the bob to one side slightly. She releases the bob and measures the time t for 20 complete oscillations. The time t is shown on the stopwatch in **Fig 2.3**. [2]

i Write down the time t shown on the stopwatch. [1]

ii The period T of the pendulum is the time for one complete oscillation. Calculate the period T of the pendulum. [2]

c To determine a value for the acceleration of free fall g, the student needs the value of T^2.
   i Calculate T^2 to a suitable number of significant figures. Give the unit. [1]
   ii Calculate the acceleration of free fall g using the equation:
   $$g = \frac{4\pi^2 l}{T^2}$$ [2]
d Suggest two improvements that the student could make to the experiment. [2]

3 A student investigates how the length l of a wire affects its resistance R.

The student sets up the circuit in **Fig. 3.1**.

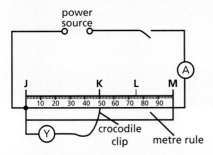

Fig. 3.1

The student measures the current I through the wire. He also measures the potential difference V across sections **JK**, **JL**, and **JM**. The student's measurements are shown in **Table 3.1**.

section of wire	l/cm	I/A	V/V	R
JK			0.95	
JL				
JM			1.9	

Table 3.1

a Using **Fig. 3.1**, complete the length column in **Table 3.1**. [1]
b i Fig. 3.2 shows the ammeter reading for all three sections of the wire.

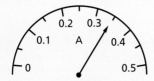

Fig. 3.2

Complete the current column for each section with the reading shown on the ammeter. [1]

ii Fig. 3.3 shows the voltmeter reading for section **JL** of the wire.

Fig. 3.3

Fill in the missing potential difference reading for section **JL** in **Table 3.1**. [1]

c i Calculate the resistance R of each section of the wire.
   Use the equation:
   $$R = \frac{V}{I}$$ [1]
   ii State the unit of resistance. [1]
d Using the results in **Table 3.1**, predict the resistance of the same piece of wire with a length of 1.5 m. [1]

4 A student investigates if the resistance of a wire depends on the metal from which the wire is made. The following equipment is available to the student:
ammeter
connecting leads
power supply
resistance wires made of different metals
switch
variable resistor
voltmeter

The resistance is calculated using the equation:
$$resistance = \frac{potential\ difference}{current}$$

Plan an experiment to investigate if the resistance of a wire depends on the metal from which the wire is made.
In your plan include:
– a diagram of the circuit you would set up to determine the resistance of each metal wire [3]
– a description of how you would perform the experiment, including any measurements you would take [3]
– key variables that you would need to control [2]
– a table to display your readings and calculated values, including column headings (you do not need to write any readings in your table). [2]

IGCSE PRACTICE QUESTIONS

5 Two students do an experiment to calculate the speed of sound in air.

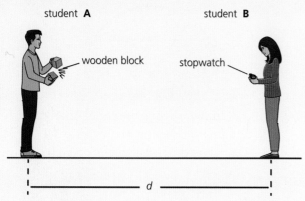

The students stand a distance d apart. Student **B** starts the stopwatch when she sees student **A** hitting the wooden blocks together. Student **B** stops the stopwatch when she hears the sound made by the wooden blocks and records the time interval t

The students repeat the experiment five times. They record the results in a table and calculate the speed of sound v.

t/s	$\dfrac{v}{(m/s)}$
1.16	344.83
1.19	336.13
1.12	357.14
1.11	360.36
1.20	333.33

a Estimate the distance d between the two students. [1]

b Suggest a suitable piece of equipment that the students use to measure the distance d. [1]

c Calculate the mean value for the speed of sound from the students' results. [2]

d In the table, the students recorded their values for the speed of sound v to five significant figures.
Explain whether this is a suitable number of significant figures for the students to use. [1]

6 A student is investigating the cooling of water using the equipment in **Fig. 6.1**.

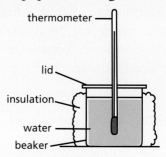

Fig. 6.1

The student pours 200 cm³ of hot water into the insulated beaker. He covers the top of the beaker with a lid.

a The thermometer in **Fig. 6.2** shows the temperature of the water in the beaker at the start of the experiment.

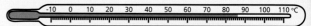

Fig. 6.2

Write down the starting temperature θ_s shown on the thermometer in **Fig. 6.2**. [1]

b The student measures the temperature θ of the water every 30 s as the water cools. He records the readings in **Table 6.1**. [2]

t/s	θ/
0.0	86
30	80
60	76
90	73
120	71

Table 6.1.

i Complete the missing unit in column two in **Table 6.1**. [1]

ii Calculate the temperature drop $\Delta\theta_1$ during the first 60 s. [1]

iii Calculate the temperature drop $\Delta\theta_2$ between 60 s and 120 s. [1]

iv Explain why $\Delta\theta_1$ is different from $\Delta\theta_2$. [1]

c The student repeats the experiment. Suggest two changes that the student could make to the equipment to make the water cool down more quickly. [2]

16
Reference

- USEFUL EQUATIONS
- UNITS AND ELEMENTS
- ELECTRICAL SYMBOLS AND RESISTOR CODES
- ANSWERS TO QUESTIONS
- INDEX

Useful equations

In most cases, the equations below are given in both word and symbol form.

$g = 10$ N/kg (Earth's gravitational field strength)
$= 10$ m/s² (acceleration of free fall)

Density, mass, and volume

$$\text{density} = \frac{\text{mass}}{\text{volume}}$$

$$\rho = \frac{m}{V}$$

Speed

$$\text{average speed} = \frac{\text{distance moved}}{\text{time taken}}$$

Acceleration

$$\text{acceleration} = \frac{\text{change in velocity}}{\text{time taken}}$$

$$a = \frac{\Delta v}{\Delta t}$$

Force, mass, and acceleration

force = mass × acceleration

$$F = ma$$

Momentum

momentum = mass × velocity

Impulse

impulse = force × time = change in momentum

Weight

weight = mass × g

$$W = mg$$

Moment of a force

moment of force about a point = force × perpendicular distance from point

Stretched spring

load = spring constant × extension

$$F = kx$$

Pressure and force

$$\text{pressure} = \frac{\text{force}}{\text{area}}$$

$$p = \frac{F}{A}$$

Pressure in a liquid

pressure = density × g × depth

$$p = \rho g h$$

Temperature

Kelvin temperature = temperature in °C + 273

Compressing gases

For a fixed mass of gas at constant temperature:

pressure₁ × volume₁ = pressure₂ × volume₂

$$p_1 V_1 = p_2 V_2$$

(Boyle's law)

Work

work done = force × distance moved in direction of force

$$W = Fd$$

Gravitational potential energy

gravitational potential energy = mass × g × height

$$PE = mgh$$

Kinetic energy

kinetic energy = ½ × mass × velocity²

$$KE = \tfrac{1}{2}mv^2$$

REFERENCE

Energy and temperature change
energy transferred = mass × specific heat capacity × temperature change

$$E = mc\Delta\theta$$

Power
$$\text{power} = \frac{\text{work done}}{\text{time taken}} = \frac{\text{energy transferred}}{\text{time taken}}$$

Efficiency
$$\text{efficiency} = \frac{\text{useful work done}}{\text{total energy input}}$$

$$= \frac{\text{useful energy output}}{\text{total energy input}}$$

$$= \frac{\text{useful power output}}{\text{total power input}}$$

Waves
speed = frequency × wavelength

$$v = f\lambda$$

Refraction of light
$$\text{refractive index} = \frac{\text{sine of angle of incidence}}{\text{sine of angle of refraction}}$$

$$n = \frac{\sin i}{\sin r}$$

Total internal reflection
$$\text{sine of critical angle} = \frac{1}{\text{refractive index}}$$

$$\sin c = \frac{1}{n}$$

Current and charge
$$\text{current} = \frac{\text{charge}}{\text{time}} \qquad I = \frac{Q}{t}$$

p.d., work done, and charge
$$\text{p.d.} = \frac{\text{work done}}{\text{charge}} \qquad V = \frac{W}{Q}$$

Resistance, p.d. (voltage), and current
$$\text{resistance} = \frac{\text{p.d.}}{\text{current}}$$

$$R = \frac{V}{I}$$

Resistors in series....
total resistance $R = R_1 + R_2$

...and in parallel
$$\frac{1}{R} = \frac{1}{R_1} + \frac{1}{R_2}$$

Electrical power
power = p.d. × current = current² × resistance

$$P = VI \qquad P = I^2R$$

Electrical energy
energy transferred = power × time
$$= \text{p.d.} \times \text{current} \times \text{time}$$
$$E = VIt$$

Transformers
$$\frac{\text{output voltage}}{\text{input voltage}} = \frac{\text{output turns}}{\text{input turns}}$$

$$\frac{V_s}{V_p} = \frac{n_s}{n_p}$$

For 100% efficient transformer:
power input = power output
$$V_p I_p = V_s I_s$$

Units and elements

SI units and prefixes

quantity	unit	symbol
mass	kilogram	kg
length	metre	m
time	second	s
area	square metre	m²
volume	cubic metre	m³
force	newton	N
weight	newton	N
pressure	pascal	Pa
energy	joule	J
work	joule	J
power	watt	W
frequency	hertz	Hz
p.d., e.m.f.	volt	V
current	ampere	A
resistance	ohm	Ω
charge	coulomb	C
capacitance	farad	F
temperature	Kelvin degree Celsius	K °C

prefix*	meaning	
G (giga)	1 000 000 000	(10^9)
M (mega)	1 000 000	(10^6)
k (kilo)	1000	(10^3)
d (deci)	$\frac{1}{10}$	(10^{-1})
c (centi)	$\frac{1}{100}$	(10^{-2})
m (milli)	$\frac{1}{1000}$	(10^{-3})
μ (micro)	$\frac{1}{1\,000\,000}$	(10^{-6})
n (nano)	$\frac{1}{1\,000\,000\,000}$	(10^{-9})
p (pico)	$\frac{1}{1\,000\,000\,000\,000}$	(10^{-12})

Examples
1 μF (microfarad) = 10^{-6} F   1 km (kilometre) = 10^3 m
1 ms (millisecond) = 10^{-3} s   1 MW (megawatt) = 10^6 W

Note: 'micro' means 'millionth'; 'milli' means 'thousandth'.

* G, μ, n, and p are not required for Cambridge IGCSE examinations.

Elements

For simplicity, many of the rarer elements have been omitted from the table below.

atomic number (proton number)	element	chemical symbol
1	hydrogen	H
2	helium	He
3	lithium	Li
4	beryllium	Be
5	boron	B
6	carbon	C
7	nitrogen	N
8	oxygen	O
9	fluorine	F
10	neon	N
11	sodium	Na
12	magnesium	Mg
13	aluminium	Al
14	silicon	Si
15	phosphorus	P
16	sulfur	S
17	chlorine	Cl
18	argon	Ar
19	potassium	K
20	calcium	Ca
22	titanium	Ti
25	manganese	Mn
26	iron	Fe
27	cobalt	Co
28	nickel	Ni
29	copper	Cu
30	zinc	Zn
35	bromine	Br
38	strontium	Sr
47	silver	Ag
48	cadmium	Cd
50	tin	Sn
53	iodine	I
55	caesium	Cs
74	tungsten	W
78	platinum	Pt
79	gold	Au
80	mercury	Hg
82	lead	Pb
86	radon	Rn
88	radium	Ra
90	thorium	Th
92	uranium	U
94	plutonium	Pu

Electrical symbols and codes

Electrical symbols

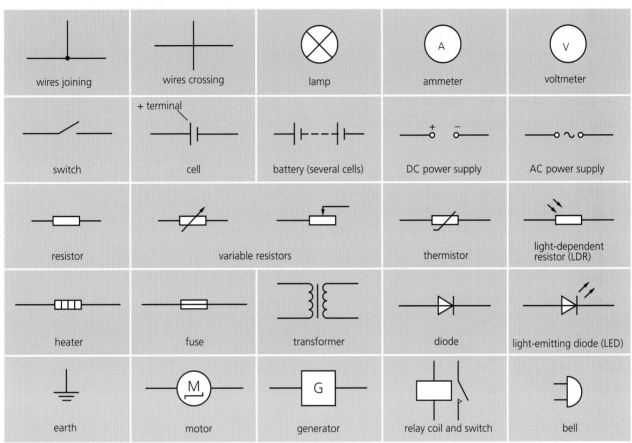

Resistor codes

The resistance of a resistor in ohms (Ω) is normally marked on it using one of these codes:

The resistor is marked with coloured rings. Each colour stands for a number:

black	0
brown	1
red	2
orange	3
yellow	4
green	5
blue	6
violet	7
grey	8
white	9

You 'read' the first three rings like this:

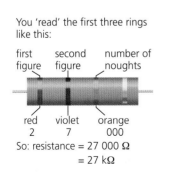

first figure — second figure — number of noughts

red — violet — orange
2 — 7 — 000

So: resistance = 27 000 Ω
= 27 kΩ

The fourth ring gives the **tolerance**. This tells you by how much the resistance may differ from the marked value:

gold ±5%   silver ±10%   no colour ±20%

The resistance is printed on the resistor:

R27 means 0.27 Ω
2R7 means 2.7 Ω
3K0 means 3000 Ω
5K6 means 5600 Ω
47K means 47 Ω
2M2 means 2.2 MΩ

So: resistance = 8.2 kΩ

The extra letter at the end gives the tolerance:

F ±1%   G ±2%   J ±5%   K ±10%   M ±20%

Answers

1.1 (page 13)
1 1000 g **2** 1000 mm **3** 10^6 μs **4** 6 m² **5** 2 km, 0.2 km, 20 km **6** 5 s, 50 s **7** 1.5×10^3 m, 1.5×10^6 m, 1.5×10^{-1} m, 1.5×10^{-2} m

1.2 (page 15)
1 m **2** kg **3** s **4** gram, milligram, tonne, micrometre, millisecond **5 a** 1.564 m **b** 1.750 kg **c** 26 000 kg (2.6×10^4 t) **d** 6.2×10^{-5} s (0.000 062 s) **e** 36.5 kg **f** 6.16×10^{-10} m **6 a** 5×10^{-3} kg **b** 5000 mg **7** mass: t, kg, g, mg, μg; length: km, m, mm, μm, nm; time: s, ms, μs, ns

1.3 (page 17)
1 a 87 mm **b** 95 mm **2** 2.3 s **b** Time more swings **3** Measure total thickness of all 336 pages, divide by 336 **4 a** Reading shown when result is known to be zero **a** Using scales to measure a weight

1.4 (page 19)
1 10^6 cm³ **2** 10^3 cm³ **3** 10^6 ml **4 a** 200 l **b** 2×10^5 cm³ **c** 2×10^5 ml **5 a** 2.7 g/cm³ **b** 54 g **c** 10 cm³ **6** Steel (stainless) **7** 39 kg **8** 4 m³ **9** 22.8×10^3 kg

1.5 (page 21)
1 Crowns: A silver, B gold, C mixture **2 a** 80 g, 100 cm³, 0.8 g/cm³ **b** 120 g, 48 cm³, 2.5 g/cm³

1.6 (page 22)
1 a Yes **b** No **2 a** 2600 kg **b** 2200 kg **c** 400 kg

Check-up questions (pages 22–24)
1 measurement: mass, time; unit: metre, second; symbol: m, kg
2 a 1000 **b** 1000 **c** 1 000 000 **d** 4 000 000 **e** 500 000
3 a 3 m **b** 0.5 kg **c** 1.5 km **d** 0.25 s **e** 500 ms **f** 750 m **g** 2500 g **h** 800 mm
4 24 cm³, 4 cm, 10 cm, 0.5 cm
5 a 2500 m **b** 2 m **c** 3000 kg **d** 2 litres
6 B and D
7 a kg **b** m, km **c** m³, cm³, ml **d** ms, s **e** g/cm³, kg/m³
8 D
9 B
10 1.25 kg/m³
11 a 0.1 m³, 0.05 m³ **b** 800 kg **c** 800 kg/m³ **d** 1000 kg/m³
12 a expanded, polystyrene, wood, ice, polythene; all less dense than water **b** expanded, polystyrene, wood **c** petrol will float on water; petrol less dense than water
13 A only is true
14 a No; too many significant figures **b** Time more swings **c** 0.93 s

2.1 (page 29)
1 20 m/s; actual speed varies **2** velocity also includes direction of travel **3 a** 64 m **b** 20 s
4 Runner 6.7 m/s, Grand Prix car 100 m/s, passenger jet 250 m/s, sound 333 m/s, International Space Station 7690 m/s
5 Increases by 2 m/s every second, velocity decreases by 2 m/s every second **6** 2.5 m/s²
7 4 m/s² **8 a** 12 m/s **b** 44 m/s **9** 17 m/s

2.2 (page 31)
1 a Not moving **b** A and B **c** B and C **d** 4 m/s **e** 60 m **f** 3 m/s **2 a** 30 m/s **b** 3 m/s² **c** 6 m/s² **d** 150 m **e** 525 m **f** 25 s **g** 21 m/s

2.3 (page 33)
1 Accelerating
2 a 0.1 s **b** 200 mm/s **c** 800 mm/s **d** 600 mm/s²
3 a 10 mm **b** 100 mm/s **c** 50 mm **d** 500 mm/s **e** 400 mm/s **f** 1000 mm/s²

2.4 (page 35)
1 a 10 m/s **b** 20 m/s **c** 50 m/s **2 a** 30 m/s **b** 40 m/s **c** 70 m/s **3 a** 10 m/s **b** 0 m/s **c** 30 m/s
4 a Downwards **b and c** B **d, e, and f** 10 m/s² **g** C

2.5 (page 37)
1 a CD **b** AB **c** DE **d** AB **e** BC, DE **f** DE
2 See below; will level off at a much lower speed than for a stone

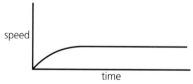

2.6 (page 39)
1 newton **2 a and b** They balance (are equal)
3 a Terminal velocity **b** Air resistance: upward force on parachute equal to weight **c** Equal **d** Lower because air resistance would be greater

2.7 (page 41)
1 a resultant force = mass × acceleration **b** 10 N, 20 N
2 a 1000 N **b** 1.25 m/s² **c** Acceleration zero (steady velocity)

2.8 (page 43)
1 a Brakes, tyres on road **b** Air resistance, engine parts **2** Lower fuel consumption **3 a** Top, so that feet grip board **b** Bottom, for faster movement over water **4 a** Static; no heating effect **b** Dynamic; heating effect

2.9 (page 45)
1 a 50 N, 100 N **b** 10 N/kg (both) **c** 10 m/s²

ANSWERS

2 a 1000 N **b** 100 kg **c** 100 kg **d** 370 N **e** 3.7 m/s²

2.10 (page 47)
1 a 500 N **b** Upward force of 500 N **2** Motion caused by equal but opposite (i.e. backward) force on gun **3** Ground is part of Earth which has a huge mass so change in motion is far too small to detect

2.11 (page 49)
1 momentum = mass × velocity
2 resultant force = change in momentum/time
3 a 48 kg m/s to right **b** 72 kg m/s to right
c 24 kg m/s to right **d** 8 kg m/s to right **e** 8 N
f +2 m/s **g** 0.67 m/s² **h** force = mass × acceleration **i** 8 N **4 a and b** 7500 N

2.12 (page 51)
1 a and b 0 **c** 12 kg m/s to left **d** 12 kg m/s to right **e** 4 m/s to right **2 a** 80 kg m/s to right
b 20 kg m/s to left **c and d** 60 kg m/s to right
e 3 m/s to right

2.13 (page 53)
1 Vector (e.g. force) has magnitude and direction, scalar (e.g. mass) has only magnitude **2 a** 17 N
b 7 N **c** 13 N at 23° to 12 N force **3 a** Horizontal component 87 N, vertical component 50 N **b** 350 N
c Force reduced (to 250 N)

2.14 (page 55)
1 Path is at a tangent to circle **2** Friction between tyres and road **3** Centripetal force
a less **b** less **c** more **4 a** gravity **b** electrostatic force **5 a** Gravity (weight) is only force on satellite, towards centre of Earth; acceleration is in same direction **b** Lower speed **c** Lower force

Check-up questions (pages 55–56)
1 a speed = distance/time **b** 100 m
2 a i 8 m **ii** 2.0 s **b** 4.0 m/s **c i** Increasing distance between positions **ii** Weight has a component down slope, force causes acceleration
3 a i More force causes more acceleration
ii force = mass × acceleration **iii** 2.0 kg
4 a 25 s **b** 1080 N **c** Resisting force (air resistance) increases with speed, so resultant force less
5 a 2000 N **b i** 1200 N
ii force = mass × acceleration
iii 1.5 m/s² **c** Total drag force will increase with speed until resultant force is again zero
6 a 5 N **b i** Both forces in same direction
ii Forces in opposite directions
7 a 5 km **b i** 10 m/s **ii** 8 min 20 s **c** 2 m/s²
8 a 4 s **b** Friction **c** On tyres, from road **d** 3000 N
9 a 20 m/s **b** Graph with speed on vertical axis and time on horizontal axis **c** 4 s; reduction in speed **d i** Weight (gravity) **ii** Air resistance **iii** Weight; air resistance; equal **e** Straight and level
f i No change **ii** Greater loss of speed
10 a See below **b i** 1.33 m/s²
ii ($\frac{1}{2} \times 20 \times 15$) + (5 × 15) = 225 m

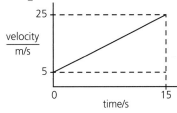

11 a and b 30 kg m/s to left **c** 10 kg **d** 3 m/s to left
12 a Centripetal force **b** Ball travels in straight line at tangent to circle **c** Gravity

3.1 (page 61)
1 Magnitude of force, perpendicular distance from point **2** For principle see p60, forces must balance
3 a 16 N m **b** 12 N m **c** No; clockwise **d** 1 N
e Downwards **4 a** 21 N **b** 10 N, 8 N, and 3 N forces; 84 N m **c** 21 N force; 84 N m **d** Yes

3.2 (page 63)

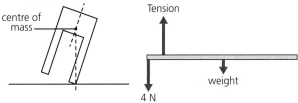

1 a See above left **b** Shorter legs, wider apart
2 a See above right **b** 1 N **3** See below:

3.3 (page 65)
1 a See below **b** 720 N m **c** 180 N **d** 600 N
e 420 N **f** zero **g** 0.5 m

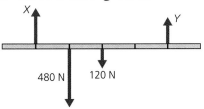

2 Yes, they are still equal

3.4 (page 67)
1 One that returns to original shape when force (load) removed **2** Point beyond which material won't return to original shape when force removed
3 No, not a straight line **4 a** 40 mm **b** extension/mm: 0, 9, 18, 27, 36, 48, 70 **d** Elastic limit is at extension of 36 mm **e** Up to 36 mm extension (end of straight section) **f** 3.9 N **g** 2.8 N

321

ANSWERS

3.5 (page 69)
1 a 50 Pa **b** 100 Pa **2 a** 200 N **b** 400 N **3** Large area of contact with soil reduces pressure on soil
4 a 300 N **b i and ii** See below **c** 7500 Pa, 500 Pa

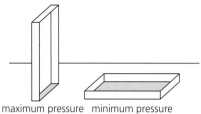

maximum pressure   minimum pressure

3.6 (page 71)
1 a Less **b** Same **c** Same **d** Less **2 a** 24 m^3
b 19 200 kg **c** 192 000 N **d** 16 000 Pa **3** 20 000 Pa

3.7 (page 73)
1 Increases with depth, acts in all direction
2 Pressure in straw reduced so greater outside air pressure pushes liquid up **3** Height of column reduced **4 a** 100 mm of mercury **b** 860 mm of mercury **c** 113 000 Pa **5 a** 96 000 Pa **b** 0.96 atm **c** 960 mb **6 a** 9810 Pa **b** 10.3 m

3.8 (page 75)
1 More molecules in each cm^3, so more collisions every second on each cm^2 of inside surface of balloon **2 a** 12 m^3 **b** 15 m^3 **3 a** Pressure × volume is constant **b** Straight line (through origin)

Check-up questions (pages 76–77)
1 a Moment due to F produces larger force at shorter distance from pivot **b** Larger force, further from pivot
2 a 0.4 N m **b** 1.6 N
3 a 100 kPa, 200 kPa, 300 kPa, 400 kPa **b** 2 m^3
4 a Downward force (weight) through M, upward force through A **b i** 30 N m **ii** 0.38 m
5 b i 10.0 cm **ii** 14.0 cm **iii** 2.5 N
6 a 100 000 Pa, pressure = force/area
b Not sufficient to exceed elastic limit of window
7 a W acts downwards from C **b** Force spread over greater area, so less pressure on heel **c** 200 N
d the perpendicular distance from the pivot increases so the moment of the force increases
8 a the same as; more than; less than **c** 2.5 N/cm^2
9 a i 0.5 m^2 **ii** 2.0 m^2 **b** 50 000 N/m^2
10 a 12 m^3 **b** 12 000 kg **c** 120 000 N **d** 20 000 Pa

4.1 (page 81)
1 60 J **2** 0.5 m **3 a** 10 000 J **b** 35 000 000 J
c 500 000 J **d** 200 J **4** 18 **5 a** kinetic, gravitational potential, chemical **b** chemical

4.2 (page 83)
1 a 50 J **b** 50 J **c** Changed to thermal energy (heat)
2 For law see p84 **3** Energy can't be made. Because of losses, generator can't deliver enough energy for motor.

4.3 (page 85)
1 a 240 J **b** 360 J **2 a** 75 J **b** 300 J **3** 20 m/s
4 a and b 25 J **c and d** 5 m

4.4 (page 87)
1 a 30% **b** Wasted as thermal energy (heat)
2 500 W **3 a** 3000 W **b** 3000 J **c** 60 000 J
d 75% **4 a** 6000 J **b** 300 J **c** 300 W **5 a** 6000 N
b 4000 W **6** 50 000 W

4.5 (page 89)
1 Coal, oil, natural gas, uranium **2 a** Turning turbines **b** Condense steam (turn it back to liquid)
3 a Turbines **b** Thermal energy (heat)
c X: 2000 MW, Y: 1500 MW **e** X is 36%, Y is 27%

4.6 (page 91)
1 Gravitational potential energy of water behind a dam **2 a** Useful energy output (as electricity) is 25% of energy in fuel **b** B **c** E **d** A **e** A **f** No fuel required or burned

4.7 (page 93)
1 Can't be replaced; oil, natural gas, coal, uranium
2 a Wind, hydroelectric solar, tidal, geo-thermal, wave **3** See p96: Sun's energy stored in ancient plants during growth (animals get energy by eating plants) → ancient remains buried and changed into crude oil over millions of years → petrol extracted from oil **4** Carbon dioxide emissions, other pollutants **5** Storage of nuclear waste, power stations expensive to decommission **6** Energy from hot rocks (or water) underground; heating, heat source for power stations **7** Energy radiated from Sun; solar panels (for hot water), solar cells
8 Hydroelectric, tidal, wave **9** Better insulation, more efficient transport, making goods last longer, being less wasteful with electricity etc.

Check-up questions (pages 96–97)
1 a Wound up spring, stretched rubber bands
b Via gearwheels so that toy gains kinetic energy
2 a i PE **ii** PE + KE **iii** PE + KE **b** Changed into thermal energy (heat)
3 a i Elastic potential energy (strain energy)
ii Changed to KE + gravitational PE **b** 0.75 J
c energy lost as thermal energy
4 a Wind, hydroelectric, tidal, solar **b** No polluting gases from sources in part a; resources don't run out **c** Output can be variable (e.g. wind)
d building, operating, and maintenance costs
5 a 225 000 J **b** 225 000 J **c** 1.25 m
6 a 3000 N **b** 180 000 J **c** 4000 W **d** 0.8 (80%)
7 a kettle 2 kW; food mixer 600 W **b** television

c food mixer **d** wasted as thermal energy
8 a Resources that can't be replaced **b** Simplest way of releasing energy, as heat (e.g as in a power station) **c** Uranium
9 a Heating water → steam → motion in turbines → turning generator → electricity **b** Compared with fossil fuels, wind power more expensive, much lower output, variable, but less polluting.
10 a wood – yes, no; uranium – no, no **b ii** Not renewable, use may be causing global warming
11 a Oil (or coal or natural gas) **b** energy, burns **c** Non-renewable fuels can't be replaced/regrown **d** (From top, example, use) petrol, cars; wood, burning for heat; wood, burning for heat; petrol, cars
12 a current, thermal **b** kinetic, thermal **c** chemical, current, light **d** sound, thermal
13 elastic; gravitational; potential

5.1 (page 101)
1 a, e and g gas **b and c** solid **d and f** liquid **2 a** Random motion of smoke particles **b** Brownian motion **c** Smoke particles light enough to be moved by collisions with individual molecules in gas **3** Move faster on average **4** Total kinetic energy of all atoms or molecules in a material

5.2 (page 103)
1 a 100 °C **b** 373 K **c** −273 °C **d** 0 K **e** 0 °C **f** 273 K
2 a Volume increase with temperature **b** Change of conducting ability (resistance) with temperature
3 a Slower on average in B **b** A to B **c** When temperatures are the same

5.3 (page 109)
1 a Particles (atoms) vibrate faster and push each other further apart **b** To allow for contraction if temperature falls **c** Aluminium expands more than concrete and would crack it **d** Metal on one side expands more than different metal on other side **e** More open arrangement of molecules in ice takes up more space than in water **2 a** Bimetal strip bends, so contacts separate **b** Right

5.4 (page 107)
1 a Particles (e.g. molecules) cause force when they collide with walls (because of momentum change) **b** Particles move faster, so force of collisions greater **2** Increases **3** Liquid; weaker attractions to hold particles together **4** Gas; very weak attractions so particles not held together

5.5 (page 109)
1 a Bottom needs to let heat (thermal energy) through, handle needs to reduce heat flow into hand **b** They trap air **c** Aluminium conducts heat away more rapidly than wood **d** Water is a much better thermal conductor than the air trapped in cloth **2** Loft insulation, mineral wool in cavity walls, insulation around hot water storage tank **3** Thicker lagging, keeping water at a lower average temperature
4 a Copper **b** Length, diameter, temperature difference same for all the metals **5** Free electrons present to move through metal and carry energy

5.6 (page 111)
1 a 'Radiator' causes convection current **b** Hot air rises by convection, carrying smoke with it **c** Cooler air flows in to replace hot air rising from bonfire **d** Cooled air sinks, setting up a convection current in 'fridge **e** Air can't circulate by convection
2 a and b For explanations, see diagrams on p114
3 B; hot water rises, so collects from top down

5.7 (page 113)
1 a and b matt black **c** silvery **2** More energy radiated per second, shorter wavelengths
3 Through metal pipes and fins **4 a** Temperature, detector distance and area same for all the surfaces **b** Plate area, distance, and radiation source same for both surfaces **5 a** Temperature of sphere will rise **b** Temperature of sphere will fall **6 a** To absorb Sun's thermal radiation **b** To carry warmed water away, into house

5.8 (page 115)
1 a Much more of the water is close to the surface where it can evaporate **b** Increase in temperature, wind across surface (or increased surface area, reduced humidity) **2** Evaporating water takes thermal energy (heat) from skin **3** Refrigerator, sweating **4 a** Evaporation from skin reduced, so less cooling **b** breeze speeds up evaporation
5 Evaporation occurs from the surface of liquid at all temperatures; boiling occurs throughout a liquid at a specific temperature **6** Humid air cooled by glass, so water vapour turns into liquid

5.9 (page 117)
1 Water used to carry thermal energy (heat) in central heating system; also in car cooling systems
2 a 400 J
b 200 000 J **c** 2 100 000 J **3 a** 8400 J
b 42 000 J **c** 5 °C

5.10 (page 119)
1 a Turning solid **1 b** 68 °C **2** Energy needed to separate particles (molecules) so that they form a liquid **3 a** 3 300 000 J **b** 23 000 000 J **4** 0.12 kg

Check-up questions (pages 120–121)
1 a Faster molecules escape from liquid surface to form gas **b** Motion over ground compresses air

ANSWERS

and warms it up. Molecules move faster, so force larger when they bounce off inside of tyres.
2 B
3 a To absorb Sun's thermal radiation **b** To prevent loss of thermal energy (heat) which should be absorbed by water **c** Pump circulates warmed water through coil in tank **d** 2 kW **e** 5 m² **f i** The energy is free/renewable **ii** it is unreliable as the Sun is not always shining.
4 a Flows equal **b** Reduces radiation flow from Earth **c** Rate of emitting radiation is slightly less than rate of absorbing it
5 a i liquid **ii** liquid **b i** 440 °C
6 a evaporation **b, c, and e** convection **d** conduction
7 a Insulation; mineral wool **b** Hot water rises by convection, so collects from the top down **c i** kilo (×1000) **ii** 3000 J **iii** 1 260 000 J **d i** 4200 J **ii** 420 000 J **iii** 3 °C
8 a Larger surface area gives increased heat transfer rate **b** 12.6 MJ (12 600 000 J)
9 a 80 °C **b** None **c** Boiling rapid (or expanding vapour bubbles in liquid)

6.1 (page 125)
1 a transverse **b** 2 m **c** 0.5 m **d i** 2 Hz **ii** 0.5 s **e** 4 m/s **f** 4 m **g** 1 Hz

6.2 (page 127)
1 b refraction **c, d, and e** diffraction **2 a** Reflect **b** Refract (bend) **c** Diffract (spread out) **d** Less diffraction (less spreading)

6.3 (page 129)
1 a Sounds can be heard across a room **b** Sounds can be heard underwater in a swimming pool **c** Sounds can be heard through walls **2 a** No medium to carry vibrations **b** Sound waves diffract **3 a** Oscillations (vibrations) backwards and forwards **b** Oscilloscope display is a graph
4 Reflected (some energy also transmitted through wall)

6.4 (page 131)
1 a Sound is much slower than light **b** 1320 m
2 a solid **b** warm air **3** refraction **4 a** 440 m **b** 1.33 s **c** 82.5 m

6.5 (page 133)
1 a C **b** B **c** A and D **2** They have different overtones **3 a** Peaks closer together **b** Peaks higher (greater amplitude) **4 a** 20 kHz **b** 16.5 m **c** 0.016 5 m

6.6 (page 135)
1 Sounds with frequency undetectable by human ear **2** Scanning the womb, breaking up gall stones
3 a Measuring depth of water **b** Depth calculated from time for reflected sound pulse to return

4 a 40 000 Hz **b** 21 m **c** 0.035 m

Check-up questions (pages 140–142)
1 a circular, transverse **b i** Oscillates up and down **ii** Transverse waves produce only up and down motion **c i** 2 Hz **ii** 0.25 m
2 a 15 s **b i** Wall **ii** 264 m **c** Sound waves are longitudinal and much faster
3 a Waves should have same spacing but higher peaks **b i** If two waves take 0.02 s, one wave takes 0.01 s, so 100 waves per second **ii** 3.3 m
4 a i A **ii** B **b i** From greater than average to less than average **ii** Repeatedly backwards and forwards **iii** One wavelength is distance from centre of one cluster of particles to centre of next
5 a Longitudinal - sound, P-waves, S-waves; Transverse - light waves, ripples, X-rays **b** By underground rock movements (earthquakes) **c** 66 m
6 a Compressions in sound waves push ball forward, then it swings back **b i** Greater amplitude (greater forwards-and-backwards motion) **ii** More vibrations per second **c** 680 Hz
7 a 0.48 ms **b** 5000 m/s
8 a B louder than A **b** C higher pitch than A **c** B **d** C **e** 1.5 m **f** 440 Hz
9 a Transverse: vibrations up and down (or side to side), at right angles to direction in which wave travels; longitudinal: vibrations backwards and forwards, parallel to direction in which wave travels **b** Safer, can distinguish between tissue layers **c** Cleaning (or metal testing)

7.1 (page 141)
1 a Sun, light bulb **b** Moon, walls in a room
2 Point light source causes sharp shadows
3 a Reflected **b** Absorbed **4 a** 1.28 s **b** 500 s
5 a red **b** violet **6** Single wavelength (and colour)

7.2 (page 143)

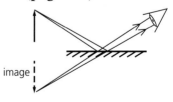

1 a See above **b** Virtual **c** See above **c** Virtual **d** No; no rays from B striking mirror will reflect into eye **2** 7.5 m

7.3 (page 145)

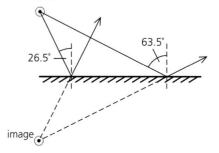

1 a, b, c, and d See above **e** 63.5°

7.4 (page 147)

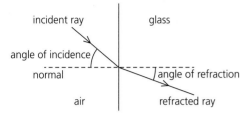

1 a See above **b** Refraction (bending) would be less in water (larger angle of refraction compared with glass) **2 a** Dispersion **b** Violet **c** Red
3 226 000 km/s

7.5 (page 149)
1 For light travelling from glass towards boundary with air, rays at angle of incidence greater than 41° are completely reflected with no refraction at all
2 a See below **b** No; angle of incidence less than critical angle

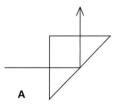

3 a Carrying telephone signals, endoscope for looking inside body **b** In periscope, binoculars (or rear reflectors)

7.6 (page 151)
1 a 17.8° **b** 36.9° **2 a** 30° **b** greater
3 a 124 000 km/s **b** 24.4°

7.7 (page 153)
1 a A **b** A **c** Point where parallel rays converge after passing through lens **d** Distance from principal focus to centre of lens **2 a** At principal focus **b** Further from lens, larger **3** Image is real, inverted, same size as object, and 2× focal length away from lens

7.8 (page 155)
1 a 12 cm from lens, height 2 cm, real and inverted **b** 15 cm from lens, height 3 cm, real and inverted **2 a** Closer than principal focus **b** At twice focal length **c** Closer than in part b, but no closer than principal focus **3** In a room, focus image of a distant window on a screen and measure distance from lens to screen

7.9 (page 157)
1 Further away **2** Rays bent inwards too much; image is formed in front of the retina **3** Rays not bent inwards enough; image would be formed behind the retina **4 a** Concave (diverging) **b** Convex (converging)

7.10 (page 159)
1 Transverse waves; can travel through vacuum; have same speed in vacuum **2** microwaves, infrared, red light, violet light, ultraviolet, X-rays **3 a** Light **b** Infrared **c** Radio waves **d** Ultraviolet **e** Microwaves **f** X-rays or gamma rays **4 a** 100 000 000 Hz **b** 3 m **c** 1500 m

7.12 (page 163)
1 Digital - represented by numbers; analogue - continuous variation **2 a** Changes electrical signals into light pulses **b** Changes light pulses into electrical signals **c** 'Cleans up' and amplifies signals **d** Easier to maintain power and quality; ideal for optical fibres and computers **e** Carry more signals; less attenuation (energy loss)
3 Contactless card reader, or security tag

Check-up questions (pages 164–165)
1 a Speed, direction **b** Totally internally reflected
2 a and b See below left **c** 20 cm **d** 18 cm

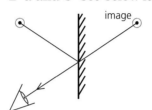 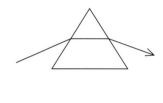

3 a See above right **b** Refraction **c** Light waves slow down
4 a and b Diagram should be similar to that at bottom of p158; size (height) of image is 3 cm
5 Larger, further from lens
6 a Diagram should be similar to that at top of p158 **b** Two of virtual, magnified, upright **c i** converging **ii** diverging **d i** concave (diverging) **ii** convex (converging)
7 a Single wavelength (single colour) **b** totally internally reflected **c** Endoscope for looking inside body **d** Represent numbers **e** Less affected by interference; easier to boost power without affecting quality **f** Contactless card reader, or security tag reader
8 a Ray travels straight through glass with no change in direction **b** Diagram similar to that at top left on page 154; 28° **c** 2×10^8 m/s
9 a Any three from radio waves, microwaves, infrared, ultraviolet, X-rays **b** Two of frequency, wavelength, penetrating power **c** 5×10^{14}
10 a i X-rays **ii** infrared
b i wave speed = wave frequency × wavelength
ii 3×10^{11} Hz **iii** microwaves
11 a 35° **b** 42° **c** Strikes KL at more than critical angle but, after reflection, strikes LM at less than critical angle
12 a i greater than **ii** the same as **iii** greater than **b i** Microwaves **ii** Ultraviolet or gamma

8.1 (page 169)
1 a repel **b** attract **c** repel **2 a** Positive (+)
b Negative (−) **c** No charge **3** Free electrons

ANSWERS

4 Polythene an insulator so charges don't move, copper a conductor so charges flow away easily **5** Carbon **6 a** Comb **b** Fewer electrons than normal so less negative charge present than positive

8.2 (page 171)
1 a Electroscope **2** coulomb **3 a** End B **b** Charges being attracted (+ and -) are closer than those being repelled (- and -) so force of attraction is stronger **4** Refuelling aircraft; earthing aircraft and tanker **5** Two of electrostatic precipitator, photocopier, laser printer

8.3 (page 173)
1 a Arrows point away from sphere **b** Away from sphere **c** Towards sphere **d** Become less (because of flow through point)

8.4 (page 175)
1 a 0.5 A **b** 2.5 A **2 a** 2000 mA **b** 100 mA **3 a and b** See below **c** 0.5 A **d** A and B; incomplete circuit **4 a** 50 C **b** 10 C

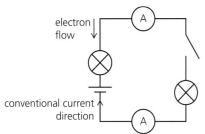

8.5 (page 177)
1 a and b volt **c** coulomb **d** ampere **e** joule **2 a** Ammeter **b** Voltmeter **c** 8 V **d** 12 J **e** 4 J **f** 2 C **g** 8 J

8.6 (page 179)
1 a 23 Ω **b** Would not heat up without resistance **2** Bulb gets brighter; less resistance in circuit, so more current **3 a** LDR **b** thermistor **c** diode

8.7 (page 181)
1 a 16 V **b** 32 V **c** 0.75 A **2** B **3 a** 2 Ω **b** 4 Ω **4** Reverse; from the graph, the current is close to zero so the value of V/I will be very high

8.8 (page 183)
1 2 Ω **2 a** 720 mm **b** 250 Ω **c** 200 mm

8.9 (page 185)
1 In series **2** All bulbs get the full battery voltage; if one breaks, others keep working **3** See below **4 a** X: 2 A, Y: 2 A **b** 6 V

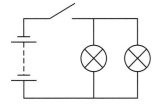

8.10 (page 187)
1 a 1.5 A **b** 6 V (both) **2 a** 3 A **b** 3 A (both) **c** 6 A **d** 2 Ω **3** D (9.9 Ω)

8.11 (page 189)
1 Allows current through in one direction only **2** Changes a.c. to d.c. **3 a** Y **b** X **4** Reduced to 2 V **5** Bring a magnet close

8.12 (page 191)
1 a 2000 W **b** 2 kW **2** 920 W **3** 2 A **4 a** 11 J **b** 660 J **5 a** 36 W **b** 21 600 J (21.6 kJ)

8.13 (page 193)
1 Thin wire which protects circuit from too high a current; wire overheats, melts, and breaks circuit if current too high **2** So that wire in cable can't still be live when switch is off **3** Safety: if there is a fault, current flowing to earth blows fuse (or trips circuit breaker) so that circuit is off **4** 3 A fuses for lamp and food mixer, 13 A fuses for hairdryer and iron **5** If there is a fault, circuits might overheat without fuse blowing **6** Switch off at socket; pull out plug

Check-up questions (pages 196–198)
1 a i Electrons pulled from hair to balloon **ii** Positive; equal but opposite to charge on balloon **b** Induces positive charge on ceiling (i.e. electrons pushed away, leaving surface of ceiling with positive charge) which is attracted to negative charge on balloon.

2 a i Like charges repel **ii** Negative **b**

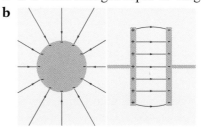

c Diagram same as on p172 bottom left **3 a** See below **b** Brighter, more current

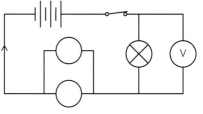

4 a 1 A **b** 3 A
5 a 4.5 J **b** 4.5 W
6 a 6 Ω **b and c** 0.2 A **d** 0.8 V **e** 0.16 W
7 a i Only for use with alternating current **ii** Frequency: current flows backwards and forwards 50 times per second **b** Two layers of insulation **c** 2.17 A **d i** Breaks circuit if current is too high for safety **ii** 3 A; nearest fuse value above actual current, otherwise faulty appliance might overheat without

ANSWERS

blowing fuse **e** Lower voltage causes lower current, so much lower power output (less heat per second)
8 a 8.7 A **b i** 840 W **ii** 2000 W greater than 840 W required but less than 3000 W needed if bulbs changed **c** All bulbs get the full generator voltage (shared if in series); if one bulbs breaks, others keep working (all stop working if in series) **d** 1890 Ω **e i** 50 Hz **ii** Graph has peaks of only half the amplitude (height) but same spacing
9 a A is ammeter, V is voltmeter, B is variable resistor **b** Use B to increase voltage in steps; measure current each time **c** 0.4 A **d** 5 Ω **e** 7.5 Ω **f** Increases
10 a 5 A **b** 2.4 Ω **c** 100 C **d** 1200 J

9.1 (page 199)
1 North-seeking pole **2 a** Unlike hard magnetic materials, soft ones easily lose magnetism **b** Steel (hard); iron (soft) **3** Iron, nickel, cobalt
4 Aluminium, copper, zinc **5** Bars 1 and 3 are permanent magnets, bar 2 is not

9.2 (page 201)
1 a N is at top (black) end **b** N pole at right-hand end of magnet; field direction is from N to S (right to left) **c** X

9.3 (page 203)
1 a As on p202 bottom right but with field direction reversed because battery is other way round in question **b** Higher current; more turns on coil **c** Reverse current direction **2** Needles form part of a circle with black ends pointing clockwise

9.4 (page 205)
1 a To increase strength of magnetic field **b** Field doesn't remain when current in coil switched off **c** Increasing current, increasing turns on coil
2 a With a relay, small current through switch can turn much larger current on/off **b** Relay core magnetized, so armature closes contacts to switch on motor **3 a** To switch off current if this is too high **b** Trips (cuts off) at lower current **4 a** To magnetize particles in a varying pattern along tape **b** Magnetism must remain but demagnetizing not be too difficult

9.5 (page 207)
1 a Higher current, stronger magnet **b** Upwards **c** Reverse current direction or turn magnet round
2 As current alternates (changes direction), force changes direction, causing vibrations
3 a Stronger turning effect (higher forces) **b** Turning effect in opposite direction

9.6 (page 209)
1 a and b Split ring (commutator) **2 a i** Coil horizontal **ii** Coil vertical **b** Stronger magnet, higher current, more turns on coil **c** Anticlockwise
3 Motor can be used with a.c.

9.7 (page 211)
1 a Current direction reversed **b and c** No current

2 a and b Greater EMF **c** Current direction reversed

9.8 (page 213)
1 Unchanged **2 a** S pole because it repels the S pole of the magnet **b** AB **3** Eddy currents induced in disc create magnetic field which opposes motion

9.9 (page 215)
1 a a.c. each side of coil reverses its direction of motion through magnetic field every half turn **b** Increase turns on coil, rotate faster, use stronger magnet **c** Horizontal; fastest motion through field lines **d** Vertical; field lines not being cut **2** Fixed coil with rotating electromagnet, more turns on coil, specially-shaped core

9.10 (page 217)
1 a Galvanometer needle flicks **b** …stays at zero **c** …flicks opposite way **2 a** Needle deflection much more **b** a.c. induced in coil (so average deflection of needle is zero) **3 a** 3 V **b** 3/1

9.11 (page 219)
1 More turns on output coil so increases voltage
2 a Magnetic field not changing **b** To reduces eddy currents which waste power by heating core **c** Because output power [voltage × current] can't be more than input power **3 a** 10 V **b and c** 23 W **d** 2.3 A

9.12 (page 221)
1 Using transformers **2 a** Transformers only work with a.c. **b** To reduce current so that less power is lost from heating effect in cables **3** In densely populated or scenic areas **4** 2640 MW
5 0.02 W **6 a** 2 kW **b** 0.002 W

Further questions (pages 224–226)
1 a Any two of: reduce turns on coil, reduce current, remove core **b** So that no magnetism remains after switch off
2 a and c a magnetic material **b** a magnet **d** a non-magnetic material
3 Current in coil creates magnetic field. Soft iron pieces attracted together which closes contacts and switches on current through motor.
4 a F is to the right, at right angles to wire **b i** Stronger **ii** Weaker **ii** Opposite direction **c** A motor
5 a i and ii Needle deflects to **right** **iii** Larger deflection to left **b** Electromagnetic induction
6 a Heating turns water into steam which pushes turbines round to turn generators **b** To reduce current so that less power is lost from heating effect in cables **c** 32 000
7 a One that reduces voltage **b** Fewer turns on output coil **c** 11.5 V **d** To reduce current in transmission lines **e** iron or Mumetal
8 a 90 C **b i** Magnetic field **ii** Become magnetized, so will repel

ANSWERS

9 a Each side of coil alternately moves up and down through magnetic field so direction of induced current keeps changing **b i** 2 **ii** 4 **c** Stronger magnet (or faster rotation)

10.1 (page 227)
1 a, c, and e electrons **b** neutrons **d** protons and neutrons **2 a and b** 13 **c** 14 **3** Different numbers of neutrons **4 a** $^{12}_{6}C$ **b** $^{16}_{8}O$ **c** $^{226}_{88}Ra$
5 X and Y are carbon, Z is nitrogen

10.2 (page 229)
1 carbon-14 **2 a, d, f, h, and i** gamma **b, e, and g** alpha **c** beta **3** Atoms of radioactive isotope have unstable nuclei which decay and emit radiation **4** Atoms are charged because electrons have been lost (or gained)

10.3 (page 231)
1 Radon gas from ground **2** Health risk if radioactive gas is absorbed by body **3 a** Gamma **b** Alpha **4 a** 2 counts per second **b** 26 counts per second **c** Gamma; it is able to penetrate the lead block

10.4 (page 233)
1 a Alpha particle **b** $A = 228, Z = 88$ **c** Radium **d** $^{232}_{90}Th \rightarrow {}^{228}_{88}Ra + {}^{4}_{2}\alpha$
e radium-228, alpha particle **2 a** 0 **b** –1 **c** Beta particle (electron)

10.5 (page 235)
1 Strontium-90 **2 a** 400 Bq **b** 200 Bq **c** 50 Bq
3 a Radioactive decay is a random process
b 1.5 hours

10.6 (page 237)
1 Emitted particles transfer energy to surrounding atoms when they collide with them **2 a** Splitting of heavy nucleus into two lighter nuclei **b** Emitted particles (neutrons) triggering further fission... and so on **3 a** Energy released in a nuclear reactor
b Explosion of nuclear weapon **4 a** Formed in reactor when U-238 is bombarded by neutrons
b Toxic, and dust can get into lungs
5 a $^{235}_{92}U + {}^{1}_{0}n \rightarrow {}^{141}_{56}Ba + {}^{92}_{36}Kr + 3{}^{1}_{0}n$
b Total mass of products is slightly less than total mass of U-235 nucleus and neutron. Loss of mass represents loss of energy.

10.7 (page 239)
1 a Joining etc. **b** Nuclear fission **2** Fuel plentiful, more energy per kg of fuel, less radioactive waste, failure is safe **3** Difficult to maintain high temperatures and pressures needed for fusion **4 a** Hydrogen **b** Nuclear fusion
c Helium **5 a** fusion **b** burning

10.8 (page 241)
1 a Radioactive isotopes **b** In nuclear reactor, when stable isotopes absorb neutrons or gamma radiation **c** Tracers, imaging **2 a** Alphas stopped, gammas pass straight through, but betas partly absorbed depending on thickness **b** Reading goes down **3 a** Any two from bulleted points on p240 **b** Safer: little radiation emitted after testing has finished **4 a** Gammas penetrate food, alphas don't **b** Sterilizing medical instruments
c X-ray-type metal testing, or medical imaging

10.9 (page 243)
1 In Thomson's model, positive and negative charges spread throughout atom **2** Rutherford–Bohr model has quantum energy levels for electrons
3 a Nucleus extremely small **b** Repelled by highly concentrated charge **4** Alphas are positive and are repelled by like charges

10.10 (page 245)
1 a Radiated as photon **b** Shorter wavelength
c Only certain energy levels exist within an atom **2** Particle not made up of other particles
3 electrons, quarks
4 Sum of fractional charges $= +\frac{2}{3} - \frac{1}{3} - \frac{1}{3} = 0$
5 Charge emitted $= +\frac{2}{3} - \left(-\frac{1}{3}\right) = +1$

Check-up questions (pages 246-248)
1 a i nuclei **ii** electrons **iii** waves **b** Alpha particles
2 a 17 electrons, 17 protons, 18 neutrons
b protons and neutrons in nucleus, electrons around it
3 a i 33 **ii** 52 **b** Atoms of same element (same atomic number/electrons) but with different mass number (or different numbers of neutrons) **c** Use fact that lead will stop alpha and beta particles but not gamma rays
4 a i Nucleus of phosphorus-32 has extra **neutron** **ii** Same electron arrangement
b i Electron **ii** 16, 32 **iii** Time taken for half radioactive atoms to decay (or activity to halve)
c i gloves/tongs, keep distance, sealed storage
ii GM tube, photographic film
5 a Too easily absorbed by tissue **b i** 12 hours
ii $\frac{1}{16}$ × original value **c** Rocks, cosmic rays
6 a Nucleus **b** Total of protons and neutrons in nucleus **c ii** 8 days **d** Much longer half-life
7 a i Only atoms with unstable nuclei are radioactive **ii** Naturally occurring radioactive materials in soil, rocks **iii** Time taken for half radioactive atoms to decay (or activity to halve) **b i** 24 counts/minute **ii** 40 hours approx
c Gamma not very ionizing **d i** beta particle
ii I 22 II 13
8 a i Unstable atoms present, emitting radiation
ii Emits radiation very close to cells and can damage/change them **b** Atoms of same element (same atomic number/electrons) but with different mass number (or different numbers of neutrons)

ANSWERS

c i and ii 86 **iii** 136
9 a Too easily absorbed by tissue **b** Tracking plant's uptake of fertilizer, or detecting leaks in underground pipes
10 a 146 **b** Nucleus

11.01 (page 251)
1 a 1 day **b** 27 days **c** 365 days **d** 27 days
2 a Moon's rotation time same as orbit time **b** We are only seeing small part of sunlit side of Moon **c** Tilt of Earth's axis is towards Sun in June so bigger fraction of each rotation is spent in sunlight **d** More hours in sunlight, also Sun's radiation comes in at higher angle so less spread out **3** 1.07×10^5 km/h

11.02 (page 253)
1 Jupiter **2** Jupiter **3** Mercury **4** Mars and Jupiter; time for one orbit lies between values for those two planets **5 a** Further from Sun, so less energy received per square metre every second **b** Venus; further from Sun than Mercury but average surface temperature is higher **6** 500 s

11.3 (page 255)
1 inner - Mercury, Venus, Earth, Mars; outer – Jupiter, Saturn, Uranus, Neptune **2** Pluto, or Ceres **3 a** Very high surface temperature, acidic atmosphere **b** No solid surface **4 a** Comet orbit is elliptical **b** Sun heats comet, particles of dust and gas stream off into space and reflect sunlight

11.4 (page 257)
1 a B **b** D **c** B **d** D **2 a** Orbit time matches Earth's rotation time **b** 24 hours **c** Sending/receiving dish on Earth can be in fixed position

11.5 (page 259)
1 a Huge cloud of gas and dust (in which stars can form) **b** Collection of billions of stars **c** The galaxy of which our Sun is a member **d** Everything – all the stars and galaxies **2 a** Slowly rotating disc around star (or planet) into which additional material is drawn **b** Because of gravitational attraction **c** Particles of matter falling inwards lost gravitational potential energy so gained kinetic energy – faster particles meant higher temperature **3 a** Nuclear fusion **b** Hydrogen **4 a** Distance travelled by light in one year **b** 4×10^{13} km **c** 8×10^8 hours (over 90 000 years)

11.6 (page 261)
1 Helium **2** Core will collapse, outer later will expand and cool (becoming red giant) leaving a hot, dense core (white dwarf) which will eventually fade **3** Fusion reactions in supernovae **4** Gigantic nuclear explosion of massive star **5** Type 1a supernovae all have the same brightness, so distances can be compared by comparing brightnesses **6** Neutron star is formed from compressed core of supernova; black hole formed when gravity is so strong that core goes on collapsing **7** Heavy elements present which could have only been made in supernova

11.7 (page 263)
1 a 4 million **b** 4 **c** 13 billion **d** 14 billion
2 a Increase in wavelength of light from very distant objects **b** Objects moving away from Earth at very high speed **3** Red shift suggest galaxies are moving apart so may have started in same place; microwaves from every direction in space may be red-shifted radiation from single event **4 a** Hubble constant = 1/age of universe **b** Rate of separation of galaxies has been constant **c** 12.7 billion years

Check-up question (pages 264-265)
1 a i A and B **ii** A **b** 24 hours
2 a i Each place spends more hours in daylight, Sun reaches higher angle in sky **ii** Autumn; winter **iii** Summer **b** Less; Uranus further from Sun so receives less radiation per m2 per second
3 a i Earth **ii** Earth **iii** Moon **b i** Earth's axis tilted towards Sun so less time in dark during each rotation **ii** Moon's surface reflects light from Sun **iii** As Moon orbits Earth, fraction of sunlit part of Moon visible from Earth varies
4 a i Venus **ii** Earth **iii** Mercury **iv** Sun **v** Neptune **vi** Circle **b i** Moon reflects Sun's light **ii** Stars give off own light **iii** Clouds and dust in atmosphere; effect of city lights **c** Solar System; galaxy; Milky Way; Universe; Big Bang; black hole
5 a i D **ii** B **b** Orbit in same direction as Earth's **c i** Ellipse **ii** Gravity
6 a Inner planets small and rock, outer planets large and gassy **b** Asteroids **c** 6.42×10^4 km/hour
7 nebula; supernova; gravity; protostar; star; accretion disc; planets
8 a Red shift shows that galaxies are moving apart **b** Red-shifted remnant of radiation from Big Bang **c** Nuclei of atoms (hydrogen for example) joining together to form nuclei of heavier elements
9 a Millions of stars **b** Reflects sun's light **c i** Motion affected by gravitational attraction between planet and Sun **ii** Venus travels at higher speed in shorter orbit **d i** Venus in new position, Milky Way in same position **ii** Venus close enough for changing position in orbit to notice, stars so far away that motion appears negligible

13.3 (page 283)
1 47 °C **2** 36 counts/second **3** 5.4 N **4** 86 kPa **5** 0.79 mV

Multichoice questions (Core) (pages 298–299)
1 A **2** C **3** C **4** D **5** A **6** B **7** D **8** A **9** A **10** A **11** D **12** C **13** C **14** B **15** C **16** B **17** B

Multichoice questions (Extended) (pages 300–301)
1 B **2** C **3** D **4** A **5** C **6** A **7** C **8** D **9** C **10** C **11** A **12** A **13** B **14** D **15** C **16** B **17** D

ANSWERS

IGCSE theory questions (pages 302–315)
All answers should be given to 2 or 3 significant figures unless stated otherwise.
All answers should include the units.
Allow error carried forward where appropriate.

1 a 0.75 s **b** 0.80 s **c** 3.8 m/s **d** to improve accuracy / reduce the effect of random errors
2 a mass is the quantity of matter in an object; weight is the gravitational force acting on an object **b** (top-pan / electronic) balance **c** measure the original volume of water in the measuring cylinder (without the piece of iron); place the piece of iron carefully into the measuring cylinder and measure the new volume of water in the measuring cylinder; subtract the original volume from the new volume to determine the volume of the piece of iron. **d** 7.9 g/cm^3
3 a i constant/steady/uniform speed **ii** stopped/not moving **b** 500 m **c** 70 s **d** 7.1 m/s **e** Section P as the gradient is steeper (accept speeds for time = 0-20s and time = 40-70s calculated)
4 a constant/steady/uniform speed **b** 1.5 m/s^2 **c** distance travelled = area under graph; distance travelled in first 40 s = 1200 m; distance travelled between t = 40 s and t = 60 s = 900 m; so total distance = 2100 m. **d** average speed = 35 m/s
5 a 1320 cm^3 or 1.32 × 10^{-3} m^3 **b** 2200 kg/m^3
6 a 0.050 kgm/s **b** (–)0.195 kgm/s **c** –0.29 m/s
7 a constant speed / terminal velocity **b** one arrow up labelled air resistance/drag; one arrow down labelled weight/gravitational force; two arrows the same size **c** accelerating, at a decreasing rate, weight is bigger than air resistance, as speed increases, air resistance increases (until it reaches terminal velocity)
8 a no resultant force, no resultant moment **b** 600 Nm **c** 450 Nm **d** 0.6 m (from pivot on right hand side)
9 a 44.1 N **b i** the limit of proportionality is where the extension stops being proportional to the load. **ii** 221 N/m or 2.2 N/cm.
10 a i kinetic, thermal **ii** chemical **b i** 2.8 m/s^2 **ii** 868 000 N **c i** towards the centre of the circle/circular path **ii** increases
11 a 200 000 kg/m^2 **b** 402 000 N
12 a measure the time for multiple (>5) swings; divide the total time by the number of swings **b** weight (downwards); tension (in string, upwards) **c** 0.060 J
13 a renewable energy will not run out / can be replenished (easily); non-renewable energy will run out / cannot be replenished (easily) **b i** no greenhouse gases/CO_2 produced, no sulfur dioxide produced/no acid rain, large amount of energy produced per kg of fuel, reliable **ii** non-renewable, risk of leak of radioactive waste, risk of accident **c** renewable – biofuel, biomass, tidal, wave, hydroelectric, geothermal, wind non-renewable – coal, oil, gas **d** 0.25 (or 25%)
14 a Milky way **b** An increase in the observed wavelength of the electromagnetic radiation emitted from stars and galaxies moving away from Earth. **c** 5 × 10^2 s
15 a 50 N **b** 765 J **c** 3.8 s **d** 17 A
16 a the drawing pin will fall off the rod of the best conductor first (as the wax on that rod will melt first). **b** copper
- has free electrons which gain kinetic energy when heated
- free electrons move through the copper and collide with atoms/ions to make them vibrate faster (and energy is transferred from hotter parts to cooler parts).
plastic
- no free electrons
- atoms vibrate faster when heated and these vibrations pass from atom to atom (transferring heat energy).
c The particles are far apart
17 T, F, F
18 a gas molecules collide with the walls (and exert a force per unit area) **b** increases; molecules have higher average speed so collide more frequently with walls of container (so bigger force per unit area) **c** 1920 cm^3
19 a nearer – small and rocky; furthest – large and gaseous **b** 4 from: dust and gas from a nebula; pulled together by gravity; accretion disc formed (from spinning dust and gas); rocks formed to make inner planets; extreme temperature caused lighter material to move away and form outer planets
20 a measure the time that the block is heated for; calculate energy supplied using: energy = power x time; measure the start and end temperatures/calculate temperature rise; use: energy = mass x specific heat capacity x temperature change (to calculate the specific heat capacity) **b** some of the energy supplied to the iron block is lost to the surroundings **c** add insulation around the iron block
21 a any 3 from: faster/ more energetic molecules escape; molecules escape from the surface/water; slower/less energetic molecules remain; less energy means lower temperature **b** 178 500 J **c** any 2 from: evaporation - occurs at any temperature; occurs only at the surface of the liquid; lowers the (kinetic) energy of the molecules remaining in the liquid / causes cooling of the liquid. (Allow reverse argument for boiling – occurs only at one temperature / boiling

point; occurs throughout the liquid; does not lower (kinetic) energy of the molecules in the liquid / cause cooling; bubbles occur)
22 a sound wave is reflected **b** transverse wave has oscillations at right angles to the direction of travel/energy transfer; longitudinal wave has oscillations parallel to the direction of travel/energy transfer **c** 412.5 Hz
23 a and b see diagram below

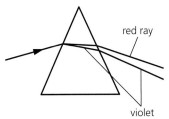

c 30°
24 a Uranus is further from the sun **b** Jupiter; largest gravitational field strength **c** 4.9×10^{12} m **d** 2.3
25 a normal **b** angle of reflection **c** they are equal **d** total internal reflection shown inside the optical fibre with a maximum of 4 reflections in total; angle of incidence = angle of reflection **e** 42°
26 a (student **B**) measures the time using the stopwatch between hearing the clap and hearing the echo; uses speed = distance/time; where d = 500 m. **b** (student **A** is incorrect) wave **X** has a lower pitch as it has a lower frequency; (student **A** is correct) wave **X** is louder as it has a larger amplitude.
27 a **Q** – microwaves; **R** – gamma rays **b** infrared **c** ultraviolet
28 a

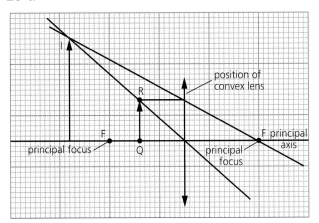

b virtual / magnified / erect / same side of lens as object **c** the image of a distant object is formed in front of the retina; a diverging lens diverges/spreads out the light rays; so the rays will focus on the retina
29 a and b see diagram

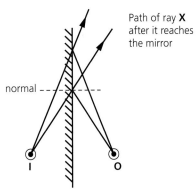

c any two from: same size as the object; upright; laterally inverted; same distance away from the mirror as the object **d** (see diagram) extend both rays behind the mirror; rays meet at a point; label image I
30 a correct symbols; ammeter, cell and fixed resistor in series; voltmeter in parallel **b** 100 Ω **c** arrow showing current from positive terminal to negative terminal **d** 14 s
31 a variable resistor **b i** (filament) lamp **b ii** as the potential difference increases, the current increases at a decreasing rate; as the current increases, the temperature increases; as the temperature increases, the resistance of the lamp increases.
32 a a region where an electric charge feels a force **b**

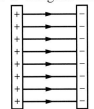

c 0.0020 A **d** 0.010 A
33 a any 3 from: place plotting compass on the cardboard/near to the wire (and switch on current); draw a dot at the head of the compass arrow; move the arrow and draw another dot at the head (idea of top-to-tail); repeat at different distance from the wire **b** circular magnetic field lines (getting further apart); anticlockwise direction **c** reverses direction of magnetic field
34 a iron core **b** alternating current in the primary coil; causes a changing magnetic field in the iron core; induces alternating p.d./current in secondary coil **c** B; C **d i** 500 V **ii** 12.5A **iii** to reduce the current; to reduce heat/energy loss to the surroundings
35 a magnetic field (lines) are cut / there is a changing magnetic field; current/pd induced; **b** one from: increase magnetic field strength, increase number of turns on coil of wire, increase speed of movement of magnet **c** one from: change direction

of magnetic field, change direction of movement of magnet **d** P repels/opposes south pole of magnet
36 a ratio of speed at which a galaxy is moving away from Earth to its distance from Earth **b** 7.3×10^{22} m **c** gradient of graph; value within 5 km/s/Mpc
37 a i circuit A **a ii** circuit B **a iii** 400Ω **a iv** 75Ω **b** 9V
38 a i 226 **a ii** 88 **a iii** 138 **b** (forms of an element with) same atomic number/number of proton; different mass number/number of neutrons
39 a i time taken for the activity to halve/ half of the radioactive nuclei to decay **a ii** 8 minutes **b** smooth curve starting at 400 counts/min; above the original curve **c** 350 – 30 = 320; 50 – 30 = 20; 4 half-lives; half-life = 20/5=5 days
40 a i alpha **a ii** alpha **b i** radiation that is all around us **b ii** any two from: radon gas, rocks/buildings, food/drink; cosmic rays **c** curved path; out of paper; at right angles to the magnetic field
41 a $^{237}_{93}\text{Np}$; $^{4}_{2}\alpha$ **b** α-particles cause air particles to ionise so a small current flows; smoke particles absorb the α-particles; so less air particles are ionised; the current is reduced (so the alarm rings) **c** curve upwards
42 a 10 million years **b** 2 years **c i** protostar **ii** red giant **iii** white dwarf **iv** red supergiant **v** neutron star **vi** black hole

Alternative-to-practical paper questions
1 a vertical ruler drawn ≤ 0.80cm from the spring **b** column 2: l or length of spring / mm; e or extension / mm **c** (in order) 3.0, 5.0, 9.0, 12.0, 15.0 (mm) **d** axes the correct way round **and** labelled with the quantity and unit; linear scales and not awkward numbers; points (small dots or crosses) plotted accurately (to the nearest ½ square); straight line of best fit drawn **e** (yes) straight line through the origin **f** correct value read from graph e.g. 6.9 mm
2 a look perpendicular to the ruler when reading the scale **b i** 26.36 s **ii** 1.32 s **c i** $1.7 s^2$ **ii** 10 m/s^2 **d** any two from: repeat using different length(s) and calculate the mean; repeat the timing and calculate the mean; use increased number of oscillations; use a fiducial marker
3 a (from top) 50, 75, 100 **b i** (all readings) 0.35 **ii** 1.4 **c i** JK 2.7 JL 4.0 JM 5.4 **ii** Ω (ohm) **d** 8.0 (or 8.1) Ω
4 circuit diagram with correct symbols for power supply, ammeter, voltmeter and resistance wire and components correctly connected with ammeter in series and voltmeter in parallel with the resistance wire (variable resistor is optional); measure the potential difference and current for each metal wire; calculate the resistance for each metal wire; repeat (with different p.d.) and calculate the mean; length, diameter, temperature (of wire); table with columns including correct units for type of wire, voltage, current and resistance
5 a 400m **b** trundle wheel, measuring tape **c** 346 m/s **d** too many significant figures used / two significant figures are more appropriate
6 a 86°C **b i** °C **ii** 10°C **iii** 5°C **iv** the (starting) temperature is closer to room temperature **c** two from: remove lid, remove insulation, increase starting temperature of water

Check-up on practical papers (page 291)
1 a correct t values for every 30 s up to 210 s; temperatures decreasing for beakers **A** and **B**; least decrease in beaker **A**. **b** °C in both columns **c i** axes correctly labelled and right way round; suitable scales (not awkward numbers); points plotted correctly to +/- half a square; smooth curve of best fit **ii** points plotted correctly to +/- half a square; smooth curve of best fit **d** adding insulation decreases the rate of cooling of the water; reference to readings showing the temperature differences for beaker **A** and beaker **B**. **e** any two from: same volume of water; same starting temperature, same material/size of beaker, both beakers must have no lid, same temperature of the surroundings **f** any one from: (avoid parallax error by) reading the thermometer perpendicular to the scale, thermometer must have stopped rising before first reading taken, thermometer should not touch the beaker, stir the water
2 a normal at the centre of **AB** and **CD** and **EX** at 30° to the normal; P_1 and P_2 at least 5 cm apart; Line through P_3 and P_4 to meet normal at **Z**; **b** (their) ϑ measured correctly to ± 2o; l measured correctly to ±2 (mm) **c** ϑ measurements of 28 – 32° ; unit of l is mm **d** statement agrees with readings (Yes or No); Yes - idea of within the limits of experimental accuracy or No - idea of outside the limits of experimental accuracy **e** any one from: pin separation should be large, view the bases of the pins, make sure pins are vertical, use thin pins/pencil lines

Index

If a page number is given in **bold**, you should look this up first.

absolute zero 103
a.c.
    changing to d.c. 188
    generators 214–215
    mains supply 192
    voltage **176**, 192, 214
acceleration 29–37, 40–41
    of free fall, g 34–35
    uniform and non-uniform 36–37
accretion disc 259
activity 234–235
action and reaction 46–47
air resistance **38**, 39, 43
alpha decay 232
alpha particles 228–229
alternating current *see* a.c.
alternators 214–215
ammeter 174
ampere, unit of current 174–175
amplitude 125, 133
analogue signals 162
asteroids 252, 254
atmospheric pressure 72–73
atomic number 226, 320
atoms 100, 168, **226**–227
    models of 226, **242**–243, 271

balance, for weighing 14, 22, 38
balance, state of 60–61
barometer 73
battery 174, 176
becquerel, unit of activity 234
beta decay **233**, 245
beta particles 228–229
Big Bang theory **262**–263, 275
bimetal strip 105
binary code 162
biofuels **93**, 94
black hole 261
bluetooth 163
Bohr, Niels 243, 271, 276
boiling 114
boiling point of water 102
Boyle's law 74–75
bulbs, filament 178, 181
Brownian motion 101

camera 156
cells (electric) 174
    in series and in parallel 176, 185
Celsius scale 102–103
centre of gravity **62**–63, 65
centre of mass **62**, 65

centripetal (and centrifugal) force 54–55
chain reaction 236–237
charge 168–171
    on electrons and protons 168, 226
    induced 170
    link with current 175
    unit of 171
circuit breakers 192, 193, **205**
circuit symbols 174, **321**
circuits **174**–177, 184–189
    mains 192–193
circular motion 54–55
cloud chamber 231
coil, magnetic field around 202–203
colour **141**, 147
comets 255–256
commutator 208
compass 200
components of a vector 53
compressions (waves) 124
concave lenses **152**, 155
condensation 115
conduction (electrical) 169
conduction (thermal) 108–109
conductors (electrical) 169
conductors (thermal) 108–109
conservation of energy 82
conservation of momentum 50–51
convection 110–111
conventional current direction 175
converging lens 152
convex lenses 152–157
CMBR (microwave background) 262
coulomb, unit of charge 171
critical angle **148**, 151
current, electric 174–175
    direction 175
    magnetic effect 202–205
    magnetic force on 206–209

dark matter and energy 263, 275
day and night 250
d.c. 192
decay, radioactive 228, **232**–235, 245
deceleration 29
demagnetizing magnets 203, 205
density **18**–21, 23
    changes in water 105
    and floating 23
    measuring 20–21
    of planets 253
    and pressure 71
diffraction
    in ripple tank 127
    of radio waves 160

digital signals 162–163
diodes **179**, 181, 188
direct current *see* d.c.
displacement can 20
dispersion 147
diverging lens 152
double insulation 192
drag 43
Earth
    rotation and orbit 250
    planetary data 253
earthing 170, **192**
echoes 131
echo-sounding 131, **134–135**
eddy currents **213**, 219
Hubble, Edwin 275, 276
Einstein, Albert 237, 268, 270, 276
efficiency 86
    of power stations 89
elastic limit 66–67
electric cells *see* cells
electric charge *see* charge
electric circuits *see* circuits
electric current *see* current
electric fields 172–173
electric motors 208–209
electrical energy 81
    cost of 191
    equation for 191
electrical power *see* power
    equation for **190**, 221
electricity
    cost of 191
    early ideas 272–273
electromagnetic induction 210–219
electromagnetic waves 122, **158**–161
    speed of 158
electromagnets 204–205
electromotive force 176
electron shells 227
electrons
    in atoms 168, **226**–227
    in circuits 174
    discovery of **242**, 271
    in electrical conductors 169
    in thermal conductors 109
    transfer by rubbing 168–169
electroscope 168
electrostatic charge 166, 169
electrostatic energy 81
electrostatic force 51
elements 232, 324
e.m.f. 174
energy 80–95
    chemical 81
    conservation law 82

early ideas 268–269
elastic (strain) 81
electrical 81, **191**
electrostatic 81
geothermal 93, 95
gravitational potential 81, **84**–85
hydroelectric **91**, 92, 95
internal **101**, 116, 269
kinetic 81, **84**–85
magnetic 81
and mass 237
non-renewable resources 92
nuclear 81, 88, 95, **236**–238
potential 81, **84**–85
renewable resources 92–93
resources 92–95
solar 93, 94
spreading 89
from Sun 92, 93, **94**–95
thermal *see* thermal energy
tidal **91**, 93, 95
transfers 82–83
wind 90–**91**, 93, 95
energy stores 81
equilibrium 61, 63
evaporation 114–115
expansion (thermal)
of gases 107
of ice and water 105
of solids and liquids 104–105
extension of spring 66
eye, human 156–157
short and long sight 157

fair test 281
Faraday's law
of electromagnetic induction 210
fibres, optical **149**, 163
fission, nuclear 88, 92, **236**–237, 239
fixed points (temperature) 106
Fleming's left-hand rule 206
Fleming's right-hand rule 212–213
floating and density 23
fluorescence 161
focal length 152, 155
force **38**–51, 56–68
and acceleration 40–41
centripetal 54–55
gravitational 44
and momentum 48
and pressure 68–69
and work 80
turning effect of 60
fossil fuels 92, 94
frequency 125
of a.c. mains 192
of light waves 141
of radio waves 163
of sound waves **132**–134

friction 42–43
fuels 90, 92–94
nuclear 92, **236**–237, 239
fundamental particles 244
fuses 192–193
fusion, latent heat of 118
fusion, nuclear 92, 94, **238**–239
in star (Sun) 259–260

g (acceleration of free fall) **34**–35, 45
g (Earth's gravitational field strength) 44–45
galaxies **258**, 262–263, 275
gamma rays
in electromagnetic spectrum 159, 161
properties and effects **228**–229, 233
uses 240–241
gases
expansion of 107
heating 106–107
particles in 100
pressure 74–75, 106–107
pressure-volume law 74–75
Geiger-Müller tube 230–231
generators 214–215
geostationary orbit 257
geothermal energy 93, 95
global warming 90, **113**
gradient of a graph 30
gravitational field strength 44–45
on planets 253
gravitational force **44**, 55
gravity 44–45
centre of **62**–63, 65
and orbits 256
Grid (electricity) 220

half-life 234–235
heat *see* thermal energy
hertz **125**, 132
Hooke's law 67
Hubble constant 263
Hubble, Edwin 275, 276
hydroelectric power **91**, 92, 95
hydrogen
atom 55, **227**
in the Sun 239, 260
hydrometer 21

ideal gas 75
image formation
by plane mirrors 142–145
by lenses 152–157
images, real and virtual **142**, 152, 154
impulse 48
induced charge 170
induced magnetism 198
induced voltage and current 210–219
direction of current 212–213

inertia 40
infrared 112, 159–**160**, 163
insulators (electrical) 169
insulators (thermal) 108–109
internal energy **101**, 218, 269
ionization 173, 228–229
ions in air 173
isotopes 226–227

jet engine 47
joule, unit of work and energy 80

Kelvin scale 103
kilogram, unit of mass 14
kilowatt 86, 190
kilowatt hour (kWh) 191
kinetic energy 81
calculating 84
kinetic theory 75, **100**

laser, light from 141
laser diode 163
latent heat 118–119
LDRs **179**, 189
LEDs 163, **189**
length 15–16
lenses 152–157
Lenz's law 212–213
light 140–160
early ideas 270
from an atom 244
in electromagnetic spectrum 159–160
speed of **141**, 150, 158
waves 141
light year 258
light-dependent resistors **179**, 189
light-emitting diodes 163, **189**
limit of proportionality 66
liquids
expansion of 104–105
particles in 100
pressure in 70–71
long sight 157
longitudinal waves 124
loudness 133
loudspeaker 207

magnetic effect of a current 202–205
magnetic field 200–203
Earth's 201
magnetic materials 199
magnetic poles **198**–201, 203
magnetic storage 205
magnetism, early ideas 272
magnets 198–201
making and demagnetizing 199, 203
magnifying glass 154
mains electricity 192–193
supply system 220–221

manometer 73
mass 14
    and acceleration 40–41
    centre of 62–63
    and density 18–19
    and energy 237
    measuring by comparing 22
    and weight 44–45
mass number 227
medium (light) 146
medium (sound) 128
melting 118
melting point of water 102
meteors and meteorites 255
metre, unit of length 15
microwaves 131, 159–**160**
    from space 262
Milky Way 258
mirrors 142–143
molecules 100
moments **60**–61, 64–65
    principle of 60
momentum 48–51
    conservation of 50–51
monochromatic light 141
Moon 250–251
    phases of 251
moons around planets 253
motion 28–37, 54–55
    circular 54–55
    early ideas 268
    graphs **30**–31, 35–37
    Newton's first law of 38
    Newton's second law of 40
    Newton's third law of 47
motors, electric 208–209
mutual induction 216–219

nebula 258–259
neutron star 261
neutrons 168, **226**
    in fission **236**–237, 238
    structure of 244–245
newton, unit of force 38, **41**
Newton, Isaac 268, 270, 274, 276
    laws of motion *see* motion
nuclear
    energy 81, 88, 95, **236**–239
    fission 88, 92, **236**–237
    fuel 92, **236**–237
    fusion 92, 94, **238**–239, 260
    power stations **88**, 236–237, 239
    radiation 228–231
    reactors **236**–239
nucleon number 227
nucleons 226
nucleus 168, **226**
    changes during decay 232–233
    evidence for **242**, 271
nuclide **227**, 232

octaves 132
ohm, unit of resistance 178
Ohm's law 181
optical fibres **149**, 163
    in cables 163
orbital speed 251
orbits **55**, 274
    of Earth and Moon 250–251
    of planets **252**–253, 256, 274
    of satellites 55, 257
oscilloscope
    displaying sounds on 129, 132–133

P (primary) waves 124
parallel circuits 184–187
parallelogram rule for vectors 52
particle accelerators 244
particles
    in atoms **226**–227, 244–245
    fundamental 244–245
    in solids, liquids, gases 100–101
pascal, unit of pressure 68
p.d. 176
    circuit rules 177, 185
    effect on current 180–181
pendulum, period of 17
penetration (radiation) 229
period
    of orbit 55, 256
    of oscillation 17, 125
periscope 148
photodiode 163
photons 141, **244**, 271
pitch 132
planets 252–255
    data on 253
    formation of 259
    orbits of **253**, 274
plugs, electric 192–193
poles, magnetic **198**–201, 203
pollution 90
potential difference *see* p.d.
potential divider 189
potential energy 81
    gravitational 81, **84**–85
power **86**, 190
    electrical 190–191
    supply system 220–221
power loss in a cable 221
power rating of appliances 190
power stations 88–91
    reactors in 236–239
pressure 68
    atmospheric 72–73
    of gas 74–75, 106–107
    in liquids 70–71
    measuring 73
pressure-volume law for gases 74–75
principal focus 152

prisms 147, 148
proportion, direct and inverse 295
proton number 226
protons 168, **226**
    structure of 244–245
protostar 259
pumped storage 91

quality (sound) 133
quantum theory 243, 244, 271
quarks **244**–245, 271, 276

radar 131, 159
radiation
    background **230**–231, 235
    dangers 230
    electromagnetic 158–161
    nuclear 228–231
    thermal (heat) 112–113
radio waves **159**–160, 163
radioactive decay 228, **232**–235, 245
radioactive waste 237
radioactivity 228–237, 240–241
    uses of 240–241
radioisotopes (radionuclides) 240
radiotherapy 240
rarefactions (waves) 124, 128
RCD (residual current device) 193
reactors, nuclear **236**–239
real image 152
recording 205
rectifier **188**, 215
red giants and supergiants 260–261
red shift 262
reed switch and relay 189
reflection
    laws of 142
    by plane mirrors 142–145
    in prisms 148
    in ripple tank 126
    of sound **131**, 134–135
    total internal **148**–149, 151
refraction
    of light 146–147, 150–151
    in ripple tank 126
    of sound 131
refractive index **137**, 150–151
refrigerator 111, **115**
relay
    magnetic 204
    reed 189
resistance 178–87
    factors affecting (for wire) 17 182–183
resistivity 187
resistors 179
    in series and parallel 185–187
    variable 179

resultant 40, 52
retardation 29
RFID 163
right-hand grip rules 202, 203
ripple tank 126–127
rocket engine 47

S (secondary) waves 124
safety
    electrical 193
    in the laboratory 231, **278–279**
    nuclear 231, 236
satellite and mobile phones 163
satellites (artificial)
    orbits of 55, **256**–257
    uses 257
scalars **52**, 85
scientific notation 13
seasons 250
second, unit of time 15
seismic waves 124
semiconductors 169
series circuits 184–187
short sight 157
SI units 14–15
    table of 320
signals, analogue and digital 162–163
slip rings 214
Snell's law 150
solar
    cells 93, 94
    energy 93, 94
    panel 93, 94, **113**
Solar System 252–255
solenoid 202
solidification 115
solids
    particles in 100
    expansion of 104–105
sound
    characteristics of 132–133
    speed of 130
    waves 128–135
specific heat capacity 116–117
specific latent heat
    of fusion 118
    of vaporization 119
ʼral lines 243
   ım
    ctromagnetic 158–161
    147, 243

    141, 150, 158
    130

speed-time graphs **31**, 36–37
spring balance 38
spring constant 67
spring, stretching 66–67
stability of balanced objects 63
stability of nucleus 235
stars 258–261
state, changes of **115**, 118
strain (elastic) energy 81
Sun
    birth of 259
    death of 260
    energy from 94–95
    fusion in 92, 94, **238**–239, 260
    and planets 252
supernova 261
sweating 115
switches 184, 192
    magnetic (relay) 204–205
    reed (relay) 189
symbols, circuit 174, **321**

telescope 155
temperature **102**–103, 269
terminal velocity 39
thermal capacity 116
thermal energy 81, 101
    capacity and storing 116–117
    and latent heat 114–115
thermal radiation 112–113
thermistors 102, **179**
thermometers 102
thermostat 105
ticker-tape experiments 32–33
tidal power **91**, 93, 95
time 15
    measuring small intervals 17
total internal reflection **148**–149, 151
tracers (radioactive) 240
transformers 217–220
    step-up and step-down 218
transverse waves 124
ultrasonic sounds 134
ultrasound 134–135
ultraviolet 159, **161**
uncertainties (in measurements) 282
units 12–15
    SI 14–15, 320
Universe 258, 274–275
    age of 262–263
    expanding 262–263

vacuum 73
vaporization, latent heat of 119

vapour, water 114–115
variables 280–281
vectors 28, **52**–53
velocity 28
    terminal 39
velocity-time graphs 31
virtual image 142
    in lenses 154, 155
    in plane mirror 142
volt, unit of p.d. and e.m.f 176
voltage 176, 214
    see also p.d. and e.m.f
    mains **192**, 220
voltmeter 176
volume 18–20

water
    density of **18**, 19
    and ice 105
    melting and boiling points
      of 102
    specific heat capacity of 116
    specific latent heats of 118–119
water power 90–**91**, 93, 95
water vapour 114–115
watt, unit of power 86, 190
wave energy 93, 95
wave equation 125, 133
wavelength 125
    of electromagnetic waves 159
    of light 141
    of sound 128
waves
    electromagnetic 158–161
    light 141
    longitudinal 124
    radio 159, **160**, 163
    in ripple tank 126–127
    sound 128–135
    transverse 124
weight 22, **44**–45
white dwarf 260
wind power 90–**91**, 93, 95
work **80**, 83

X-rays 159, **161**

zero error 17